Understanding and Managing Tourism Impacts

An integrated approach

C. Michael Hall
and Alan A. Lew

Routledge
Taylor & Francis Group

LONDON AND NEW YORK

First published 2009
by Routledge
2 Park Square, Milton Park, Abingdon, Oxon, OX14 4RN

Simultaneously published in the USA and Canada
by Routledge
270 Madison Avenue, New York, NY 10016

Routledge is an imprint of the Taylor & Francis Group

Typeset in Times New Roman by
Swales & Willis Ltd, Exeter, Devon
Printed and bound in Great Britain by
Antony Rowe, Chippenham, Wiltshire

British Library Cataloguing in Publication Data
A catalogue record for this book is available from the British Library

Library of Congress Cataloging in Publication Data
Hall, Colin Michael, 1961–
   Understanding and managing tourism impacts: an integrated
   approach/C. Michael Hall and Alan Lew.
      p. cm.
   Includes bibliographical references and index.
   1. Tourism.   I. Lew, Alan A.   II. Title.
   G155.A1H3495 2009
   910.68′4–dc22   2008054415

ISBN 13: 978–0–415–77132–0 (hbk)
ISBN 13: 978–0–415–77133–7 (pbk)
ISBN 13: 978–0–203–87587–7 (ebk)

ISBN 10: 0–415–77132–3 (hbk)
ISBN 10: 0–415–77133–1 (pbk)
ISBN 10: 0–203–87587–7 (ebk)

Alan A. Lew dedicates this book to Jessie Wee and Polly Wee, who always provide such a warm welcome and support when he is in Singapore.

C. Michael Hall dedicates the book to the Coopers and the Jameses, and to all those friends that sustain him.

Contents

Photos

Figures

Tables

Case studies

Preface and acknowledgements

One of the books that Alan Lew used in the first tourism class that he taught as a graduate student at the University of Oregon was Alister Mathieson and Geoff Wall's *Tourism: Economic, Physical and Social Impacts* (1982), which was then hot off the press. It hit just the right balance of timeliness, content and comprehensiveness for its time, which no other book on tourism impacts has quite been able to do since. Despite the lack of a single source that meets his needs, Alan has continued to teach tourism impacts over the years, and has made it a focus area in his tourism geography and sustainable tourism classes at Northern Arizona University. The gestation of his perspective in this book came out of the material and content that he had collected through those classes. Thus, building on his urban planning background, he was interested in a book that reflected both an understanding of tourism impacts and a community planning approach to addressing those impacts. This was then combined with C. Michael Hall's background in public policy and global perspectives on environmental, economic and social change to create an outline that attempts to (1) encompass tourism-related impacts at different geographic scales, from the global to the individual, and (2) recognizes the complexity and interrelatedness of tourism-related change-inducing forces. Yes, tourism development impacts destinations, but so do larger economic, social and environmental changes impact tourism. And although we continue to use the traditional division of economic, socio-cultural and environmental realms in this book, we also recognize that these are mostly artificial and mask the true complexity of tourism and the changing world we live in. This is partly why four of the chapters provide broader perspectives that are not constrained by the 'triple bottom-line' issues.

Many colleagues, students and friends have contributed to the evolution of the ideas that have culminated in this book. Michael would also like to acknowledge the debt owed to Mathieson and Wall (1982) which he used when taking a grad class from Geoff Wall when undertaking his Master's in geography at the University of Waterloo and which provided one of the springboards to teach and research in tourism studies. Much of his recent interest in issues of impacts and tourism-related environmental, regional and political change has been influenced by working with colleagues in the Nordic countries. In particular he would like to thank Stefan Gössling at Lund University Helsingborg, Dieter Müller at Umeå

University and Jarkko Saarinen at the University of Oulu for their accommodation, collaboration and hospitality. In addition, the stimulation of working with Tim Coles, Stephen Page, Murray Simpson and Allan Williams in the UK; Harvey Lemelin and Daniel Scott in Canada; David Duval, Hannah Jean Blythe, Nicola van Tiel and Sandra Wilson in New Zealand; T.C. Chang in Singapore; Gustav Visser and Robert Preston-Whyte in South Africa; and Tim Winter in Australia has also greatly contributed to his understanding of tourism and change. Finally, he would like to thank Jody, JC and Cooper for coping with time spent on the Mac writing and otherwise being on the road and in the field.

Even though tourists were the last people he wanted to be associated with as an undergraduate student at the University of Hawaii at Hilo, Alan credits the geography faculty there with establishing his academic interests in natural and cultural landscapes and places. His tourism specializations were later formed under the direction of tourism planners Dean Runyan and David Povey in the Urban Planning master's program at the University of Oregon, and later with geographer Ev Smith, his PhD dissertation chair, also at the University of Oregon. His students, colleagues and support staff in the Geography and Public Planning programs at Northern Arizona University, along with professional planners, including former students, in the state of Arizona, have all helped to maintain his interests and expand his knowledge of the field of tourism planning. Alan also wishes to thank the many colleagues who have contributed to the success of *Tourism Geographies*, which recently completed its first decade of publication, with many of that journal's contributing editors and authors informing this volume. He is indebted to the staff, especially T.C. Chang and Shirlena Huang, and Honours students in the Geography Department at the National University of Singapore, where he spent a recent sabbatical during which a portion of this book was written. And last but not least, Alan thanks his wife, Mable, and his three children for their support of his career and scholarship activities, which often involves dragging them off to various corners of the globe.

Some acronyms

ACR	Accommodation, Cafes and Restaurant industrial category
CBD	Convention on Biological Diversity
COHRE	Centre on Housing Rights and Evictions
CPC	Central Product Classification
EEA	European Economic Area
EU	European Union
GATS	General Agreement on Trade in Services
GDP	gross domestic product
GHG	greenhouse gas
IATA	International Air Transport Association
ICT	Information and Communication Technology
ISIC	International Standard Industrial Classification
MNC	multinational corporation
NTO	National Tourism Organization
OECD	Organization for Economic Co-operation and Development
Pkm	Passenger kilometres
R&D	research and development
SAR	Special Autonomous Region
SME	small and medium-sized enterprise
SNA	System of National Accounts
TNC	transnational corporation
TSA	Tourism Satellite Account
UNWTO	United Nations World Tourism Organization
VFR	visiting friends and relations
WTO	World Trade Organization
WTTC	World Travel and Tourism Council

1 Introduction

Conceptualizing tourism

Learning objectives

By the end of this chapter, students will:

- Be able to understand definitions of tourism from academic and lay pespectives.
- Understand the issues involved in defining tourism as an industry.
- Be able to identify the different categories of UNWTO tourism and travel statistics.
- Appreciate the service dimensions of tourism operations.
- Be able to identify the different modes of international trade in tourism services.
- Understand the concept of *impact* and its relationship to change.

This first chapter provides an overview and understanding of core tourism concepts and the importance of their definition for evaluating impacts and change.

Introduction

The impacts of tourism are receiving more public attention than ever before. Issues in the media as varied as climate change, coastal urbanization, demand for water by resorts and golf courses, the loss of agricultural land for development, the spread of exotic pests and diseases (see Case 1.1), economic and industrial change, fossil fuel consumption, increased cost of energy, changes in housing and communities, and sex tourism have all focused on the more controversial roles of tourism in contemporary society. Prince Philip in a royal tour of Slovenia in 2008 reportedly branded tourism 'national prostitution', going on to say 'We don't need any more tourists. They ruin cities' (Royal Watch News 2008). However, to what extent is tourism 'guilty' of these? And how are we to understand them?

This book seeks to address such issues and provide students of tourism with an improved understanding of what the effects of tourism might be, how they can be evaluated and how they can be managed. One of the first concepts we wish to address is that of an *impact*. The way the term is used implies that tourism has an effect on something, be it a place, person, environment or economy. The term also often suggests that this is a unidimensional or 'one-way' effect (Figure 1.1). However, tourism impacts are very rarely, if ever, a one-way relationship. In fact, tourism both affects people and things and, in turn, is affected by them. Furthermore, tourism impacts are rarely, if ever, just an issue of environmental, social, economic or political impact. Instead, at least two, if not all of these dimensions emerge to varying degrees when the effects of tourism activities are

How tourism impacts are often conceived as a one-way effect or unidimensional relationship

Tourism 'impacts' as a two-way relationship. Realization that tourism not only affects something but is also affected itself

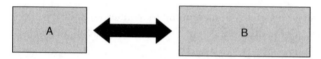

Recognition of tourism 'impacts' as change over time. A and B are constantly influencing and affecting each other (co-evolving) with each observation being a different state of the relationship between them

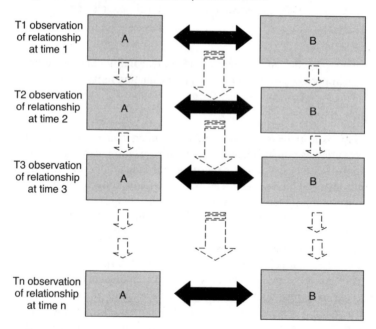

Figure 1.1 The nature of impacts.

being studied. This is because tourism affects the physical environment; it affects people, communities and the broader social environment; it has economic effects; and it can be very political, especially with respect to how places both attract and manage tourism (see Cases 1.2 and 1.4 as examples of this). Therefore, management of the impacts of tourism requires an integrated approach that aims to bring these various dimensions of tourism together.

The term *tourism impacts* is usually used as a kind of shorthand – and a poor one at that – to describe changes in the state of something related to tourism over time. A term such as *tourism-related change* would be a much better way of describing what people mean when they say tourism impacts, but unfortunately people tend to be lazy, and apart from discussions between a few tourism researchers the term *impact* is the one in common use, and the one we are stuck with! Therefore, throughout this book when we use the term impact please note that we are also using it as shorthand for change. In fact, this concept of impact as 'changes in a given state over time' is one of the key messages of this book, which starts to provide new insights into the role of tourism in cultural, economic and environmental change.

Before going into greater detail on how impacts can be understood, assessed and managed, this first chapter will provide some context with respect to the core concepts of tourism, illustrating why the nature of tourism makes it inherently difficult to monitor and manage.

Case 1.1: Ecotourism and the introduction of pests and disease

Ecotourism is generally defined as tourism that is friendly towards the environment and local people, and helps conservation. However, tourism is increasingly being connected to the introduction of exotic pests and diseases that are endangering some of the very species and environments that ecotourism is trying to protect (see also Chapter 5). This case study demonstrates the way tourism, and for that matter any human intervention in the natural environment, can have unforeseeable impacts.

Great ape tourism, to see gorillas and orangutans in their native habitats, is an extremely popular form of ecotourism to travellers who are willing to spend very high prices for the experience. Tens of thousands of visitors each year pay to see the great apes, which are only experienced by most people in developed countries either in a zoo or on television programmes such as *Animal Planet*. However, scientists became alarmed following the publication of evidence that great apes are dying from respiratory viruses directly transmitted to them by humans. They fear that existing safety measures to protect the apes, who are genetically very similar to humans and therefore subject to many of the same diseases and illnesses, do not go far enough and are calling for stricter precautions, including the mandatory wearing of face masks for all who come into relatively close contact with gorillas, orangutans and chimpanzees (Köndgen et al. 2008).

Photo 1.1 An orangutan in a rehabilitation enclosure in Balikpapan in Indonesia on the island of Borneo. Many of these orangutans were kept as pets when small, but became too strong to keep as adults. Rehabilitation centres work to prepare them to live in the wild. Most of these centres are open to tourists who help fund their efforts through entrance fees and longer-term donation programmes. (Photo by Alan A. Lew)

Their concern follows the first evidence that chimpanzees in the Ivory Coast (Côte d'Ivoire), in West Africa, died from HRSV (human respiratory syncytial virus) and HMPV (human metapneumovirus) during outbreaks at the Taï chimpanzee research station. Viral strains sampled from chimpanzees were closely related to strains circulating in contemporaneous, worldwide human epidemics. The study used a multidisciplinary approach involving behavioural ecology, veterinary medicine, virology and population biology to track human disease introduction into two chimpanzee communities in the country's Taï National Park, where researchers first began to habituate chimpanzees to human presence in 1982 (Max Planck Society 2008).

Twenty-four years of mortality data from observed chimpanzees reveal that such respiratory outbreaks could have been occurring from research activities for a much longer time than had been known. At the same time, survey data also show that the presence of researchers has had a strong positive effect in suppressing poaching around the research site (Köndgen et al. 2008).

The findings pose a major problem for those protecting the declining populations of gorillas in Uganda, Rwanda and the Democratic Republic of Congo, now numbering less than 650, as well as orangutans on Borneo and Sumatra in Indonesia, thought to number around 15,000. The tourist dollar is essential for protecting the endangered apes from poachers and funds vital work aimed at halting their decline from commercial hunting and habitat loss. However, the new research illustrates the challenge of maximizing the benefit of research and tourism to great apes while minimizing the negative side effects.

In an *Observer* newspaper report Dr Jo Setchell, a primatologist at Durham University in the UK and member of the Primate Society of Great Britain, was quoted as backing the calls for more stringent precautions, saying:

> It is very concerning. It is something that has been raised before, but this is the first report that really demonstrates concretely that these viruses are transmitted by humans . . . I think the masks are essential . . . One of the major problems is if you go on a fairly short holiday to, say, Uganda, and you have paid a lot for your permit, if you have a slight cold many people will not forgo that money. They'll take medication to hide their symptoms – because it's a big tourist experience, they have waited a long time for it, and it is very expensive.
>
> (in Davies 2008)

Source: Max Planck Society Press Release: http://www.mpg.de/english/ illustrationsDocumentation/documentation/pressReleases/2008/pressRelease 20080125/index.html

---

Defining tourism

Definitions are fundamental to any subject. Each area or domain of research has, as one of its first tasks, the identification of the things that comprise the foci of study. In tourism studies we are faced with four interrelated concepts – tourism, tourist, tourism industry and tourism resources – which provide the core, in one form or another, for the subject that we study (Hall 2005a; Hall and Page 2006). Such issues are not abstract. They are fundamental to being able to understand impacts and how to manage them, no matter how wearisome or academic it might seem. For how can you possibly assess the impacts of tourism unless you can define what tourism is? How can you manage or regulate tourism unless you can define what it is you are trying to regulate?

Tourism is a *slippery* (see MacNab 1985; Eden 2000; Wincott 2003, for other slippery concepts) and a *fuzzy* concept (Markusen 1999). It is relatively easy to visualize yet difficult to define with precision because it changes meaning depending on the context of its analysis, purpose and use.

Tourism is, therefore, a concept that, while initially looking very easy to define, is actually quite complex, with a substantial literature written specifically on the issue of definition (see Smith 2004, 2007). Much of the problem with considering the concept of tourism is that most people think of tourism in terms of leisure travel or being on a holiday or vacation. However, the concept is much wider than that (Figure 1.2) and can be interpreted from a number of academic or technical perspectives.

In distinguishing tourism from other forms of human movement several ideas become significant. First, tourism is voluntary and does not include the forced movement of people for political or environmental reasons, that is tourists are not refugees. In fact, the more impoverished someone is the less likely it is that they travel for leisure; and if they have to travel across a border, they are less likely to be welcomed. Rich people, by contrast, are usually welcomed and given far more privileges in crossing international borders than are the poor.

Second, tourism can be distinguished from migration because a tourist is making a return trip while the migrant is moving permanently from what was their home environment. While migration represents a form of voluntary one-way human mobility, tourism can be referred to as a type of voluntary return mobility.

Third, the distinction between tourism and migration sometimes becomes blurred because some people travel away from their home environment for a long time (such as on a gap year or 'overseas experience', although they are still intending to return) or because some are working, if only temporarily, in another country. In these types of situations, time (how long they are away from their normal or permanent place of residence) and distance (how far they have travelled or whether they have crossed jurisdictional borders) become determining factors in defining tourism and migration.

Based on these factors tourism includes those forms of voluntary travel in which people travel from their usual home environment to another location and then

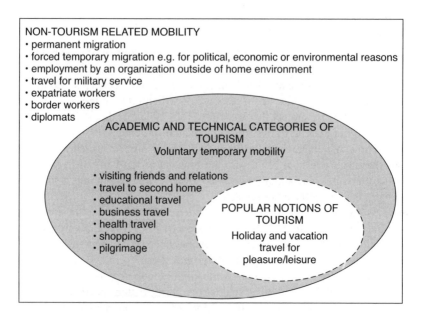

Figure 1.2 Popular and academic conceptions of tourism.

return in a manner that is shorter in time and longer in distance than non-tourism forms of similar human mobility (Figure 1.3). Using this broad definition of tourism, and given academic, statistical and research considerations, there are a range of different types of travel that are included in addition to the often narrow focus of tourism on leisure holidays or vacations. These include:

- visiting friends and relations (VFR)
- business travel
- travel to second homes
- health- and medical-related travel
- education-related travel
- religious travel and pilgrimage
- travel for shopping and retail
- volunteer tourism.

Confusion over the definition of tourism does not end here. The same word *tourism* is used to describe *tourists* (the people who engage in voluntary return mobility), as well as the *tourism industry* (which is the term used to describe those firms, organizations and individuals that enable tourists to travel). And, to complicate matters further, *tourism* is also used to refer to the whole social and economic phenomenon of tourism, including tourists, the tourism industry, and the people and places that comprise tourism destinations and landscape.

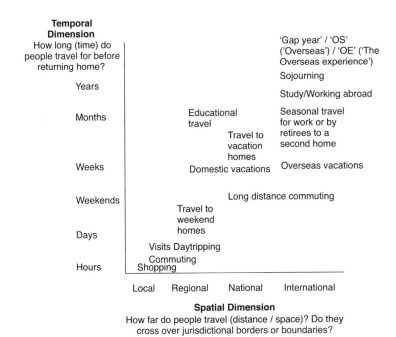

Figure 1.3 Temporary mobility in space and time (after Hall 2003).

Tourism products and their consumption present further definitional challenges because many of them serve both tourists and non-tourists (Smith 2004). In particular:

- visitors consume both tourism and non-tourism commodities and services;
- locals (non-visitors) consume both tourism and non-tourism commodities and services;
- tourism industries produce (and often consume) both tourism and non-tourism commodities services; and
- non-tourism industries produce (and often consume) tourism and non-tourism commodities and services.

In addition to the basic concepts of tourist and tourism, several other seemingly simple terms require surprisingly complicated definitions to understand the positive and negative impacts of tourism on destinations and societies.

The concept of *the home* or *usual environment* of an individual tourist is an important dimension of tourism definitions and statistics. It refers to the geographical (spatial or jurisdictional) boundaries within which an individual routinely moves in their regular daily life. According to the United Nations (UN) and the United Nations World Tourism Organization (UNWTO) (2007: 16):

> the usual environment of an individual includes the place of usual residence of the household to which he/she belongs, his/her own place of work or study and any other place that he/she visits regularly and frequently within his/her current routine of life, even when this place is located far away from the place of usual residence.

The term *trip* is also used extensively in tourism studies and refers to the movement of an individual outside their home environment until they return. The term, therefore, actually refers to a 'roundtrip'. The trip concept and its implications for understanding impacts can be understood through what academics refer to as the *tourism system* which is usually conceptualized as a geographical or spatial system (Hall and Page 2006). The *tourism system* includes the various elements that make up a trip: the *generation* or *origin region* (or place) of the tourist, the *transit region* that the tourist travels through, the *destination place* where the tourist is going, and the *overall environment* in which these exist (Figure 1.4).

A trip may also be made up of visits to different destination places. To simplify statistical data collection, countries will usually define a multi-destination trip in one of two ways:

- Based on the *point of first arrival* after departing the home environment. For example on an international flight this would be the city where you get off the aircraft, even though you may be going on to somewhere else.
- Based on the *main destination* of the trip, which is defined as either the location outside of the home environment where the most time was spent, or

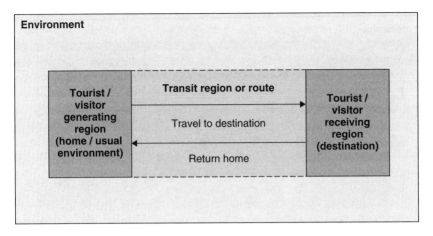

Figure 1.4 Geographical elements of a geographical tourism system.

as the place that most influenced the decision to take the trip (Figure 1.5). If the same amount of time was spent in two or more places during the trip, then the main destination is usually defined as the one that is the farthest from the place of usual residence (UN and UNWTO 2007).

The trip concept, and its representation via a tourism systems model, is important as it suggests that tourism may not just have impacts on a destination but also on the transit route, the wider environment and the tourist's home generating region. An insight which clearly has substantial implications for measuring and understanding the scale of the impacts of tourism.

An *international trip* is one in which the main destination is outside the country of residence of the traveller. A *domestic trip* is one in which the main destination is within the country of residence of the traveller (UN and UNWTO 2007). However, given how international and domestic trips are defined, an international trip could include visits to places within the country of residence in the same way as a domestic trip might include the crossing of international borders and visits outside the country of residence of the traveller.

Based on generally accepted international agreements for collecting and comparing tourism statistics, the term *tourism trip* has come to refer to a trip of not more than 12 months and for a main purpose other than being employed at the destination (UN and UNWTO 2007). However, despite UN and UNWTO recommendations there are substantial differences between countries with respect to the length of time that they use to define a tourist, as well as how employment is defined (Lennon 2003; Hall and Page 2006).

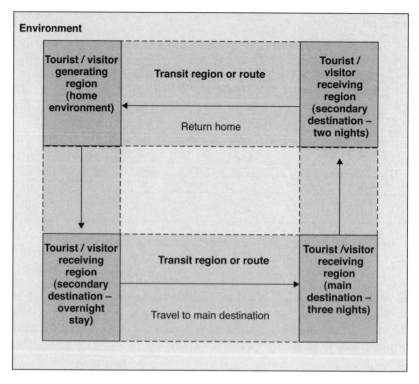

Figure 1.5 Representation of a geographical tourism system with multiple destinations in a seven-day trip.

Three types of tourism are usually recognized:

1 *domestic tourism*, which includes the activities of resident visitors within their home country or economy of reference, either as part of a domestic or an international trip;

2 *inbound tourism*, which includes the activities of non-resident visitors within the destination country or economy of reference, either as part of a domestic or an international trip (from the perspective of the traveller's country of residence); and

3 *outbound tourism*, which includes the activities of resident visitors outside their home country or economy of reference, either as part of a domestic or an international trip.

The term 'economy of reference' is used for special conditions, such as Hong Kong, which is part of the country of China, but is a separate economy of reference. Travel between China and Hong Kong is, therefore, considered international travel. However, for statistical purposes the UN and UNWTO (2007:

Photo 1.2 Local vendors selling their wares on the beach at Los Cabos, Baja California
Sur, Mexico, to tourists from north of the border. Note the rope that
demarcates the boundary between vendor and tourist (resort) territory. (Photo
by Alan A. Lew)

21–2) have recommended the adoption of the following concepts, though they
have not been widely adopted:

Internal tourism, which comprises both domestic tourism and international
inbound tourism, including the activities of resident and non-resident visitors
within the economy of reference as part of a domestic or an international trip;
this is all of the tourism expenditures in a country or economy.

National tourism, which comprises both domestic tourism and international
outbound tourism, including the activities of resident visitors within and
outside the economy of reference, either as part of a domestic or an
international trip; this is all of the tourism expenditures of domestic tourists
both inside and outside their home country or economy.

International tourism, which comprises inbound tourism and outbound
tourism, including the activities of resident visitors outside the economy of
reference, either as part of a domestic or an international trip and the activities
of non-resident visitors within the economy of reference as part of a domestic
or an international trip (from the perspective of their country of residence);
this is all tourist expenditures that are made by tourists outside of their home
country or economy.

The term *visit* refers to the stay (overnight or same-day) in a place away from home during a trip. The stay need not be overnight to qualify as a visit. Nevertheless, the notion of stay supposes that there is a stop of some undefined duration. Entering a geographical area, such as a county or town, without stopping usually does not qualify as a visit to that place (UN and UNWTO 2007).

Distinguishing such a feature can be extremely important for assessing the numbers of tourists a place receives. For example, you can imagine a situation in which a location, such as a small country town on a main road, has substantial numbers of people travelling through in their cars, but only a small proportion of those travellers actually stop in the town for anything more than refilling their fuel tank. Research that describes all those travelling through such a town as tourists or visitors, as opposed to only those who actually stop and visit, could be extremely misleading.

In order to improve statistical collection and improve understanding of tourism, the UN and the UNWTO have, for many years, recommended differentiating among visitors, tourists and excursionists. For example, the UNWTO (1991) recommended that an *international tourist* be defined as:

> a visitor who travels to a country other than that in which he or she has his or her usual residence for at least one night but not more than one year, and whose main purpose of visit is other than the exercise of an activity remunerated from within the country visited;

and that an *international excursionist* (for example, a cruise-ship visitor) be defined as:

> a visitor residing in a country who travels the same day to a different country for less than 24 hours without spending the night in the country visited, and whose main purpose of visit is other than the exercise of an activity remunerated from within the country visited.

International excursionists are also known as *day-trippers*. This type of travel at the domestic level is often considered a form of recreation, rather than tourism – though this too confuses the definition of who is a tourist and who is not, especially as some government authorities include day-trippers in their tourism data (Hall 2005b).

In order to try to clarify an increasingly complex series of travel categories the UN and UNWTO (2007) have recommended use of the term *visitor* rather than *tourist* per se, with a number of criteria needing to be satisfied for an international traveller to qualify as an *international visitor*, although the terms international visitor and international tourist tend to be used interchangeably in everyday usage:

- The place of destination within the country visited is outside the traveller's usual environment.
- The stay, or intention of stay, in the country visited should last no more than 12 months, beyond which this place in the country visited would become part

of his/her usual environment. At which point this would lead to a classification as migrant or permanent resident. The UN and UNWTO recommend that this criterion should be applied to also cover long-term students and patients, even though their stay might be interrupted by short stays in their country of origin or elsewhere.

- The main purpose of the trip is other than being employed by an organization or person in the country visited.
- The traveller is not engaged in travel for military service or is a member of the diplomatic services.
- The traveller is not a nomad or refugee. According to the UN and UNWTO (2007: 21), 'For nomads, by convention, all places they visit are part of their usual environment so that beyond the difficulty in certain cases to determine their country of residence.. . . For refugees or displaced persons, they have no longer any place of usual residence to which to refer, so that their place of stay is considered to be their usual environment.'

Domestic visitors can be similarly classified. Therefore, for a traveller to be considered a *domestic visitor* to a place in the country he or she is resident, the following conditions should be met:

- The place (or region) visited should be outside the visitor's usual environment which would exclude frequent trips, although the UN and UNWTO (2007) recommend that trips to vacation homes should always be considered as tourism trips.
- The stay, or intention of stay, in the place (or region) visited should last no more than 12 months, beyond which this place would become part of his/her usual environment. As with the international visitor classification the UN and UNWTO recommend that this criterion should be applied to also cover long-term students and patients, even though their stay might be interrupted by short stays in their place of origin or elsewhere.
- The main purpose of the visit should be other than being employed by an organization or person in the place visited.

Nevertheless, it remains debatable as to how successful the recommendations for the change of terminology will be given that the term *tourist* is in such widespread use. In addition, in the longer term if the change of terminology is accepted it may well have implications for how tourism is conceptualized as an industrial sector and how public policy is developed.

Understanding growth in international tourist arrivals

International tourism has demonstrated significant growth since statistics began and is expected to continue to grow into the foreseeable future (Table 1.1). Although slowing in some years, such as following the 2001 attack on the World Trade Center in New York or the increase in the price of oil and associated

Photo 1.3 Airplanes in line for takeoff at John F. Kennedy International Airport in New
York. Overloaded airports and an ageing air traffic control system in the
United States, combined with occasional inclement weather, can cause hours-
long delays for the travelling public. For some, the resulting discomfort is a
significant barrier to travel. (Photo by Alan A. Lew)

Table 1.1 International tourism arrivals and forecasts, 1950–2020

Year	World	Africa	Americas	Asia & Pacific	Europe	Middle East
1950	25.3	0.5	7.5	0.2	16.8	0.2
1960	69.3	0.8	16.7	0.9	50.4	0.6
1965	112.9	1.4	23.2	2.1	83.7	2.4
1970	165.8	2.4	42.3	6.2	113.0	1.9
1975	222.3	4.7	50.0	10.2	153.9	3.5
1980	278.1	7.2	62.3	23.0	178.5	7.1
1985	320.1	9.7	65.1	32.9	204.3	8.1
1990	439.5	15.2	92.8	56.2	265.8	9.6
1995	540.6	20.4	109.0	82.4	315.0	13.7
2000	687.0	28.3	128.1	110.5	395.9	24.2
2005	806.8	37.3	133.5	155.4	441.5	39.0
forecast						
2010	1006	47	190	195	527	36
2020	1561	77	282	397	717	69

Source: UNWTO 1997, 2006.

recession in developed countries in the early 1970s, the trajectory for international
tourism has generally been one of steady growth for all regions of the world since
the end of the Second World War. But how are these international tourism
numbers derived, and how accurate are they?

The basic definition for the international tourist arrivals in Table 1.1 is someone
who travels to a place outside of their home country and stays away for at least one

night, but no more than one year (as already discussed). The UNWTO requires that its member states use this definition to collect the international arrival statistics. Unfortunately, different countries around the world collect these data in different ways. In much of Western Europe, borders between countries are virtually meaningless from a traveller's perspective because of the European Union (EU)–European Economic Area (EEA) Schengen Agreement, which allowed for the abolition of systematic border controls between participating countries (Hall 2008a). For the EU, international tourist numbers can only be estimated from survey data. Other countries do not have the technology or manpower to collate the data they collect. Border crossings in remote areas where borders are not monitored, such as in much of central Asia, and almost all illegal crossings are either estimated or are not included in the numbers that countries report.

Other data collection and comparison issues include places like Hong Kong and Macau, that are Special Autonomous Regions (SAR) of China, yet still count mainland Chinese as international tourists. Almost half of the international tourists to Hong Kong are from China, while the majority of China's international tourists are from Hong Kong. There are also regions such as West Africa and Europe where the geography consists of many small countries, thereby allowing a tourist to easily pass through several in one day. This situation contributes to the very high international tourist numbers found in Europe. Alternatively, someone in Australia, Canada, the United States and other large countries could travel for several days in a straight line and not cross an international border.

In addition, some countries use different international arrival numbers than those preferred by the UNWTO. China, for example, prefers to include day-trippers in the numbers that it uses widely in Chinese media outlets, which almost triples its number of international visitors (about 125 million with day-trippers versus about 42 million without) (Lew 2007a). All the considerations delineated here must be taken into account whenever you see international tourist arrival numbers published by the UNWTO, a national tourism organization (NTO), or any other tourism organization, including a local chamber of commerce. The *fuzziness* in defining who is and who is not a tourist has major implications for under-standing what activities are and what activities are not part of the tourism industry, and what is and what is not a tourism-related change or impact.

Case 1.2: The Galápagos

The Galápagos Islands in the eastern Pacific Ocean are among the most famous islands in the world and have developed a thriving tourism industry. The volcanic archipelago has a unique flora and fauna, including different species of finches on different islands, flightless cormorants, giant tortoises, blue-footed boobies and marine iguanas that inspired Charles Darwin's theory of natural selection. Growth of interest in the environment, the showing of numerous documentaries on the islands on television, and improved air and sea access has led to substantial international tourism growth with the islands being promoted as a leading ecotourism destination. This case study shows how an increase in the number of tourists is often associated with an increase in the size of the local resident population, with concomitant impacts on the natural environment.

In 1959, the centenary year of Charles Darwin's publication of *Origin of Species*, the Ecuadorian government declared 97 per cent of the archipelago's land area a national park (the Galápagos National Park). The remaining 3 per cent is distributed between the inhabited areas of Santa Cruz, San Cristobal, Floreana and Isabela. In 1978 the islands became the first site to be placed on the UNESCO World Heritage List. The Galápagos Marine Reserve, comprising the surrounding 70,000 km^2 (43,496 m^2) of ocean, was declared a marine reserve in 1986 and is one of the world's largest (along with such sites as Australia's Great Barrier Reef and the Phoenix Islands Protected Area of Kiribati (which is in turn part of the World Heritage Central Pacific project that aims to establish a transboundary and serial World Heritage site)). In 2001 the World Heritage Site was expanded to include the marine reserve. However, in 2007 the Galápagos World Heritage Site was added to the list of *World Heritage in Danger* owing to threats posed by invasive species, growing tourism activities and immigration.

The Galápagos provide a good example of why the impacts of tourism require an integrated approach. Tourism growth has been significant with, for example, the number of days spent by passengers of cruise ships increasing by 150 per cent from 1992 to 2007. This growth in tourism has created economic opportunities, which has fuelled increased immigration from the Ecuador mainland to the islands. Annual revenues from Galápagos tourism are now estimated at $200m, although much of this is taken by airlines and tour operators on the mainland.

In 1972 the island group's population was 3,488. By the 1980s, this number had risen to more than 15,000, and in 2006 there was an estimated 40,000 people. As Carroll (2008) reported, 'For decades, thousands flocked from the impoverished Ecuadorean mainland and found jobs in the tourist industry as maids, waiters, cleaners and shop assistants'. The growth in population has also led to land-use competition between conservation activities and activities associated with food production, including growing vegetables, raising animals (chickens, cows, pigs and goats) and fishing. There has also been a number of incidents of poaching and killing of the protected native wildlife. The growth in people and farm animals, along with greater inter-island mobility, has raised fears about the introduction and spread of pests, diseases and yet more invasive species.

However, the problems on the Galápagos Islands are not just environmental, they are also political, social and economic. In 2008 the Ecuadorean government started a crackdown on illegal immigration. 'It is a policy to send home all those who do not have legal status or the proper documentation.. . . We are enforcing the law' said Carlos Macias, a spokesperson for Ingala, the regional planning agency (in Carroll 2008). One thousand people had been returned to the mainland in the 12 months to October 2008, while at that time another 2,000 had been told to leave within 12 months. However, according to Carroll, 'there are no plans to curb the soaring number of tourists (mostly well-heeled Europeans and Americans who visit for a few days) which this year is set to reach 180,000', to which Macias commented 'Of course the tourist numbers have an environmental impact, but we cannot forfeit the economic opportunity' (in Carroll 2008).

Nevertheless, the Ecuadorean government is seeking to develop a new *tourism model* for the islands that seeks to reconcile the economic benefits of tourism and environmental concerns, with the removal of illegal migrants back to the Ecuador mainland being a significant part of the strategy. This is happening despite Ecuador's Environment Minister, Marcela Aguiñaga, telling the *Los Angeles Times* newspaper in September 2008 that there was no sign that tourism was 'oversaturated'. The idea is to maintain the economic impacts of tourism but reduce its environmental footprint by scaling back ancillary activities that require imported labour – 'a process of weeding out the poor to make room for the rich by government, as opposed to natural, selection' (Carroll 2008). Environmentalists welcomed the initiative, but were worried it did not go far enough. According to Johannah Barry, president of the Galápagos Conservancy 'The system is currently broken, or certainly strained.. . . The problem is not so much the number of tourists as the ancillary economy that's going up around it. It makes sense to limit the strain.' One way of doing this is to change the tourism products that are provided to visitors, including tourist packages which offered kayaking, horse-riding, scuba diving, deep-sea fishing and other activities that disrupted the islands' ecology. According to Barry, 'You can do those things in Hawaii; there is no reason to do them in the Galápagos'.

Sources: Carroll, R. (2008) Tourism curbed in bid to save Galapagos haven. *The Observer*, Sunday October 12, http://www.guardian.co.uk/travel/2008/oct/12/galapagosislands-travelnews

Galapagos Conservancy (formerly the Charles Darwin Foundation): http://www.galapagos.org/

World Heritage List, Galápagos Islands: http://whc.unesco.org/en/list/1/

Defining tourism industry and trade

As noted previously, the term tourism is often equated with that of the tourism industry. This definition is complicated by the service industry characteristics of tourism, which means that it tends to be defined by consumption rather than commodity-producing activities. Defining service industries, such as tourism, can be an extremely difficult exercise. The essential characteristic of service industries is that they cannot be produced without the agreement and cooperation of the consumer. In addition, the outputs produced (such as adventure experiences) are not separate entities that exist independently of the producers or consumers.

In the case of tourism such consumption and production usually occurs outside of the home environment of the consumer. The recent growth in information and communication technologies (ICTs) has changed the nature of tourism distribution (or sales) channels so that it is possible to purchase tourism products from a home computer over the Internet, although final consumption will still occur outside of the home environment.

International and national government tourism agencies and industry organizations have sought since the 1980s to develop a production or supply-side approach so as to compare the economic dimensions of tourism with other sectors and industries. This approach focuses on tangible tourism products in the destination, such as hotels and restaurants, and has been developed particularly through the concept of Tourism Satellite Accounts (TSAs). TSAs have been developed in a number of countries, including Canada, New Zealand, Norway and Australia (see Chapter 3 for a discussion of TSAs in more detail). Like other international tourism statistics, the numbers produced in international TSAs are not standardized, though they usually contain one or more of the following accounts:

- tourism consumption by commodity
- tourism consumption impact on supply
- production accounts of the tourism industry
- tourism-related gross fixed capital formation
- employment related to tourism
- stocks and flows of fixed assets related to tourism
- imports and exports of goods and services generated by tourism
- tourism balance of payments and
- tourism value added (contribution to GDP) by commodity and activity.

From a supply-side perspective, the tourism industry can be defined as the aggregate of all businesses that directly provide goods or services to facilitate business, pleasure and leisure activities to people who are voluntarily away from their home environment. According to Smith (2004: 31), when utilizing a supply-side approach 'A tourism industry is any industry that produces a tourism commodity'. Three key features emerge from this definition:

- The tourism industry produces tourism commodities that can be defined as 'any good or service for which a significant portion of demand comes from persons engaged in tourism as consumers' (Smith 2004: 30).
- The inclusion of business, pleasure and leisure activities emphasizes the nature of the goods and services a traveller requires to undertake the trip.
- The definition includes the notion of a 'home environment' (as already discussed), which requires the delineation of a distance or time threshold, such as an overnight stay (Hall 2005a).

Smith (2004) argues that these three elements can be combined to conceptualize and measure tourism in a manner that is consistent with that of other industries. However, while Smith's approach, which is utilized in TSAs, assists with understanding the economic impacts of tourism consumption it does not provide much help with understanding the strategies that firms use with respect to tourism (or not, in the case of those businesses which do not see tourists as a market even though they may sell commodities to them). Nor does it necessarily help in the formulation of policies and strategies used by places with respect to economic development, although it is often used to justify pro-tourism development policies (Hall 2007b).

In addition, there are some firms that supply products to tourists and which explicitly recognize the tourist as either a direct (tourist as end consumer) or indirect (business-to-business) market for their goods and services. These are a subset within the larger realm of firms that provide products that tourists use and buy, and are therefore captured within TSA methodology (Figure 1.6). The former are *tourism firms* as these businesses consciously recognize that all or part of their business is comprised of tourists and consequently they develop strategies to meet the needs of this market. It is these firms that are usually perceived by most people as comprising the tourism industry, and it is these firms that are usually active within tourism industry organizations. Businesses that would usually be recognized as being tourism firms, and hence are regarded as the primary contributors to the tourism industry include:

- international and domestic transport operators and carriers;
- accommodation operators, including hotels, motels, caravan parks and camping grounds;

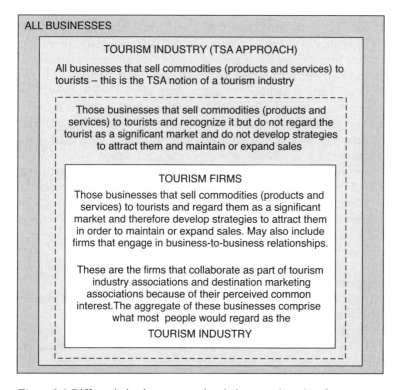

Figure 1.6 Differentiating between tourism industry and tourism firms.

- restaurants and other catering establishments;
- tour operators, wholesalers, travel and booking agents;
- attraction, entertainment and event facility operators;
- national parks;
- manufacturers of souvenirs;
- specialist travel information suppliers;
- specialist event, convention and meeting centre operators; and
- specialist retailers, such as souvenir shops.

Further distinction between different businesses is also important as, for example, some businesses, such as restaurants, in some locations may regard themselves as part of the tourism industry, while in another location they do not. Recognition of having the tourist as a market will, therefore, play a significant role in influencing how individual firms cooperate, network and develop business strategies, as well as broader public tourism policy.

Hall and Coles (2008a) further conceptualize international tourism by highlighting the international business dimensions of tourism within the context of the World Trade Organization (WTO) framework, and in particular the General Agreement on Trade in Services (GATS) (UN et al. 2002). GATS recognizes four different modes in the international supply of services (illustrated in Figure 1.7).

- *Cross-border supply* (Mode 1 under GATS): from the territory of one into the territory of another. This mode is similar to the traditional notion of trade in goods where both the consumer and the supplier remain in their respective countries when the product is delivered, while the service or product crosses jurisdictional borders. Delivery of the service can be made by various telecommunication technologies (ICTs), traditional mail or freight delivery. Online ticket purchases are sometimes purchased and consumed in this way, but it is generally not a significant part of tourism retail sales because most international tourism is partially or wholly consumed abroad.
- *Consumption abroad* (Mode 2): a consumer moves outside his or her home country and consumes services and products in another country. International tourism provides the classic example of consumption abroad although, as well as leisure consumption, it also includes medical-related travel and travel for education and language courses. Under GATS statistical guidelines, consumption abroad refers to trips under 12 months in duration, although countries have a range of classification schemes for such consumption depending on regulatory and visa requirements.
- *Commercial presence* (Mode 3): the service is provided by a service supplier of a country through commercial presence in the territory of another country at the various stages of production and delivery, as well as after delivery. Under GATS, the 'supply of a service' includes production, distribution, marketing, promotion, sale and delivery. Examples include transportation ticketing available via a foreign-owned travel agency or accommodation in a foreign-owned hotel.

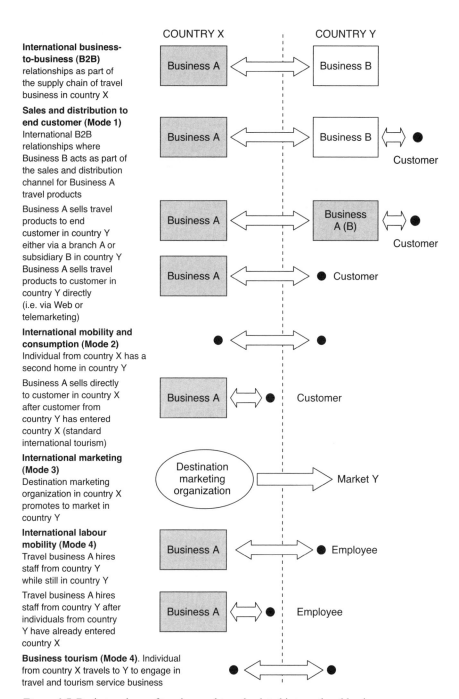

Figure 1.7 Basic typology of tourism and travel-related international business dimensions (source: Hall and Coles 2008a: 9).

Photo 1.4 Approaching the Canadian border customs and immigration station from northern Michigan in the United States. Friendly political relations and special treaty agreements between countries can significantly enhance travel between them. Poor relations can have the opposite effect. (Photo by Alan A. Lew)

- *Presence of natural persons* (Mode 4): occurs when an individual (not a company or organization) has moved into the territory of the consumer to provide a service, whether on his or her own behalf or on behalf of his or her employer. This includes both self-employed persons as well as employees. This category covers only non-permanent employment in the country of the consumer. However, to complicate matters there is no agreed definition of 'non-permanent' employment under GATS or other international agreements. Although under each country's GATS commitments the temporary status generally covers a period of two to five years, with it being different for different categories of natural persons. Mode 4 is becoming increasingly important for international tourism in both an empirical sense, with respect to the growth in seasonal international workers and other short-term labour migrations, and in a conceptual sense, with the growth of the concept of working holidays.

Tourism mobilities

As discussed previously, tourism is just one form of human mobility. However, since the late 1990s increasing attention has been given to the various concepts of mobility and the extent to which they are interrelated. These have been particularly highlighted as a result of globalization processes, which include the impacts of transport and communication technologies and the expanding international labour and leisure markets. Travel that once took two or three days to accomplish may now be completed as a day trip – a factor that clearly challenges how we conceive of the notion of *home* or the *daily routine environment.*

This is further complicated when combined with the extent to which a person may have multiple homes at one time, as well as many homes over their lifetime as they move for education, employment and relationships (Hall 2005a). There is also a form of existential tourism in which individuals have special life and identity relationships with certain places where they have had life-changing experiences or diasporic ancestral roots (Lew and Wong 2003, 2005). Such changes in mobility and nuanced relationships with different places have implications not only for tourism but also for a wide range of human activities, as well as ideas of accessibility, distance, extensibility, home, identity and networks (cf. Frändberg and Vilhelmson 2003; Coles and Timothy 2004; Coles et al. 2004, 2005; Hall 2005b).

Figure 1.8 adds a third dimension, number of trips, to the diagram presented in Figure 1.3 to present a model for describing different forms of temporary mobility, such as short-term travel for work, gap years, travel for education or health, as well

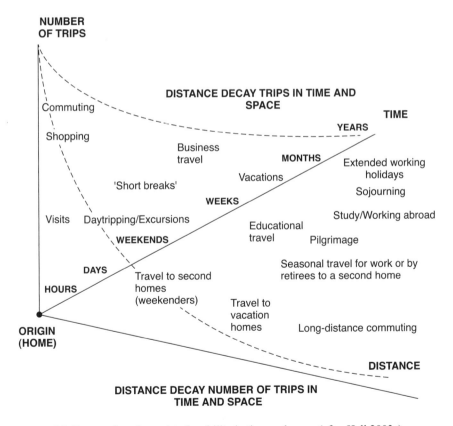

Figure 1.8 Extent of tourism-related mobility in time and space (after Hall 2003a).

as travel to second homes, return migration and travel among diasporic communities, in terms of three dimensions of space, time and number of trips (Hall 2003). Figure 1.8 illustrates the decline in the overall number of trips or movements with time and distance away from a central generating point which would often be termed as *home*. The fact that the number of movements declines the further one travels in time and space away from the point of origin is well recognized in the study of travel behaviour and has a number of implications for the impacts of tourism (McKercher and Lew 2003). For example, with respect to the greenhouse gas (GHG) emissions from tourist travel that are connected to climate change (see Case 1.3). The 'distance decay' relationship represented in Figure 1.8 holds whether one is describing the totality of movements of an individual over their lifespan from a central point (home), or whether one is describing the total overall behavior of a larger aggregate population (Hall 2005a).

Case 1.3: Tourist trips and CO_2 emissions

Carbon dioxide (CO_2) is the most significant greenhouse gas (GHG), accounting for 77 per cent of global anthropogenic (human-caused) global warming (Stern 2006). CO_2 emissions from tourism have grown steadily since the 1950s to the current estimated level of about 5.0 per cent of all anthropogenic emissions of CO_2. Globally, an average tourist trip lasts 4.15 days (average length for all international and domestic tourist trips) and generates average emissions of 0.25 tonnes (t) of CO_2 per traveller. This case study shows how transportation, which is fundamental to all tourism, is overall the single most serious contributor to global climate change today.

The vast majority of trips produce lower emissions, but a small share are highly emission-intense. For instance, a 14-day holiday from Europe to Thailand may cause emissions of 2.4 t of CO_2 per person and a typical fly-and-cruise trip from the Netherlands to Antarctica produces some 9 t CO_2 (Lamers and Amelung 2007). Even holidays said to be eco-friendly, such as dive holidays, will cause high emissions in the range of 1.2 to 6.8 t CO_2 owing to emission from air, automobile and boat travel (Gössling et al. 2007). Some trips, therefore, generate per person emissions in a single holiday that vastly exceed annual per capita emissions of the average world citizen (4.3 t CO_2), or even the higher average of an EU citizen (9 t CO_2) (see Peeters et al. 2007 for a discussion of the environmental impacts of European tourism transport).

Transport generates the largest proportion of tourism-related CO_2 emissions (75 per cent), followed by accommodation (21 per cent) and tourist activities (3 per cent). In terms of radiative forcing, which is the change in the balance between radiation coming into the atmosphere and radiation going out, the contribution of transportation is significantly larger, ranging from 82 per cent to 90 per cent, with air transport alone accounting for 54 per cent to 75 per cent of the total change. Variation in emissions from different types of tourist trips is large. While the average trip generates 0.25 t of CO_2, as already mentioned, long-haul trips and high-end luxury cruises can generate up to 9 t CO_2 per person per trip (35 times the emissions caused by an average trip) (UNWTO and UNEP 2008).

The majority of tourist trips cause only small amounts of emissions, though. For instance, trips by coach (bus) and rail account for 34 per cent of all trips worldwide, but comprise only 13 per cent of global CO_2 emissions (including emissions from accommodation and activities). In contrast, a small number of energy-intense trips is responsible for the majority of emissions. While air-based trips comprise 17 per cent of all tourist trips, they cause about 40 per cent of all tourism-related CO_2 emissions, and 54–75 per cent of radiative forcing. Likewise, long-haul travel between the five world regions the UNWTO use for statistical aggregation purposes (Africa, Americas, Asia-Pacific, Europe, and the Middle East) accounts for only 2.7 per cent of all tourist trips, but contributes 17 per cent of global tourist emissions (UNWTO and UNEP 2008).

Travel between and within the more economically developed regions of the world (Europe plus parts of the Americas and the Asia-Pacific) is the most significant contribution to emissions, comprising 67 per cent of international trips worldwide and 50 per cent of all passenger kilometres (pkm) travelled globally. In contrast, Africa is the least important destination for travellers from the developed countries, receiving just 2.2 per cent of all trips from Europe and the Americas, which represents 3.3 per cent of CO_2 emissions from tourism. Long-haul (interregional) tourism from Europe represents 18 per cent of all international trips, and 49 per cent of all CO_2 emissions from travel. Long-haul travel within and between the world's most industrialized countries, therefore, causes most of tourism's greenhouse gas emissions (UNWTO and UNEP 2008). Because of such issues there is enormous controversy about how international transport emissions can and should be incorporated into climate change mitigation and adaptation schemes, who should pay the associated costs, and how future transport infra-structure should be developed (see Chapters 5 and 7 for further discussion on tourism and climate change).

Tourism life course

Figure 1.8 reinforces an important issue that has already been raised with respect to assessing the impacts of tourism, and that is the role of time. The geographical tourism system, outlined in Figures 1.4 and 1.5, shows the progression of a tourist trip over space and over time. Each stage of the tourism system can be matched with the psychology of a consumer's travel experiences, because at each stage the tourist will be in a different psychological state in different environments and sites, and with a growing biography of new experiences and understandings that change their motivations and expectations (Hall 2005a; Lew and McKercher 2002).

Tourist travel within the basic geographical tourist system can therefore be regarded as consisting of five stages (Fridgen 1984):

- travel decision-making and anticipation
- travel to a tourism destination or attraction

- the on-site or at destination experience
- return travel and
- recollection of the experience and influence on future decision-making.

Given that travel behaviours and motivations are, at least in part, determined by prior experiences, it is then possible to talk of a 'travel career' (Pearce 1988, 2005; Pearce and Lee 2005) that expresses the development of an individual's travel experiences over time. Oppermann (1995) developed the travel career notion to connect travel with the life-cycle concept to describe changes in travel behaviour across someone's lifespan. As it was originally conceived, the notion of a 'life cycle' referred to individuals moving through certain stages of life (such as school, university, work, marriage, children, retirement) at certain ages, and that these stages would then influence their patterns of consumption as well as overall well-being.

However, the generalized life-cycle concept has been criticized for being too time and space specific (post-Second World War, white, United States), as well as posing the more philosophical issue of whether life actually follows a cycle in the sense that when you die you then start at the same point again (which is what a cycle implies). Human life paths are not constituted by the endless repetition of orderly sequences, 'the deterministic implication that life is irreversibly leading something back to where it came from' (Bryman et al. 1987: 2). Personal time, like historical time, is linear not cyclical. Instead, an alternative approach to describing an individual's life path is the concept of the 'life course', which is gaining recognition in the social sciences for its greater appropriateness to contemporary society, as well as to describing mobility.

From a life-course approach, the unit of analysis is the individual situated in geographical, social, historical and political space and time. The study of the individual, household or family becomes the study of conjoined or interdependent life courses or paths. This recognizes that social interactions influence decisions with respect to mobility behaviour, rather than being undertaken in isolation. A life-course approach does not impose a 'normal' or ideal life path with stages. Instead, what is central to the concept of the life course is that of the event or transition (Boyle et al. 1998).

Early transitions have implications for later ones with transitions occurring in 'personal time', 'historical time' and 'family time'. The life-course paradigm, therefore, emphasizes that changes in one dimension of the household-ageing process, for example, are necessarily linked to changes in other dimensions. Such an analysis clearly has implications for understanding tourism behaviour (Figure 1.9).

Rather than examine tourism motivations and activities in isolation, a life course approach suggests that tourist behaviour and demand should not only be seen in a wider context of previous travel experiences – which is the 'travel career' concept – but also in terms of other social, economic, political and even environmental situation factors, such as quality of life (see Chapter 3), that affect individual decision-making. Therefore, the life-course approach identifies the importance of events or transitions that influence life paths. Warnes (1992) identifies several life-course transitions that influence travel careers:

- leaving parental home
- sexual union
- career
- family
- children
- career promotion
- divorce or separation
- cohabitation and second marriage
- retirement
- migration
- bereavement or income collapse
- frailty or chronic ill-health.

In terms of understanding tourism and travel behaviour over the lifespan of the individual, Hall (2005a) uses a life-course approach that sees travel and movement within three domains of:

- 'lifestyle': including cohort and demographic characteristics and different types of careers with respect to employment, relationships, leisure and consumption;

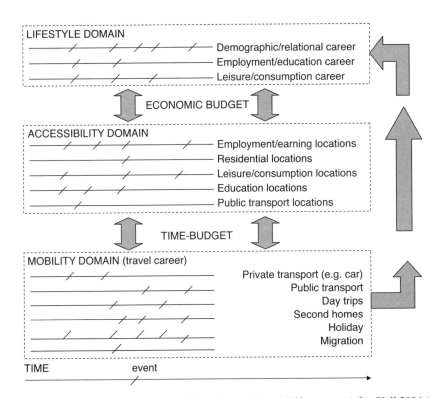

Figure 1.9 The construction of mobility biographies and life courses (after Hall 2004a).

- 'accessibility': referring to the places to which accessibility is required from where you live; and
- 'mobility': what is otherwise often referred to as travel career and identifies travel capacities and patterns.

All of these domains are constrained by an economic budget and a time budget, which refers to the time available to undertake travel.

According to Schafer (2000: 22): 'Aggregate travel behavior is determined largely by two budgets: the share of monetary expenditure and the amount of time that individuals allocate to transportation. However, neither budget is unique.' Given that a travel money budget represents the fraction of disposable income devoted to travel, a fixed travel money budget establishes a direct relationship between disposable income and distance travelled, provided average user costs of transport remain constant (Schafer and Victor 2000).

If people are on a fixed time budget then those that are willing to pay the increased costs will shift from one mode of transport to another so as to increase speed and, therefore, reduce the amount of time engaged in travelling relative to other activities within the constraints of the overall time budget (Hall 2005a). These concepts can be illustrated through the notion of a time–space prism with different potential path spaces depending on modes of transport used and the route of such transport (Figure 1.10).

As Hall (2005a) argued, for most people in societies around the world mobility is relatively constricted within such a prism as not only does one need money to

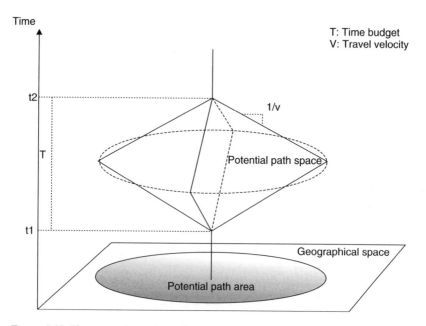

Figure 1.10 The space–time prism of mobility.

access transport to enhance mobility beyond that of walking, but one also needs access to time resources in order to undertake particular tasks. This is why one can argue that tourism, and the long-distance travel associated with international tourism in particular, is the domain of the wealthier members of our planet – because they are the ones who are time and money rich and can, therefore, engage in substantial leisure-related mobility.

The life-course approach is useful because it highlights the constraints that act on tourism and travel at particular points in time and over longer periods of time. For example, travel decision-making is often a joint process engaged in with partners, family and friends. Even if someone decides to travel alone, they will often talk about it with other people beforehand – yet many studies of travel decision-making seem to suggest that such decision-making occurs in isolation. Also such an approach focuses attention on constraints such as money and time budgets, as well as aspects such as culture, religion, race, language, physical capability, prior experience and issues of access. What constrains us in travel is, therefore, as important as what motivates us.

Photo 1.5 Families enjoy a boat ride on the canals of the historic city of Bruges, Belgium. Architectural heritage is carefully conserved, and sometimes carefully recreated, making it among the major attractions of many European cities. Recreational and leisure activities associated with heritage destinations are not only tourist attractions, but enhance the quality of life for local residents, as well. (Photo by Alan A. Lew)

Tourism products and services

As already noted, tourism is usually categorized as a service industry. This means that tourism services have some distinctive characteristics in terms of their consumption and production, as well as the nature of the product. Some of the commonly understood dimensions of services are that they are:

- *intangible*: they are experiences; although people can keep reminders of the experience through souvenirs and photographs;
- *inseparable*: production and consumption of tourism services (experiences) occur simultaneously;
- *variable* (also referred to as *inconsistent* or *heterogeneous*): because tourism businesses are geared towards the selling of experiences, they vary substantially from one service experience to another; and
- *perishable*: the 'product' cannot usually be stored from one day to the next; if a room has not been sold tonight, the opportunity for that sale is lost forever; experiences can be stored only in people's heads!

Pride and Ferrell (2003: 325) suggest that services also imply *client-based relationships* and *customer contacts* (see Table 1.2). However, substantial doubts have been expressed about the extent to which the characteristics of services can actually be generalized, especially given the growth of ICT and the so-called knowledge economy (Boden and Miles 2000; Lovelock and Gummesson 2004). Table 1.3 illustrates the applicability of the characteristics of services to tourism.

According to Hill (1999: 426) 'the distinction between goods and services has become erroneously and unnecessarily confused with quite a different one, namely that between tangible and intangible products'. Hill distinguished services as different from both tangible and intangible goods. Tangible goods are entities that exist independently of their owner and preserve their identity over time with their ownership being able to be exchanged and transferred. Intangible goods are described by Hill (1999) as intangible entities that have all the economic characteristics of tangible goods and were originally produced as outputs by persons or enterprises engaged in creative or innovative activities such as arts and entertainment.

In contrast, services are separate entities that do not exist independently of the direct interaction between consumers and producers, and they therefore cannot be stocked or have their ownership transferred. Hill (1999) regarded such characteristics as extremely important as they identified a 'fundamental divide' in the ways and means by which processes of production and distribution are organized. Instead, he argued that the essential characteristics of services are (1) that they cannot be produced without the agreement, cooperation and possibly active cooperation of the consumer, and (2) that the outputs are not separate entities that exist independently of the producers or consumers. Services are, therefore, co-produced or co-created (cf. Prahalad and Ramaswamy 2002, 2004; Prahalad and Hammond 2004; Vargo and Lusch 2004).

Table 1.2 Consumption, market, product and production characteristics of services

Service consumption	
Delivery of product	Consumption and production are coterminous in time and space, often requiring consumer or supplier to move to meet the other party.
Role of consumer	Services are consumer-intensive.
Organization of consumption	Often hard to separate production from consumption as they often occur simultaneously.
Service production	
Technology and plant	Low levels of capital equipment, heavy investment in buildings.
Labour	Some services are highly professional, often requiring high level of interpersonal skills; others relatively unskilled, with often a high degree of flexibility in terms of casual and part-time labour. Specialist knowledge may be important but rarely technological skills.
Organization of labour process	High degree of variability with some workers often engaged in craft-like production while others have high degree of management control of details of work.
Features of production	Production is often non-continuous and economies of scale are limited and often involve client-based relationships and customer contact.
Organization of industry	Some services are state-run public services, large multinational firms operate in trade in international services and domestic-based firms are often small scale with high preponderence of family businesses and self-employed.
Service product	
Nature of product	Immaterial, hard to store and transport with process and product hard to distinguish.
Features of product	Often customized to consumer and client requirements.
Intellectual property	Hard to protect and it is easy to copy many service innovations. Reputation and brand is often crucial.
Service markets	
Organization of markets	Some services are delivered via public sector provision. Some costs are invisibly bundled with goods, e.g. retail sector.
Regulation	Professional regulation and accreditation programmes in some services, government regulation of some service standards.
Marketing	Difficult to demonstrate product in advance.

Source: after Boden and Miles 2000; Pride and Ferrell 2003; Lovelock and Gummesson 2004; Hall 2005a, 2007a.

Table 1.3 The applicability of the characteristics of services to tourism

Characteristic	Service category involving			
	Physical acts with customers' bodies, e.g. lodging, passenger transport	*Physical acts with owned objects, e.g. freight transport, cleaning and laundry, food*	*Non-physical acts with customers' minds, e.g. entertainment, interpretation*	*Processing of information, e.g. Internet booking, travel insurance, tourism research*
Intangibility	Misleading, performance is ephemeral, but experience may be highly tangible	Misleading, performance is ephemeral but may physically transform object in tangible ways	Yes	Yes
Inseparability	Yes	No, customer usually absent during production	Only when performance is delivered live	Many exceptions, customers often absent during production
Variability	Yes, often hard to standardize because of direct labour and customer involvement	Numerous exceptions, can often be standardized	Numerous exceptions, can often be standardized	Numerous exceptions, can often be standardized
Perishability	Yes	Yes	Numerous exceptions, performance can often be stored in electronic or print form	Numerous exceptions, performance can often be stored in electronic or print form
Client-based and customer contact	Yes	No, customer usually absent during production	Numerous exceptions because of storage	Strong client orientation though numerous exceptions

Source: after Lovelock and Gummesson 2004; Hall 2005a, 2007a.

The interrelationship of consumption and production that exists in tourism services and the significance of co-creation or production further illustrate the difficulties of defining a single tourism industry, as discussed earlier in this chapter. However, just as significantly, it forces us to recognize that the notion of what comprises a tourism product is also complicated. Unlike most commodity products, tourism consists of multiple products.

The tourism product is an amalgam of tangible and intangible elements, as well as experiences, including physical resources, people, environments, infrastructure, materials, goods and services. Combined, they provide the tourist experience within a specific location. Hall (2005a) argues that most of the time a tourist is simultaneously consuming at least four embedded products, as illustrated in Figure 1.11. These include: (1) individual service encounters and experiences (this could

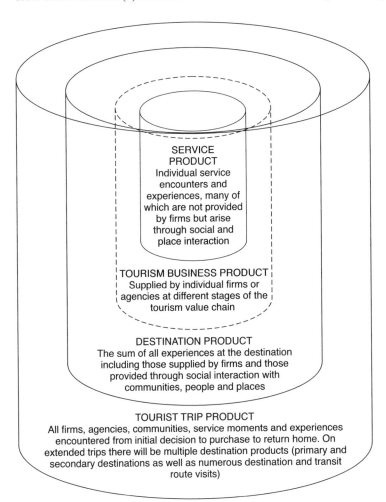

Figure 1.11 The multiple and embedded nature of tourism products.

be conceptualized as two products rather than one, but in a commercial sense they coexist through co-creation and co-production), (2) firm or organization-level products (also note that firms offer multiple product packages, each of which can be regarded as embedded within a business's overall product), (3) destination products, and (4) the total tourist trip product.

Hall (2007a) also illustrated that it is possible to divide the nature of the tourism product even further, given what we know about the nature of services (see Table 1.3 for the various categories). For example, it is possible to distinguish between the physical acts with owned objects (e.g. the purchase of food at a restaurant), the physical acts with customers' bodies (e.g. the physical quality of the restaurant in terms of layout, chair, tables, table setting) and the non-physical acts with customers' minds (e.g. the entertainment, atmosphere and the information regarding the menu and wine list from the waiter) – all of which can occur as simultaneous acts of consumption. Furthermore, they can also all happen while the credit card payment is being made (processing of information) (Hall 2007a).

Such a situation reinforces the understanding that when defining the impacts of tourism it is extremely important to know what it actually is we are seeking to

Photo 1.6 Tourist services in the Thamel District of Kathmandu, Nepal. Kathmandu has been a popular youth or backpack tourist destination since the 1960s. These alternative budget tourists, however, require the same types of travel services as mass tourists, though they may be able to find them at somewhat lower prices. (Photo by Alan A. Lew)

assess: is it individual encounters or experiences? tourism consumption or production? a firm or a location? a destination, the transit route or the entire trip? Furthermore, the conceptualization of co-creation again stresses that tourism impacts are relational, and as such the simultaneous consumption and production of tourism products – as well as the preparation for co-production by consumer and producers – also has impacts. Therefore, the scale of analysis is clearly significant in assessing impacts, a point which we will discuss in more detail in the next chapter. However, before we do, we will briefly address the final core concept, that of 'tourism resources.'

Tourism resources and attractions

A tourism resource is 'that component of the environment (physical or social) which either attracts the tourist and/or provides the infrastructure necessary for the tourist experience' (Hall 2007a: 34). From an economic perspective tourism resources can be categorized as scarce, private good (e.g., capital, labour and most land) or free, public good (e.g. climate, culture and most roads) (Bull 1994). Yet resources are an entirely subjective, relative and functional concept that are co-created and produced.

What actually constitutes a tourism resource from the perspective of the market depends on the motivations, desires and interests of consumers, and the cultural, social, economic and technological context within which those interests occur. Importantly, a tourism resource only exists when it provides, either directly or indirectly, the experiences being sought by the consumer. Producers do have a part in making the resource available, but it is the consumer who is the final arbiter of what constitutes a resource.

To repeat Zimmermann's (1951: 15) often-quoted seminal words on resources: 'Resources are not, they become; they are not static but expand and contract in response to human wants and human actions.' A tourism resource therefore becomes a resource only if it is seen as having utility value, and different cultures and nationalities can have different perceptions of the tourism value of the same commodity. What may be a resource in one culture may be 'neutral stuff' in another. 'Or in other words, what may be a tourist attraction in one culture or location may not be recognized as an attraction in another' (Cooper and Hall 2008: 118).

A tourist attraction is a specific type of tourism resource. 'An attraction is a resource that tourists are prepared to experience for a purpose other than to support their travel, e.g. provision of accommodation, transport or other hospitality services' (Cooper and Hall 2008: 118). In some cases an attraction may also serve to simultaneously support travel, such as a cruise ship or tourist canal boat, or, in some cases, historic hotels. Classification of attractions is sometimes determined along the lines of:

- cultural or built, e.g. townscapes, museums, monuments
- natural, e.g. wilderness areas and national parks.

Although such a division between natural and cultural attractions is extremely artificial, given that all tourism resources are inherently cultural. For example, a decision to set aside an area as a national park is as much of a reflection of cultural interpretation of the environment as it would be to turn the same area of land into a forest plantation or farm (see Chapter 5).

Destinations are also determined by the tourist because 'if people from outside a location do not visit a place it is not a destination' (Cooper and Hall 2008: 112). A destination is, therefore, a spatial or geographical concept that is primarily defined by visitors from outside the location although many places obviously seek to promote themselves as destinations to visitors. However, as will be discussed in Chapter 2, the perceptual destination of a visitor may not match administrative or jurisdictional boundaries, nor even a clear environmental boundary, therefore making it difficult to assess or manage the effects of tourism.

Case 1.4: Growth in Tibetan tourism and concerns over culture and politics

In 2007 just over 4 million tourists visited Tibet, the thinly populated Himalayan region of 2.8 million people. This represented a 64 per cent increase over the previous year, which was brought about by the opening of the Qinghai–Tibet Railroad in July 2006. The railroad is the world's highest rail line and is almost 2000 kilometres long with over half of the railway laid on permafrost soils. This case study demonstrates how tourism impacts and is impacted by the often shifting relationship between contestations over environmental conservation, cultural ownership, economic development and political power.

Fears have been raised about rising pressures on Tibet's attractions, infrastructure and the environment. However, tourism is integral to the Chinese government's policies for the economic development of Tibet and the integration of Tibet with China. Tourism accounts for about 9 per cent of Tibet's GDP and would generate about 4.8bn yuan (£340m) for 2007, up 73 per cent from 2006 (BBC News 2007a).

> There are concerns that the region's culture is being 'swamped' by China's majority Han population. The Dalai Lama, Tibet's exiled spiritual leader, has accused the Chinese government of committing 'cultural genocide', and warned that the fragile environment of Tibet is being put under threat, causing problems not just in Tibet, but in India, Bangladesh and China itself, which depend on the Himalayan plateaus for their water supplies.
>
> (Stanway 2008)

According to Matt Whitticase of the Free Tibet Campaign 'This is the first time that the number of tourist arrivals exceeded the total population.. . . Tourism is obviously a pillar of China's western development strategy but it is putting unacceptable strains on Tibet's fragile environment' (quoted in Stanway 2008).

The Chinese government claims that tourist growth will help generate the income necessary to protect Tibet's ancient monuments and culture. This agenda has been supported by the designation of three sites in Tibet as UNESCO World Heritage Sites. These include the Potala Palace, Jokhang Temple and Norbulinka Temple. But Whitticase claims that the Tibetans themselves were not benefiting. 'Tibetans are being left behind and the tourist industry is being run by Han Chinese companies not domiciled in Tibet' (quoted in Stanway 2008). Furthermore, while the World Heritage Site designation recognizes these as unique symbols of Tibetan culture, the requirements for such a designation is managed through the nation of China, which also defines them in terms of Chinese nationalism.

Despite the large influx of Chinese workers and business people, Tibetan culture remains strong and monasteries have experienced a revival as China overall has gradually opened its economy since the 1980s. Social change is particularly noticeable among younger Tibetans, who are more closely tied to both the national (Chinese) economy and the global economy than are the older generations.

However, tensions between Chinese newcomers and Tibetan old-timers exist due to both cultural differences and economic concerns. These were brought into sharp focus by pro-Tibetan protestors in Tibet and worldwide during the Olympic torch relay prior to the 2008 Beijing Summer Olympics. As a result of anti-Chinese protests in Tibet in March 2008 as well as Tibetan supporters using the torch relay as a means to gain media attention for the Tibetan cause, the Chinese government closed Tibet to both domestic and international tourist travel for almost half a year, resulting in a dramatic drop in visitor arrivals in 2008 (Ang 2008), and major economic hardships for the province's tourism economy.

Sources: Free Tibet Campaign: www.freetibet.org/

International Campaign for Tibet: http://www.savetibet.org/

PRC China Tibet Information Center: http://info.tibet.cn/en/

Summary and conclusions

This chapter has identified some of the key concepts of tourism in order to better understand the various ways in which tourism may be said to have impacts.

It has stressed that definitions of tourism tend to share several common elements:

- tourism is the temporary, short-term travel of people (non-residents) along transit routes to and from a destination that is outside of their normal home environment;
- tourism can have a wide variety of impacts on the destination, the transit route and the source points of tourists;
- tourism can influence the attitudes and behaviours of the tourist as well as the people that provide tourism experiences; and
- tourism is voluntary and therefore encompasses not only travel for leisure but also other aspects of mobility such as short-term travel for work, working

holidays, travel to second homes, gap years, and travel in relation to health, education and visiting friends and relations.

As will be seen in forthcoming chapters this broad approach to tourism is important as places are often seeking to attract mobile people both for the short and long term (Hall 2005a). The chapter has also provided an account of the way terms such as *international tourism* and *domestic tourism*, and *trip* and *visit* are used in a technical sense (UN and UNWTO 2007), as well as discussing the concepts of *tourism industry*, *tourism services*, *tourism resources* and *tourist attractions*.

We hope that one of the most important insights gained from this chapter is that tourism is an extremely complex subject and does not accurately reflect the popular understandings of, and attitudes towards, tourism by people who travel – 'if I can travel then surely it cannot be that difficult to understand!' Indeed, as the next chapter stresses, the inherently complex nature of tourism creates significant challenges for understanding and assessing its impacts.

Chapter 2 provides a discussion of what constitutes impacts and the role of impact assessment. It also illustrates the importance of scale in assessing and managing impacts and provides an integrated framework to help understand such tasks. Chapters 3, 4 and 5 provide overviews of the economic, cultural and physical dimensions of tourism-related impacts and associated change. Each chapter provides an account of such change with respect to different scales of analysis, ranging from the global to the local and, in some cases, to individuals. As noted in the next chapter, while such divisions are in one sense artificial, these standard categories and scales provide a useful means to gain a sense of the range of tourism-related impacts.

Chapter 6 offers a response to the challenges raised in the earlier chapters with respect to tourism-related impacts and change and discusses the role of planning in providing an integrated approach to tourism impacts. It identifies issues associated with destination management, as well as issues of tourism planning in different jurisdictions and environments. The final chapter discusses some of the emerging issues with respect to tourism-related change in the future, as well as the limits of management and planning approaches.

This book aims to provide an introduction to the field of tourism impacts and change, and how they might be assessed, managed and ameliorated. The field is so big that this volume cannot pretend to be able to cover all the changes that tourism has been related to over the many years that tourism has existed, nor can it chart all of the responses to those changes. However, it does seek to provide the reader with a foundation in the main themes and issues with respect to tourism and change so as to enable students of tourism to better assist industry, destinations and planners to respond to the challenges that tourism provides.

Self-review questions

1   What are the four ways in which tourism is categorized with respect to international trade in services?
2   What is the importance of differentiating between academic and lay definitions of tourism?
3   How do time and space influence our understanding and definitions of tourism-related mobility?
4   The UN and UNWTO (2007) have recommended use of the term *visitor* rather than *tourist*. What might be the potential implications for how tourism is conceptualized? For example, should we be referring to the 'visitor industry' rather than tourism industry, and visitor studies rather than tourism studies?

Recommended further reading

Mathieson, A. and Wall, G. (1982) *Tourism: Economic, Physical and Social Impacts.* Harlow: Longman Scientific and Technical.
Arguably the seminal work with respect to tourism impacts. The growth of the field can be demonstrated by noting that this book sought to provide a comprehensive overview of all the tourism impact literature available at the time and was based on a Master's thesis. Such a task would be almost impossible to undertake now and would require many volumes.

Leiper, N. (1983) 'An etymology of "tourism"', *Annals of Tourism Research*, 10: 277–81. Early work that provides a good account of the historical development of the tourism concept.

Shaw, G. and Williams, A.M. (2002) *Critical Issues in Tourism: A Geographical Perspective*, 2nd edn, Oxford: Blackwell.
One of the leading textbooks on tourism by geographers.

Lew, A.A., Hall, C.M. and Williams, A.M. (eds) (2004) *A Companion to Tourism*, Oxford: Blackwell.
A substantial volume with over 50 review chapters on various dimensions of contemporary tourism research.

Hall, C.M. (2005) *Tourism: Rethinking the Social Science of Mobility*, Harlow: Prentice-Hall.
Provides a comprehensive account of tourism from a mobilities perspective with considerable emphasis on some of the spatial dimensions.

Hall, C.M. and Page, S. (2006) *The Geography of Tourism and Recreation*, 3rd edn, London: Routledge.
One of the leading texts on the geography of tourism that also includes discussion of the relationship to recreation and leisure.

Coles, T.E. and Hall, C.M. (eds) (2008) *International Business and Tourism: Global Issues, Contemporary Interactions*, London: Routledge.
Provides a comprehensive account of the various ways in which tourism is part of international business and trade in services.

Cooper, C. and Hall, C.M. (2008) *Contemporary Tourism: An International Approach*, Oxford: Butterworth Heinemann.

Provides a service marketing-oriented understanding of contemporary tourism.

Web resources

United Nations World Tourism Organization (UNWTO): http://www.unwto.org/
Website of the leading intergovernmental body with respect to tourism. Contains information on statistics as well as on significant issues facing the global tourism industry.

World Travel and Tourism Council (WTTC): http://wttc.org/
The WTTC is the leading international organization for travel and tourism businesses, providing research and educational outreach, and advocating for their interests.

Adjectival Tourism and Tourism Issues: http://hubpages.com/hub/Adjectival-Tourism
This website lists the wide variety of niche and special interest travel and tourism, most with some definitions, along with contemporary academic research interest in tourism.

Key concepts and definitions

Domestic tourism Includes the activities of resident visitors within their home country or economy of reference, either as part of a domestic or an international trip.

Home environment (also called) *usual environment* Refers to the geographical (spatial or jurisdictional) boundaries within which an individual routinely moves in their regular daily life.

Inbound tourism Includes the activities of non-resident visitors within the destination country or economy of reference, either as part of a domestic or an international trip (from the perspective of the traveller's country of residence).

Internal tourism Comprises both domestic tourism and international inbound tourism, including the activities of resident and non-resident visitors within the economy of reference as part of a domestic or an international trip; this is all of the tourism expenditures in a country or economy.

International tourism Comprises inbound tourism and outbound tourism, including the activities of resident visitors outside the economy of reference, either as part of a domestic or an international trip and the activities of non-resident visitors within the economy of reference as part of a domestic or an international trip (from the perspective of their country of residence); this is all tourist expenditures that are made by tourists outside of their home country or economy.

National tourism Comprises both domestic tourism and international outbound tourism, including the activities of resident visitors within and outside the economy of reference, either as part of a domestic or an international trip; this is all of the tourism expenditures of domestic tourists both inside and outside their home country or economy.

Natural person A human perceptible through the senses and subject to physical laws, in contrast to an artificial or juridical person, such as a corporation or an organization that the law treats for some purposes as if it were a person distinct from its owner or members.

Outbound tourism Includes the activities of resident visitors outside their home country or economy of reference, either as part of a domestic or an international trip.

Services Commodities that cannot be produced without the agreement, cooperation and possibly active participation of the consumer. The outputs are not separate entities that exist independently of the producers or consumers.

Tourism A form of voluntary human mobility associated with the temporary movement of persons from their usual home environment and subsequent return. Usual examples include: leisure holiday or vacation, visiting friends and relations (VFR), business travel, travel to second homes, health- and medical-related travel, education-related travel, religious travel and pilgrimage, travel for shopping and retail, volunteer tourism. A tourist is therefore someone who engages in tourism mobility.

Tourism firm Any firm (business) that produces a commodity (product or service) that is consumed by tourists and which recognizes tourists as a market.

Tourism impact The change in the state of something related to tourism over time.

Tourism industry The aggregate of all businesses that produce a commodity (good or service) that is consumed by tourists.

Tourism resource That component of the environment (physical or social) which either attracts the tourist and/or provides the infrastructure necessary for the tourist experience.

Tourism trip Generally accepted under international agreements on collection of technical tourism and travel statistics to refer to a trip of not more than 12 months and for a main purpose other than being employed at the destination.

Tourist attraction A specific type of tourism resource that tourists are prepared to experience for a purpose other than to support their travel, e.g. provision of accommodation, transport or other hospitality services.

Trip The movement of an individual outside their home environment until they return.

Visit Refers to the stay (overnight or same-day) in a place away from home during a trip.

2 Understanding impacts

Learning objectives

By the end of this chapter, students will:

- Be able to define the concept of impact assessment.
- Understand the objectives of impact assessment.
- Understand the development of impact studies in relation to ideas such as conservation and sustainability.
- Understand the elements of a tourism system.

This second chapter provides a greater understanding of impacts, their nature and their assessment. It first provides an introduction to the development of impact studies, followed by a discussion of impacts and their assessment. It then goes on to examine the tourism system concept and its relevance to understanding impacts and change.

The development of impact studies and management

Impact studies are a relatively recent phenomenon. Although the undesirable effects of economic and industrial development have been noted for several hundreds of years, they have often been taken for granted as part of the 'price' of development (Goudie 2005). Impact studies emerged in the nineteenth century in response to the industrial revolution and the rapid growth in population that accompanied it. The industrial revolution began in England in the late 1700s, and expanded across Europe and to North America in the early to mid-1800s. In the United States, for example, railroad mileage grew from 23 miles in 1830 to 2,818 miles in 1840, and as a result of industrial revolution-driven European migration, the US population grew 17-fold from 1800 to 1900. That century of economic revolution and social turmoil brought environmental degradation and urban squalor in its wake and prompted the first major impact management responses as many cities had undergone rapid urbanization as a result of the new work opportunities created by industrialization. To make them more liveable, cities had to build new roads, public parks, public transportation systems, water supply and sewerage services, and gradually implement controls on polluting industries, including zoning land areas for specific industrial uses away from higher value residential and amenity areas (Brimblecombe 1987; Hassan 2008). These developments helped to lay the intellectual foundations for present-day urban and regional planning (Hall 2002), which has also proven to be hugely influential on tourism planning (Hall 2008c).

It was the publication in 1864 of George Perkins Marsh's book *Man and Nature; Or, Physical Geography as Modified by Human Action* (1965) that provides a starting point for impact studies. The book had an enormous impact when it was first published, as it was the first comprehensive critique of the extent to which inappropriate development had damaged the physical environment and hence human well-being. According to Marsh, 'Even now, we are breaking up the floor and wainscoting and doors and window frames of our dwelling, for fuel to warm our bodies and seethe our pottage, and the world cannot afford to wait till the slow and sure progress of exact science has taught it a better economy' (March 1965 [1864]: 52). More specifically, he stated:

> The earth is fast becoming an unfit home for its noblest inhabitant, and another era of equal human crime and human improvidence, and of like duration with that through which traces of that crime and that improvidence extend, would reduce it to such a condition of that impoverished productiveness, of shattered surface, of climatic excess, as to threaten the depravation, barbarism, and perhaps even extinction of the [human] species.
>
> (Marsh 1965 [1864]: 43)

Such words strike a distinct chord when compared to contemporary concern over biodiversity loss, deforestation, declining quality of agricultural lands, desertification and climate change. These are all issues, with the exception of climate change, that Marsh himself addressed during the industrial revolution (although even then observations with respect to change were being made at a microclimatic level).

Marsh's work had an international impact, leading to a growing awareness of the limits of resources. He focused on natural resources such as biodiversity, forests, water and dune systems, and the value of science and evaluation in understanding and managing such change. His writing, along with the closing of the American frontier in 1890, led to the development of the 'progressive conservation' movement (Hays 1957, 1959). The progressive conservation movement represented a 'wise use' approach to the management of natural resources, and its conservation motives were economic rather than aesthetic in intent; in the United States it led to the establishment of the Bureau of Reclamation, the National Park Service and the United States Forest Service. The latter two organizations in particular have a very strong influence on how tourism is managed in natural areas and the way natural areas are promoted and experienced as tourist attractions.

The progressive conservation movement can be contrasted with the 'Romantic ecology' (Worster 1977) of John Muir, the 'grandfather of National Parks', which stressed the spiritual values of wilderness and can be broadly categorized as 'preservation'. The creation of the US Forest Service shed significant insights into this relationship.

Although the US Forest Service was not founded until 1905, momentum for its creation had been building in the two prior decades. Several bills relating to timber on public land had been introduced into the US Congress from the 1870s onwards, but in 1891 the president was given the power to set aside areas of the public domain as forest reserves (Clarke and McCool 1985). This protected the land from homesteading, and thus privatization. Both preservationists and progressive conservationists saw the *Forests Reserves Act* of 1891 as a means to protect wilderness areas, most of which had limited agricultural value for homesteaders anyway.

Preservationists led by John Muir wanted wilderness to contain no human activity that would be unsympathetic to the primitive nature of a wilderness area. However, progressive conservationists led by noted forester Gifford Pinchot and supported by Theodore Roosevelt, who would become the twenty-sixth President of the United States, wanted forest lands to be managed on a sustained yield basis. They were therefore in favour of timber harvesting, the building of dams for water supplies, and selective mining and grazing, all in the name of conservation. In a statement which recalls much of current debates over sustainability, Gifford Pinchot stated in 1910 that

> The first great fact about conservation is that it stands for development. There has been a fundamental misconception that conservation means nothing but the husbanding of resources for future generations. There could be no more serious mistake. Conservation does mean provision for the future, but it means also and first of all the recognition of the right of the present generation to the fullest necessary use of all the resources with which this country is so abundantly blessed. Conservation demands the welfare of the country first, and afterward the welfare of the generations to follow.
>
> (Pinchot 1968 [1910]: 9)

Initially, there was a reasonable degree of correspondence in the views of romantic ecologists and economic or progressive conservationists. Muir, for example, wrote in 1895 that 'it is impossible in the nature of things to stop at preservation. The forests must be, and will be, not only preserved, but used, and . . . like perennial fountains . . . be made to yield a sure harvest of timber, while at the same time all their far-reaching [aesthetic and spiritual] uses may be maintained unimpaired' (in Nash 1967: 133–5). However, over time a split occurred between the parties as to how conservation reserves should be managed.

Pinchot and the progressive conservationists advocated the 'wise use' of natural resources, while the preservationists continued to focus on the aesthetic and spiritual qualities of forest wilderness. As Fernow (1896 in Nash 1967: 137) wrote in *The Forester*, 'the main service, the principal object of the forest has nothing to do with beauty or pleasure. It is not, except incidentally, an object of esthetics, but an object of economics'. Such a viewpoint was anathema to the preservationists. Muir believed that 'government protection should be thrown around every wild grove and forest on the mountains' in order to preserve the 'higher' uses of wilderness (in Nash 1963: 9). The problem that faced Muir, and that exists to this day, is that the existence of 'undisturbed' wild nature is incompatible with plantation forest management. As with many conservationists to the current day, Muir regarded tourism as a less evil form of economic development than grazing or commercial clear-cutting of forests (Mark and Hall 2009). In the 1870s his writing suggests 'that a growing tourist business might drive the more exploitative users' (Cohen 1984: 206) and especially sheep interests, out of the Sierra Nevada mountains, and Yosemite in particular. Accordingly, 'Despite his suspicion that the path of moderation was not the best way to a true vision of Nature, he attempted to write moderate articles which would bring urban tourists' (Cohen 1984: 206).

Indeed, Cohen goes on to argue that 'in a sense, all of Muir's writings were for the tourist, since they involved the question of how to see. Most tourists did not want to hear philosophy, but wanted to know exactly where to stop and look' (1984: 207). In some of Muir's writing this is extremely clear. For example, in John Muir's *The Yosemite*, originally published in 1914, chapter 12 is entitled 'How Best to Spend One's Yosemite Time' providing instructions for two one-day excursions, two two-day excursions, a three-day excursion and longer routes. With the Upper Tuolumne excursion being 'the grandest of all the Yosemite excursions, one that requires at least two to three weeks' (Muir 1914: 155). These excursions, along with other advice on visiting the park, are used by travellers to the park to the present day.

The passing of the US *Forest Management Act* in 1897 and the subsequent creation of the Forest Service in 1905, with Pinchot at its head, marked the institutionalization of progressive conservation in the US government (Richardson 1962). Government forestry, and wider involvement in the management of natural resources, in the United States was founded upon Marsh's and Pinchot's vision of academic forestry – 'that is, the scientific management of the timber resource according to the principles of wise use and sustained yield' (in Clarke and McCool 1985: 36). However, as a reader can immediately see, such ideas have strongly influenced not only forestry practices but also the broader field of resource management with respect to the concept of environmental usage and sustainability, including tourism.

Similar trends with respect to 'wise use' were also occurring in urban areas in the late 1800s and early 1900s. Efforts to improve the quality of life in the newly industrialized cities began in England with the efforts of social reformers and unions to improve working conditions so as to include provision of time for leisure. As early as 1817 Robert Owen had coined the slogan used by the campaign for an 8-hour working day: 'Eight hours labour, Eight hours recreation, Eight hours rest'. The growth of the middle classes and increased leisure time allowed for the first developments of mass tourism, as well as improved urban park and recreation planning. Significant amongst this change was the Garden City Movement, which was primarily associated with Sir Ebenezer Howard. The goals of this movement were to bring nature to the city through the creation of green belts and urban parks in older cities, and the creation of new towns that included these elements. London's green belt, New York City's Central Park and Vancouver's Stanley Park all resulted from the Garden City Movement. Central Park was designed by Frederick Law Olmstead, who also helped draft the Yosemite National Park bill and was that park's first supervisor under President Lincoln.

However, greening the city proved to have limited impacts on addressing urban ills – although it continues to influence urban design up to the present day. By the time of the Chicago World's Fair in 1893, the slums of the industrial cities of the northeastern United States were among the most densely crowded in the world and epidemics of typhoid, cholera and yellow fever were common. Poor housing conditions were cited as the primary cause of these social ills and 'housing crusades' to address the worst problems started to develop in the last part of the nineteenth century. The first laws to regulate housing in the United States were passed by New York City in 1860 to require light and air access to the interior of buildings, but more significant regulations and the birth of city planning did not occur until the early 1900s.

In 1906 the *Antiquities Act* was passed, which was the first US Federal law to protect archaeological sites, allowing for the establishment of National Monument areas from public domain lands that contained 'historic landmarks, historic and prehistoric structures, and objects of historic or scientific interest'. The US National Park Service was established in 1916, and the first urban historic preservation commission, the Vieux Carre Commission, was established in New Orleans in 1921. Such organizations have served as models throughout the world with respect to attempts to combine use, and tourism in particular, and conservation.

The Vieux Carre Commission evolved from the City Beautiful Movement, which grew out of the Garden City Movement in the later 1800s. This approach to urban problems focused on 'civic art' in the form of monumental structures, large open plazas with fountains and wide boulevards. The approach was based on European urban design and public art symbols and is probably best seen in Washington, DC, which was redesigned in the early 1900s by Daniel Burnham, considered the founder of American city planning, as well as cities such as Canberra in Australia.

Like the earlier Garden City Movement, the City Beautiful Movement had relatively little impact on the actual health and safety of city residents. What it did do, however, was to get people to develop city planning documents for the first time, though they initially focused simply on the physical design and layout of cities. These plans gradually expanded in the early 1900s to include health and safety controls, and in 1926 the US Supreme Court voted for the first time to support a zoning ordinance from Euclid, Ohio, based on the grounds that the forced separation of industrial, residential and retail land uses furthered the health, safety and general welfare of the community. Such zoning approaches have since been applied in planning legislation throughout the world and is the most common method to separate 'desirable' and 'undesirable' land uses.

Photo 2.1 The sea wall, Stanley Park, Vancouver, with North Vancouver in the background. In 1886 Vancouver's city council petitioned the dominion government who used the land for military purposes to lease the reserve for use as a public park. On 27 September 1888 the park was officially opened, named after Lord Stanley, then Canada's Governor General. Although the park has approximately 2.5 million users a year it is a Canadian National Historic Site with most of it remaining forested. Construction of the 8.8-km (5.5-mile) sea wall around the park commenced in 1917 but was not finished until 1971. Stanley Park's design and planning was heavily influenced by urban parks in the UK and the USA, especially Central Park. (Photo by C. Michael Hall)

Zoning ordinances, following the 'Euclidian' zoning model, were quickly adopted throughout the United States following the Supreme Court ruling, as were college degree programmes in city planning. The Great Depression, starting in 1929, caused massive unemployment throughout the whole world and pushed city planners to change their emphasis to economic and social planning, with a focus on employment, housing and transportation issues, into the 1950s.

The post-Second World War era brought increases in population, urbanization and prosperity in the United States. The guiding philosophy in urban planning at the time was what is sometimes called the *City Efficient* movement. Like the 'wise use' approach in natural resource management, it assumed that scientific rationality would produce the best social outcomes. Cities developed master plans that eliminated old, dilapidated and poverty-stricken neighbourhoods by condemning them through their eminent domain powers. This legal process forced building and home owners to sell their properties, whether they wanted to or not. Cities would then level the land and sell it as a whole to a private developer or use it for public purposes, such as the new Interstate Highway system that started in the 1950s. This was known as *urban renewal*, and it resulted in large numbers of people being displaced out of their older, inner-city homes and moving to 'cookie cutter' suburban developments.

A large-scale reaction against urban renewal and the so-called 'bulldozer' ethic occurred in the 1960s, when some planners turned to social advocacy and transactive planning approaches that more explicitly recognized the political complexity of social change (Lew 2007b). Advocacy planning actively addresses the needs of traditionally under-represented groups in the government decision-making process, while transactive planning works from the assumption that local residents have equal knowledge and an equal role with development professionals in the planning process. These approaches were closely tied to the civil rights

Photo 2.2 The old and new parliament buildings of the Commonwealth of Australia viewed along ANZAC Parade from the Australian War Memorial, Canberra. Designed by American architect Walter Burley Griffin, the capital of Australia is a planned city that incorporates many aspects of the City Beautiful Movement. The city's geography not only reflects its capital status but the provision of substantial public parkland, wide boulevards and emphasis on access to the natural environment also indicate the debt owed to the planning trends of the early twentieth century, which greatly emphasized parks and recreation opportunities. (Photo by C. Michael Hall)

movements of the 1960s, and still have significant roles to play in community-based tourism and pro-poor tourism (PPT), especially in less developed economies.

Another response against urban renewal came in the rise of historic preservation efforts. Although historic preservation efforts began in the United States in the immediate post-Second World War years, it was not until the 1960s that national legislation to support building preservation prompted a nationwide movement. Similar efforts became widespread in the 1960s in Canada and other 'New World' countries, such as Australia and New Zealand. Some of these were efforts to preserve the architectural creations of the earlier Garden City and City Beautiful movements. The tourism value of these historic districts was soon recognized, and tourism-based urban design approaches became commonplace in the 1970s (Lew 1989) (see Case 2.1).

The US Congress established the National Trust for Historic Preservation in 1949 as a non-profit/non-governmental entity that monitors, educates and advocates for historic properties and districts. In addition to identifying and registering historic sites and structures, its National Main Street Program works to revitalize older retail districts and has been a key player in reshaping the inner urban landscape of many American cities. In many instances, this has turned an older retail district into an historic-themed entertainment centre.

Case 2.1: From decay to eatertainment: the example of downtown Flagstaff, Arizona

The Interstate Highway system was introduced in the United States in the mid-1950s and expanded across the country in the 1960s and early 1970s. The new 'freeways' typically bypassed smaller cities and towns that were reliant on the older highways that brought travellers through their city centres. The result devastated the economies of many communities, some of which were never able to recover. Older downtowns, in particular, saw the movement of major retailers out of the city to freeway-based shopping malls that were more accessible to suburban housing projects (Lew 1989).

Downtown Flagstaff, Arizona, experienced this, being impacted from being bypassed by the Interstate freeway and competition from a suburban shopping mall, both of which were built in the 1970s. In the mid-1980s, the last large department store closed in downtown Flagstaff, leaving behind a large, empty and ugly structure in the middle of this community of 45,000 (1980s population). Many other shops were empty or sold second-hand (used) clothing, though there were a few well-established restaurants, along with tourist-oriented gift shops on the old Route 66 highway.

In 1987 the city government, working with local business people, funded the establishment of a National Trust for Historic Preservation Main Street Program. This involved hiring a Main Street coordinator who treated the district as a shopping mall manager – organizing events, finding the best mix of retail establishments, and coordinating building designs and upgrades.

Photo 2.3   Classic cars at the Route 66 Festival in downtown Flagstaff, Arizona.
Restaurants, book stores and speciality arts stores line this street, which leads
to the old US Route 66 in the distance. The streets are closed to vehicle traffic,
as they often are on summer weekends for special events that draw tourists
from around the state and region. (Photo by Alan A. Lew)

The empty department store was demolished, and underneath it was found the
original structure that the store eventually evolved from – the Babbit Brothers
Trading Store, which was preserved with its red brick and original signage still
visible. A new public plaza was also built on the site, along with a new building
designed to fit into the architectural style of the district. Sidewalks were widened,
new street furniture (benches, lamp posts and bicycle racks) was purchased, and
all of the businesses adopted uniform sign and awning designs.

A decade later, all of the retail space in downtown Flagstaff was full with
restaurants, outdoor outfitter stores (fitting with the town's mountain setting), art
galleries and speciality shops. Activities close the streets of downtown Flagstaff
throughout the year, but especially in the summer months when some special
event takes place every weekend, from arts festivals to classic car shows. The art
galleries host a popular Friday night art walk though the summer, and free movies
and concerts in the public plaza draw locals and tourists alike to the downtown.
The Main Street Program was dropped in the late 1990s, as it was no longer
deemed necessary.

The story of downtown Flagstaff's transformation, from a declining retail district
unable to compete with shopping malls and warehouse department stores, to a
major shopping, eating and entertainment centre offering products and experiences

not found in modern exurban shopping venues, has been repeated in many other communities in the United States where architectural resources combine with potential tourist markets. Not all communities are as successful as Flagstaff. For some communities, issues of historic authenticity and economic imperatives can politicize and complicate their development efforts. For other communities, their location or building resources create challenges to successful redevelopment. The National Trust's Main Street Program continues to be important in working with these communities to find the best mix of conservation and economic development.

This case study shows how public and private development projects brought about social change and decline in what was once the core retail centre of a city. However, combined public and private planning efforts, along with broader social changes (an increasing interest in historic districts), were ultimately successful in addressing the problems of an earlier era. The example of Flagstaff also shows how some of the best tourist attractions and tourism districts are also very popular among local residents.

See also: Thomas W. Paradis, 'Theming, tourism, and fantasy city'. In Lew et al. (2004: 195–209).

_____

Environmental impact legislation

The 1960s were also a transformative decade in the management of natural resources, including forests and other wildlands. Broader public concern over the impacts of natural resource management philosophies came about in 1962 with the publication of Rachel Carson's *Silent Spring* and a growing awareness of the environmental effects of industrial pollution. *Silent Spring* was important as it transformed the environmental debate from one that was primarily based in peripheral areas (national forest areas and national parks) to one that was urban based, where the vast majority of people lived, and also provided renewed impetus to the historic preservation movement in the United States as well as elsewhere in the developed world.

It was, therefore, primarily a result of urban, and to a lesser extent, rural environmental concerns in the 1960s that led to the passage of the first environmental impact legislation and the creation of national environmental protection agencies. For example, in the United States the *National Environmental Policy Act* (NEPA) was enacted in 1969, which made environmental impact statements required for any 'major Federal action significantly affecting the quality of the human environment'. In 1970 the US Environmental Protection Agency (EPA) was established as a result of a reorganization of previous government efforts with respect to the environment.

The American initiatives had international implications with legislation and organizations being established in Australia, Canada and Europe in the 1970s, with respect to environmental protection and assessment. Examples of the introduction of formal *environmental impact assessment* (EIA) requirements include Japan (1972), Hong Kong (1972), Canada (1973), Australia (1974), Germany (1975),

France (1976), the Philippines (1977), Taiwan (1979) and the People's Republic of China (1979). The European Community made EIA requirements applicable to all EU member states in 1988 (Gilpin 1995).

In one sense EIAs have been revolutionary with respect to solving impact issues. Russell Train, former chairman of the US Council for Environmental Quality and administrator of the Environmental Protection Agency stated:

> I can think of no other initiative in our history that has had such a broad outreach, that has cut across so many functions of government, and that has had such a fundamental impact on the way government does business? I am qualified to characterise that process as truly a revolution in government policy and decision making.
>
> (cited in Bartlett and Kurian 1999: 416)

However, concepts of appropriate use of resources were further transformed in the 1980s with the *World Conservation Strategy* (WCS) and the development of the concept of *sustainable development*. The WCS (1980) was prepared by the International Union for the Conservation of Nature and Natural Resources (IUCN) with the assistance of the United Nations Environment Programme (UNEP), the World Wildlife Fund/Worldwide Fund for Nature (WWF), the Food and Agricultural Organization of the United Nations (FAO) and the United Nations Educational, Scientific and Cultural Organization (UNESCO). The WCS is a strategy for the conservation of the Earth's living resources in the face of major international environmental problems, such as deforestation, desertification, ecosystem degradation and destruction, extinction of species and loss of genetic diversity, loss of cropland, pollution and soil erosion.

The WCS defined *conservation* as 'the management of human use of the biosphere so that it may yield the greatest sustainable benefit to present generations while maintaining its potential to meet the needs and aspirations of future generations' (IUCN 1980: s.1.6). The WCS had three specific objectives (IUCN 1980: s.1.7):

- to maintain essential ecological processes and life-support systems (such as soil regeneration and protection, the recycling of nutrients, and the cleansing of waters), on which human survival and development depend;
- to preserve genetic diversity (the range of genetic material found in the world's organisms), on which depend the breeding programmes necessary for the protection and improvement of cultivated plants and domesticated animals, as well as much scientific advance, technical innovation, and the security of the many industries that use living resources;
- to ensure the sustainable utilization of species and ecosystems (notably fish and other wildlife, forest and grazing lands), which support millions of rural communities as well as major industries.

The notion of sustainable development espoused in the WCS emphasized the relationship between economic development and the conservation and sustenance of natural resources. In many ways there was nothing new in this idea as it had been at the core of much of the conservation debate for many years (see also Chapter 5). However, what was significant was the manner in which the report highlighted the global nature of environmental problems, emphasized the significance of the environmental–economic development relationship, placed these issues in the context of the relationship between developed and less developed countries (the north–south debate), and provided a basis for some government and private sector response, albeit limited, to the problems and issues identified in the report.

The Brundtland Report and sustainable development

In a review of the 1972 United Nations Stockholm Conference on the Human Environment, UNEP recommended the creation of a World Commission on Environment and Development (WCED). In 1983 the Commission was created as an independent body reporting directly to the United Nations Assembly, and Gro Harlem Brundtland, then parliamentary leader of the Norwegian Labour Party, was appointed as its chair. Although the term *sustainability* was used by Brown (1981), Myers (Myers and Gaia Ltd staff 1984) and Clark and Munn (1986), it was not until the publication of the report of the WCED in 1987, *Our Common Future* (commonly referred to as the Brundtland Report) that *sustainable development* entered the public imagination. As we have already noted, its roots had actually been around for well over a hundred years.

According to the WCED (1987: 43), 'sustainable development' is development that 'meets the needs of the present without compromising the ability of future generations to meet their own needs'. Five basic principles of sustainability were identified:

- the idea of holistic planning and strategy-making;
- the importance of preserving essential ecological processes;
- the need to protect both human heritage and biodiversity;
- the effort to develop in such a way that productivity can be sustained over the long term for future generations; and
- the goal of achieving a better balance of fairness and opportunity between nations.

The report has been incredibly influential with respect to global resource management and development, as well as with tourism – although it should be noted that tourism was hardly mentioned at all in the original report. In fact concern over the negative impacts of tourism only emerged as a significant issue in developed countries in the 1970s and 1980s (Hall and Page 2006). Nevertheless, there is now a wealth of literature on sustainable tourism development, and concern over sustainable development aspects of tourism, usually referred to as *sustainable tourism*, is a major focus of impact research and destination management.

According to Hall (2008c: 27) '*Sustainable tourism* is a sub-set of both tourism and sustainable development'. Sustainable tourism development is not the same as sustainable development, although the principles of sustainable development do inform sustainable tourism (Figure 2.1). The key difference between the two concepts is one of scale. Sustainable tourism only refers to the application of sustainability concepts at the level of the tourism industry and related social, environmental and economic change (Figure 2.2). Sustainable development, on the other hand, operates at a broader scale that incorporates all aspects of human interaction with the Earth's environment. The implications of this scale difference are important because, for example, a tourism enterprise's operations may meet the criteria of being sustainable at the business level, but in a community context it may cause the community as a whole to be rendered unsustainable as a result of other development options not being able to be pursued (Hall 2008c). And the greenhouse gas consequences of long-haul travel, as discussed in Case 1.3, brings into question the sustainability of the entire tourism phenomenon.

The attention to the concept of sustainable development is also important because the understanding of impacts and their assessment has shifted to take into account concepts of sustainability. Although not written into the original environmental review requirements, sustainable development concepts guide much of the application of EIA techniques, especially when they involve social impact analysis (see Chapter 6).

Impacts and their assessment

As noted in Chapter 1, an *impact* is a change in a given state over time as the result of an external stimulus. This is often considered in relation to specific environmental, economic or social impacts. However, increasingly, and prompted by the insights of sustainable development, impacts are being approached in a combined fashion, taking into account the interrelationship of two, or even all

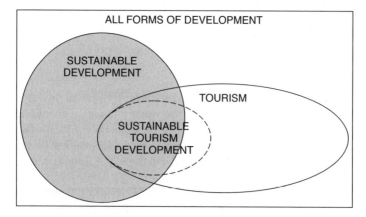

Figure 2.1 Sustainable tourism and sustainable development.

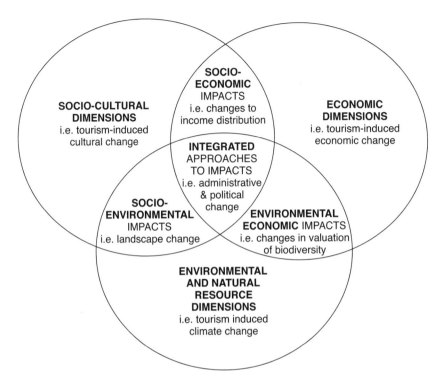

Figure 2.2 Dimensions of the impacts of tourism.

three, of the impact types. Indeed the Brundtland Report emphasized the role of impact assessments as an important tool for sustainable development, noting 'When the environmental impact of a proposed project is particularly high, public scrutiny of the case should be mandatory and, wherever feasible, the decision should be subject to earlier public approval, perhaps by referendum' (WCED 1987: 64).

The International Association for Impact Assessment (IAIA) defines *environmental impact assessment* (EIA) as, 'The process of identifying, predicting, evaluating and mitigating the biophysical, social, and other relevant effects of development proposals prior to major decisions being taken and commitments made' (IAIA 1999: 2). With the objectives of an EIA being:

- to ensure that environmental considerations are explicitly addressed and incorporated into the development decision-making process;
- to anticipate and avoid, minimize or offset the adverse significant biophysical, social and other relevant effects of development proposals;
- to protect the productivity and capacity of natural systems and the ecological processes which maintain their functions; and

- to promote development that is sustainable and optimizes resource use and management opportunities.

Within the European Union (Directive 85/377/EEC as amended by Directives 97/11/EC and 2001/42/EC) member states are required to ensure that public or private projects that are likely to have 'significant effects' on the environment should not be given development consent until their likely environmental impacts have been fully assessed by an EIA undertaken under the competent national authority.

> The environmental impact assessment must identify the direct and indirect effects of a project on the following factors: man, the fauna, the flora, the soil, water, air, the climate, the landscape, the material assets and cultural heritage, and the interaction between these various elements.
>
> (EU 2006)

Directive 85/337/EEC covers construction work and other installations or schemes, as well as other measures affecting the natural environment or landscape. Although a distinction is drawn between projects that are subject to a mandatory assessment process and those that are subject to an assessment only if the EU member state believes that they will have serious environmental impacts.

The types of projects subject to EIA in the EU are listed in annexes to the directive and affect a range of tourism projects. For example, assessment is obligatory for transport infrastructure developments, such as railways, airports, motorways, inland waterways and ports, when the infrastructure exceeds certain specific thresholds. Other projects are not automatically assessed, as member states can decide to subject them to assessment on a case-by-case basis or according to thresholds, certain criteria (e.g. size), locations (e.g. near sensitive ecological areas) and potential impacts (e.g. surface affected, duration). The list of projects in this category is much longer and includes such things as coastal land reclamation, shopping centres and car parks, motorway service areas, ski runs and lifts, golf courses, holiday villages, caravan/recreational vehicle sites and theme parks.

The annex to the directive also generally specifies minimum size thresholds. For example, golf courses and caravan sites of less than one hectare are usually exempt from the EIA process (Watson 2003). Directive 2001/42/EC also extended the assessment progress further by introducing a system of prior environmental assessment at the planning stage, while 2003/35/EEC provided for greater public access to information, public participation and legal access in order to encourage greater public participation in environmental decision-making.

The European approach to impact assessment bears substantial similarity with other assessment regimes around the world (see Chapter 6). The EU EIA requirements also cover aspects of social and economic impacts, and there is now a whole range of specific impact assessment areas that have developed out of the EIA. For example, the IAIA provide best practice guidelines on social impact

assessment (SIA) (Case 2.2), biodiversity in impact assessment, public participation, health, corporate EIA and strategic environmental assessment. Many of
these are relevant to tourism and are discussed in more detail in Chapter 6.

Social impact assessment (SIA) is an extension of the broader environmental
impact assessment methodology that focuses on the analysis, management and
monitoring of the social consequences of development. Becker (2000: 311) defines
SIA as 'the process of identifying the future consequences of a current or proposed
action which are related to individuals, organizations and social macro-systems'.
Although there are different interpretations of the SIA concept (Becker and
Vanclay 2003), it is regarded by most as the process of analysing, managing and
monitoring the positive and negative intended and unintended consequences on
the human environment of planned interventions (policies, programmes, plans)
and specific developments (projects) and their associated social changes, so as to
bring about a more sustainable and equitable biophysical and human environment
(Becker and Vanclay 2003; IAIA 2003). Some of the key features of this
understanding of SIA are:

- The goal of impact assessment is to bring about a more ecologically, socio-
  culturally and economically sustainable and equitable environment. Impact
  assessment, therefore, promotes community development and empowerment,
  builds capacity, and develops social capital (social networks and trust).
- The focus of concern of SIA is a proactive stance to development and better
  development outcomes, not just the identification or amelioration of negative or unintended outcomes. Assisting communities and other stakeholders
  to identify development goals, and ensuring that positive outcomes are
  maximised, can be more important than minimising harm from negative
  impacts.
- SIA builds on local knowledge and utilises participatory processes to analyse
  the concerns of interested and affected parties. It involves stakeholders in the
  assessment of social impacts, the analysis of alternatives, and monitoring of
  the planned intervention.
- The good practice of SIA accepts that social, economic and biophysical
  impacts are inherently and inextricably interconnected.

(IAIA 2003: 2)

The *social impact of tourism* refers to the manner in which tourism and travel
effect changes in collective and individual value systems, behaviour patterns,
community structures, lifestyle and the quality of life. Broadly conceived, *social
impacts* refer to all changes in the structure and functioning of patterned social
ordering that occur in conjunction with an environmental, technological or social
innovation or alteration. Factors and associated social indicators that contribute to
social well-being and quality of life include economic security, employment,
health, personal safety, housing conditions, physical environment and recreational
opportunities. The scale of the social impact of tourism is typically measured by
the ratio of tourists to the host population, by the economic level of the host

community as compared with that of the tourists, and cultural differences and perceptions between the two groups. However, to this should also be added other factors such as the proximity of residential spaces to tourism development, education levels and socio-economic status of local populations, variations in the roles of different neighbourhoods, political and social boundaries within the destination place, and the overall visibility and prominence of tourists and the tourism industry.

Whether or not a social impact assessment study is conducted for a proposed tourism development will depend on the legislative requirements of each national or, in some cases, regional jurisdiction and is usually dependent on planning and development legislation. Very few private developers will undertake a social impact assessment of a tourism development because they believe that it is an essential part of corporate citizenship. The externalities of economic development (impacts beyond the development site itself) are costs that business-oriented developers would prefer not to bear. Legal compulsion, therefore, is a necessary part of the broader impact assessment requirements. However, in general, SIA has been a lesser sibling in the EIA process, and is not a widely required component for project or policy appraisal (Burdge 2002).

Case 2.2: Social impacts of mega-events

One area that has become increasingly controversial with respect to the need to conduct a social impact assessment is that of large events, such as the Olympic Games and World Expos (World Fairs), and the social effects they have on the communities that host them. These mega-events are directly related to the City Beautiful Movement and urban renewal traditions of the last century in terms of their goals (redevelopment to serve a higher public good), methods (condemning and repurposing large tracts of 'undesirable' land uses) and impacts on the urban poor. There have been several instances that have demonstrated that the hosting of large events 'have a tendency to displace groups of citizens located in the poorer sections of cities' (Wilkinson 1994: 29). The people who are often most impacted by hallmark events are those who are least able to form community groups to protect their interests. At worst, this tends to lead to a situation in which residents are forced to relocate because of construction or not being able to afford rent increases (Hall 1994; Baade and Matheson 2002).

The 1986 World Exposition on Transportation and Communication, better known as Expo '86, was a World's Fair held in Vancouver, British Columbia, Canada, from 2 May until 13 October 1986. The fair, the theme of which was 'Transportation and Communication: World in Motion – World in Touch', featured pavilions from 54 countries as well as a number of corporations. It coincided with Vancouver's centennial and was held on the north shore of False Creek to the south and east of Vancouver's downtown. It was the second time that Canada had hosted a World's Fair, the first being Expo '67 in Montreal held during the Canadian Centennial of Confederation.

The study by Olds (1988) of the 1986 Expo's effects on the residents of Vancouver's 'Downtown Eastside', which was situated next to the Expo site, was the first comprehensive account of the housing and social dislocation impacts of a mega event (see Ley and Olds 1988; and Olds 1998, for more accessible public documents). During the event, Olds observed that about 600 tenants were evicted, many of whom were long-term, low-income residents, from areas near the site. Although there was no direct on-site housing impact of the 1986 Expo, there was a substantial post-announcement speculative impact. The redevelopment of the area through the hosting of Expo had three substantial, interrelated effects (Olds 1998). First was the alteration of physical space through the construction of new buildings and the refurbishment of old. Second was the displacement of short- and, more importantly, long-term residents through physical restructuring and rent increases. And third was the effect on the longer-term processes of *gentrification* and urban redevelopment in the inner Vancouver area. Interestingly, the same area is now receiving a second round of event-related gentrification, some 20 years after the Vancouver Expo, as a result of the city being the host of the 2010 Winter Olympic Games. However, substantial questions have been raised about this mega-event's broader economic and social contribution (Whitson and Horne 2006).

The Olympic Games has also been recognized as having a substantial impact on housing, with the impacts of hosting such a large event, and its related redevelopment projects, usually being felt well before the games are actually held. In a study of the potential impacts of the 2000 Olympic Games in Sydney, Australia, on low-income housing, Cox et al. (1994) concluded that previous mega-events often had a detrimental effect on low-income people who were disadvantaged by a localized boom in rent and real estate prices, thereby creating significant dislocation in extreme cases. The same rise in prices is considered beneficial to most home owners and developers, who can cash in on the increased equity of their properties. In such situations, public and private lower-cost housing are pushed out of areas designated for Olympic-related development as a result of increased land prices, higher property taxes and construction costs (Cox et al., 1994; COHRE 2007). The study by the Centre on Housing Rights and Evictions (COHRE 2007) is especially significant because, as it found, 'forced evictions, discrimination against racial minorities, targeting homeless persons, and the many other effects we noted, are in complete contradiction to the very spirit and ideals of the Olympic Movement, which aims to foster peace, solidarity and respect of universal fundamental principles' (2007: 9).

In the case of the 2000 Sydney Olympics, no social impact study was undertaken by the Sydney bid team during the bidding process. After the bid was won, a comprehensive housing and social impact study was carried out by Cox et al. (1994) on behalf of low-income housing interests – not for the state government or the Sydney Olympic organization. The report also presented a number of recommendations that could be implemented in a proactive strategy to ameliorate the negative impacts stemming from the preparation and hosting of the Sydney Games. The main recommendations included:

- the establishment of a housing impact monitoring committee;
- development of an Olympics accommodation strategy;
- tougher legislation to protect tenants and prevent arbitrary evictions;
- provision of public housing and emergency accommodation for disabled people; and
- a form of rent control.

None of these policies were adopted for the Sydney Olympics, even though research makes it abundantly clear that the Olympic Games has major impacts on the poorer elements of society. As the COHRE study observed,

> [the] project analysed examples of thousands of people forcibly evicted from their shelters in Seoul in 1987–8. We looked at the effects of 'street cleaning operations' in Atlanta in which thousands of homeless were effectively criminalised. And we studied the way in which the Roma where disproportionately affected by evictions for Olympic-related construction in Athens. Disturbingly, as our study progressed, we realised that the issue was even bigger than we had at first imagined; and getting worse. For example, since the commencement of this project, the number of people displaced due to Olympics-related development in Beijing has risen from 400,000 (a figure we reported in 2005) to a staggering 1.25 million (the number of people our research shows have been displaced as of early 2007, with another 250,000

Photo 2.4 The Telstra Stadium, one of the main venues of the 2000 Olympic Games in Sydney. The city of Sydney, the largest city in Australia, experienced a major increase in international tourist visits immediately following the games, which were credited with significantly increasing the global image and awareness of the city, as well as Australia overall. However, the Olympic tourism effect had declined substantially by 2002. (Photo by C. Michael Hall/Sarah Pollmann)

more expected to be displaced over the next year).. . . Past errors should be used to improve future conduct of host cities. It is worrying that, in spite of publicity of the violations committed by the City of Beijing, the City of London is already failing to sufficiently prioritise housing concerns in their preparations for staging the Olympic Games.

(COHRE 2007: 9)

An example of an attempt to include greater consideration of social impacts in sports tourism planning was the campaign of the non-profit public interest coalition Bread Not Circuses (BNC) in Toronto's bid to host the 2008 Olympic Games, which was eventually unsuccessful. BNC argues that given the cost of both bidding for and hosting the Olympics, the bidding process must be subject to public scrutiny. 'Any Olympic bid worth its salt will not only withstand public scrutiny, but will be improved by a rigorous and open public process' (Bread Not Circuses 1998a). They also argued that the Toronto City Council should have made its support for an Olympic bid conditional on:

- the development and execution of a suitable process that addresses financial, social and environmental concerns, ensuring an effective public participation process, and including a commitment to the development of a detailed series of Olympic social and environmental standards. A timeframe of one year from the date of the vote to support the bid should be set to ensure that the plans for the participation process are taken seriously;
- a full and open independent accounting of the financial costs of bidding and staging the Games; and
- a full and open independent social impact assessment of the Games.

Other key elements of the proposed public participation process included:

- a full, fair and democratic process to involve all of the people of Toronto in the development and review of the Olympic bid;
- an Olympic Intervenor Fund, similar to the fund established by the City of Toronto in 1989, to fund interested groups to participate effectively in the public scrutiny of the Toronto bid;
- an independent environmental assessment of the 2008 Games, and strategies should be developed to resolve specific concerns; and
- the development of a series of financial, social and environmental standards governing the 2008 Games, similar to the Toronto Olympic Commitment adopted by City Council in September of 1989.

(Bread Not Circuses 1998a)

A similar BNC proposal was sent in a letter to the International Olympic Committee (IOC) president, requesting 'that the IOC, which sets the rules for the bidding process, take an active responsibility in ensuring that the local processes in the bidding stage are effective and democratic' and specifically address concerns regarding the 'financial and social costs of the Olympic Games'. They further proposed:

- an international network be created that includes COHRE, the Habitat International Coalition (HIC), Housing Rights Subcommittee, academics, NGOs (including local groups in cities that have bid for and hosted the Games);
- a set of standards regarding forced evictions, etc., would be developed and adopted by the network;
- a plan to build international support for the standards, including identification of sympathetic IOC, National Olympic Committee (NOC) and other sports officials, would be developed and implemented;
- the IOC would be approached with the request that the standards be incorporated into the Olympic Charter, Host City Contracts and other documents of the IOC.

(Bread Not Circuses 1998b)

Such a social charter for the Olympics would undoubtedly greatly assist in making the Olympics more place-friendly and perhaps even improve the sometimes controversial image of the IOC. Unfortunately, the books of the Toronto bid were never opened for full public scrutiny, and neither was there any response to the proposal for creation of a set of social standards for the Olympics (Hall 2001). Indeed, such a situation had led authors such as Whitson and Horne (2006) to note that a full and transparent account of an Olympics will likely show a financial loss on the public investment, as most unbiased evidence suggests. The questions, therefore, become whether the event and the facilities will lead to other desirable economic, social and environmental outcomes. As Chernushenko (1994: 28) noted, 'The challenge for any host city . . . is "to make the Olympics fit the city" and not the city fit the Olympics'. What this points to, as Whitson and Horne (2006) suggest, is that environmental and social impact assessments, as well as full public consultation before submitting bids, are necessary if major sports events are ever to become democratically accountable (Flyvbjerg et al. 2003).

Sources: Bureau International des Expositions (BIE): http://www.bie-paris.org/main/

Centre on Housing Rights and Evictions (COHRE): http://www.cohre.org/mega-events

International Association for Impact Assessment: http://www.iaia.org/

Impact statements and impact studies

There are several significant differences in examining tourism-related impact assessment and impact studies. One is that impact assessment is project or programme based, whereas impact studies is a more theoretical approach, concerned with the broader aspects of change and the factors leading to change. In fact many of the largest concerns with respect to tourism-related change, such as climate change (see Chapters 5 and 7) or cultural change (see Chapter 4), cannot be easily associated with a single facility or development that would be subject to an EIA. It is likely, however, that the potential impacts on emissions and carbon budgets will be increasingly incorporated into EIAs as these start to be associated with carbon trading schemes or environmental taxes.

As Mathieson and Wall (1982: 14) reported: 'Impacts of tourism are viewed as being more than the results of a specific tourist event or facility. Impacts emerge in the form of altered human behaviour, which stems from the interactions between the agents of change and the sub-systems they impinge.' This situation has been long recognized:

> In assessing the impact of tourism developments one is not only faced with the tangible effects of major developments, situated in specific locations, but also the need to consider the more spatially diffuse, but no less considerable, impact of tourism as an economic and social activity. Indeed, even those staying in tourist centres or resorts exert a considerable influence on the surrounding locality. The impacts of tourism developments, therefore, are not confined solely to the structural changes associated with such developments but are also related to what Doxey (1975) calls the dimensional changes, i.e. those impacts that occur as a result of ever increasing numbers of tourists within an area. Such a distinction recognizes the dynamic nature of tourism as an agent of change.
>
> (Duffield and Walker 1984: 479)

Individual EIAs are, therefore, very significant in terms of the particular development they are assessing and can even contribute to more general environmental, social and other knowledge with respect to change. However, they can only provide a very partial account of the bigger picture with respect to tourism and its relationships to economic, socio-cultural and environmental change.

Another significant issue is that EIA is usually conducted under legislation that is not tourism specific. We are unaware of any EIA legislation developed solely to examine tourism anywhere in the world, although, of course, EIAs are applied to many tourism developments (e.g. Warnken and Buckley 1998; Warnken 2000; Gielen et al. 2002; Warnken et al. 2002; Mandelik et al. 2005).

Similarly, most tourism planning is conducted under general planning legislation and associated regulations and, therefore, the majority of planning for tourism is conducted by non-tourism specialist agencies. Exceptions to this occur when the planning is with respect to a particular designated land use with its own planning authority, for example in national parks and reserves or in an identified tourist zone where a planning body may have a tourism development specific mandate (Hall 2008c).

Point and non-point impacts

In seeking to provide a framework to understand the impacts of tourism we need to recognize the diffuse nature of tourism's effects throughout the economy, society and the environment. In attempting to do this we can distinguish between point and non-point sources of tourism-related change.

- *A point source* of change is a specific tourism-related facility, project or object. This could be a hotel, airport, attraction, event or some other built form of tourism development. This is a tangible aspect of tourism that will have a clear finite lifespan, and to which EIA principles and methods can be relatively easily applied. It is usually site-specific and does not move from that location – the exception potentially being ships and planes that are mobile point sources of change.
- *A non-point source* of change in tourism terms is human activity. Or, to be more precise, the interaction of the mobile population we refer to as tourists with providers of the service experience. Although both consumer and producer are required for the co-creation of the tourist experience (see Chapter 1) the non-point source of change in tourism impact studies is usually inferred as the tourist.

The analogy with the pollution literature here is deliberate. Water pollution is usually described as occurring with respect to point (a single identified source) and non-point sources. The latter refers to the many different, hard-to-trace sources with no obvious point of discharge. Many everyday human activities, such as using fertilizer or pesticides in the garden, contribute to non-point source water pollution. In the same way it is the many everyday activities of mobile people at leisure that contribute to the non-point effects of tourism. Many of the resulting impacts take years to recognize (see Case 2.3). Such a situation obviously creates challenges for understanding and managing such impacts, and due to this we must utilize the concept of the tourism system.

Case 2.3: Moose mothers and tourist roads

Pregnant moose in the Yellowstone and Grand Teton national parks have developed an innovative behaviour to avoid one of their main predators, grizzly bears. They drop their calves close to the roads that run through the parks. Each year, a team from the US Wildlife Conservation Society puts radio transmitters on between 18 and 25 female moose, about three-quarters of which are usually pregnant. They also monitor the behaviour of the moose by following them on the ground. Over the years they have noticed that the moose were giving birth to their calves closer and closer to the roads (Kaplan 2007).

In the south of the Greater Yellowstone system, where bears are still scarce, moose birthing patterns have not changed. In the north, however, where bears are more common, pregnant moose were moving an average of 122m (400ft) closer

Photo 2.5 West Thumb Geyser Basin at Yellowstone National Park. This is one of the most active geyser areas and is right on the shore, and under the waters, of Yellowstone Lake. Established as the world's first national park by US President Ulysses S. Grant in 1872, Yellowstone is the second largest US national park in the 48 conterminous states. This has made the Greater Yellowstone ecosystem the largest mostly intact temperate ecosystem in the northern hemisphere. (Photo by Alan A. Lew)

to roads each year to give birth. Given that bears seldom venture within 500m (1,600ft) of roads, moose were effectively protecting their offspring from attack. Studies carried out in Alaska suggest that grizzly bears are responsible for up to 90 per cent of young moose deaths (Berger 2007; Kaplan 2007).

This research illustrates a substantive change in how both prey and predators change their behaviour in response to human development. As Berger (2007) reported, the findings offer rigorous support that mammals can use humans to shield against carnivores and raise the possibility that redistribution has occurred in other mammalian taxa due to human presence in ways that have not yet been detected. The interpretation of pristine ecologically functioning systems within parks must therefore now also account for indirect anthropogenic effects on species distributions and behaviour.

Sources: Wildlife Conservation Society: http://www.wcs.org/

Yellowstone National Park (US National Park Service): http://www.nps.gov/yell/

The systems approach

The close relationship between the development of impact assessment to environmental concerns has led to a substantial emphasis on the interconnectedness of the factors that cause change, including economic, environmental and social change. This has meant that 'systems' approaches are integral to conceptualizing the causality that exists between external factors and the subject of any study of change (Glasson et al. 2005).

The concept of a 'tourism system' is one that has existed since the late 1960s in various forms (Hall 2008c). A system is a group of elements organized such that each element is, in some way, either directly or indirectly interdependent with every other element.

A system therefore comprises:

- a set of elements (also called entities)
- the set of relationships between the elements
- the set of relationships between those elements and their larger environment
- a definition or identification of the system's boundaries
- for some analysts, identification of the system's function, goal or purpose, even if that only means the ongoing maintenance of the system.

An example of a tourism system was seen in Figure 1.4, variants of which are widely used to illustrate basic elements of tourist flows.

Studies of systems have to address four main issues (Hay 2000):

Whether the system is open or closed. A closed system has no links to or from any environment that is external to it. Open systems interact with elements or environments outside of their system boundary. In the case of tourism an open system would almost certainly be warranted (Carlsen 1999), although this then raises the issue of defining the boundary of a tourism system.

Whether the system can be divided into sub-systems, clusters or interdependent elements that are only weakly linked to the remainder of the system. In the case of tourism, depending on the definition of the overall tourism system, it is possible to identify a number of sub-systems, although the relative strength of their relationship to the broader system will depend on the structure and dynamics of the system. For example, particular firm networks (also termed industrial districts or clusters) can be identified as sub-systems within a larger economic system.

The systems nature of tourism has even been recognized under Italian law (Legge Quadro sul Turismo, No. 135/2001) through the creation of an economic policy instrument called the 'Local Tourist System':

> We call local tourist systems, homogeneous or integrated tourist environments, which comprise territories also belonging to different regions, and which are characterized by the integrated supply of cultural, environmental

goods and tourist attractions, including typical agricultural and local handicraft products, or those characterized by a widespread presence of individual or associated tourist firms.

(National Tourist Law Reform, Law 29 March 2001 No. 135, translation Candela et al. 2005: 2)

Examples of other subsystems that have been identified in the literature on tourism, and defined at various scales, include

- tourist *attraction* systems (e.g. McKercher and Lau 2007);
- tourism *production* systems (Roehl 1998), which 'represents the mix of businesses and other organizations that provide tourism services' (Roehl 1998: 53–4); and
- tourism *consumption* systems (TCS), which is defined by Woodside and Dubelaar (2002) as the set of related travel thoughts, decisions and behaviours by a discretionary traveller prior to, during and following a trip. The central proposition of TCS is that the thoughts, decisions and behaviors regarding one's activity influence the thoughts, decisions and behaviours for a number of other activities.

Whether the links involve flows, causal relationships or 'black-box' relationships (the latter refers to when the consequences of a linkage are known but the causal factors are not). In a tourism system, the flows of tourists can clearly be identified, along with, in some cases, the flows of capital and energy. However, some of the causal relationships that result from those flows may not be so well understood.

Whether there is feedback in the system, such that a change in x may stimulate a change in y, and this in turn will have either a positive or negative impact on x. This is well recognized in the case of the impacts of tourism associated with the interaction between markets in tourism generating regions and in destinations, so that a change in the destination (x) created in part by visitors (y), leads to changed behaviour of visitors (y) at a later point in time (cf. Figure 2.3 a, b and c).

Representations of the relationships between generating regions and destinations has also been influential with respect to tourism area life-cycle research (Butler 2006a, b) (see also Chapter 7). Perhaps appropriately for understanding tourism systems, the interval between a disturbance to the system and the return to an equilibrium state is known in life-cycle literature as the *relaxation time*. However, in many instances in life-cycle studies of tourism, the characteristics and conditions of equilibrium have not been identified and nor has the timescale of analysis been such that a return to an equilibrium state has been observed, or, if it has, it has not been recognized.

Some researchers, such as McKercher (1999), argue that tourism essentially functions as a chaotic, non-linear, non-deterministic system. As such, many existing tourism frameworks and models fail to explain fully the complex relationships

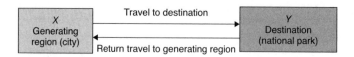

Figure 2.3 Feedback in the tourism system; (a) Systems diagram of hypothetical
example of the interaction between markets in tourism generating regions *x*
and destination *y*. This could be imagined to be the relationship between an
urban area that generates visitors (*x*) and a national park (*y*).

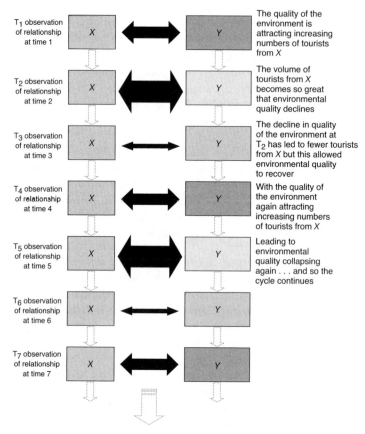

2.3(b) Hypothetical example of the interaction between markets in tourism generating
regions *x* and destination *y* so that change in *y* is affected by *x*. Size of arrow
illustrates strength of relationship between *x* and *y* in terms of numbers of tourists
travelling to and returning from *x*. Number of tourists is regarded as affecting the
quality of the environment in destination *y* with the feedback in the system also
indicating that the quality of the environment also influences the number of
visitors that come from *x* to *y* in the future which is observed at the next point of
time. Quality of environment in *y* is also indicated by shading, with the darker the
shading the higher the quality of environment.

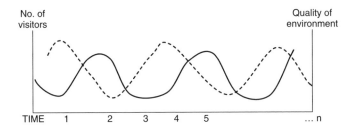

2.3(c) Relationship between visitors from *x* and environmental quality at destination *y* illustrated via hypothetical data. Dashed line indicates relative environmental quality. Solid line numbers of visitors.

that occur between and among the various elements that constitute a tourism system. One of the main reasons for this is the difference between using the notion of a system as a metaphor for a kind of explanatory or educational device (e.g. Leiper 1989; Mill and Morrison 1985; Cornelissen 2005) versus using it as a more concrete framework to statistically explain positive and negative relationships between variables and to conduct a formal systems analysis (e.g. Preobrazhensky et al. 1976; Lazanski and Kljajic 2006). In tourism studies the metaphor has dominated, which has often meant that metaphors with respect to such systems characteristics as 'complexity' and 'chaos' have been used, often with little appreciation of the concrete analytical and predictive dimensions that such concepts provide to other fields of systems research (e.g. Faulkner and Russell 2003).

A system is regarded as *complex* 'if its parts interact in a non-linear manner. Simple cause and effect relationships among the elements rarely exist and instead a very little stimulus may cause unpredictably large effects or no effect at all' (Baggio 2007: 5). Furthermore, a complex system 'can be understood only by considering it as a whole, almost independently by the number of parts composing it'. In contrast, a *complicated* system is 'a collection of an often high number of elements whose collective action is the cumulative sum of the individual ones' (Baggio 2007: 6). Tourism is usually described as a complex system (Walker et al. 1998; Ndou and Petti 2006).

Tourism as a complex adaptive system

A special type of complex system is one referred to as a *complex adaptive system* (CAS). A CAS is a complex, self-organizing and self-similar (it will look like itself at different scales) collection of interacting adaptive agents (also referred to as elements). A CAS can be used to refer to both natural and social systems, and has also been defined as:

> a dynamic network of many agents (which may represent cells, species, individuals, firms, nations) acting in parallel, constantly acting and reacting to what the other agents are doing. The control of a CAS tends to be highly

dispersed and decentralized. If there is to be any coherent behavior in the system, it has to arise from competition and cooperation among the agents themselves. The overall behavior of the system is the result of a huge number of decisions made every moment by many individual agents.

(Holland in Waldrop 1992: 144)

An ecosystem is a widely used example of natural CAS, while examples of social systems include communities and the stock market. A CAS behaves according to three key principles: (1) order is emergent and self-organized, as opposed to predetermined, (2) the system's history is irreversible, as the future behaviour of a system depends on previous behaviour, and (3) the system's future is sometimes unpredictable (Dooley 1997). The tourism sector, as an economic activity, is regarded as sharing many of these characteristics (Hall 2005a; Farrell and Twining-Ward 2004; Baggio 2007), with, for example, a destination comprising many different agents in the form of firms, organizations, and individual decision-makers. Figure 2.4 illustrates a tourism CAS.

Several important points emerge from thinking about tourism impacts and change in the context of systems and CAS. These are summarized under the broader concepts of scale, networks and behaviour.

patterns

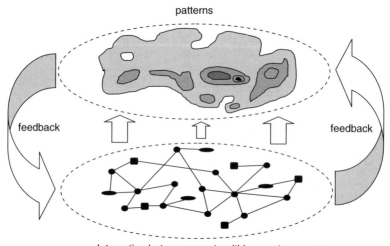

Interaction between agents within a system

Figure 2.4 Complex adaptive systems. Diagrammatic attempt to represent a CAS. From the interaction between agents within a tourism system, with the different agents representing firms from different sectors, regularities emerge which create patterns (patterns here represent overnight stay distribution of tourists within the tourism system) which, in turn, feeds back on the system and informs the interactions of agents, i.e. firms recognize the distribution of visitors and then take action to affect the pattern for their own advantage. If we were to follow process to see what happens at another point in time (as in the case of Figure 2.3) we would be able to observe a new pattern of overnight stay.

Scale Scale affects the definition of the system and hence the behaviours observed, and the feedback processes that can occur between system levels. For example, how we define an element depends on the scale at which we conceive the system's boundaries, otherwise referred to as the 'resolution level' (Hall 2008c). In CAS, emergence is regarded as a bottom-up rather than a top-down process, that is it goes from lower to higher scales in the sense that the sum of micro-behaviour produces a macro-behaviour. The resulting higher-level behaviour feeds back to individual units at the lower level.

With respect to scale, selection of the level of analysis of a system, including its composite sub-systems, raises a number of significant questions in understanding impacts. For example, Figures 2.5 and 2.6 illustrate that within research on tourism and global environmental change some scales of analysis have been well studied while others are almost completely unknown (Hall 2004b; Gössling and Hall 2006a).

Significantly Hall (2004b) found that much of the research undertaken in tourism tended to be at the destination, omitting other elements of the geographical tourism system or trip, such as the transit region and the origin area. Much of the research also tended to be highly localized in one site or place, thereby potentially limiting our capacities to generalize about the tourism system as a whole (see also Gössling and Hall 2006a).

Hall (2008c: 81) suggested that there were three basic questions with respect to scale:

• *Scale coverage*: do we have regular and comprehensive monitoring of the world at all relevant scales from local to global? This issue is clearly important

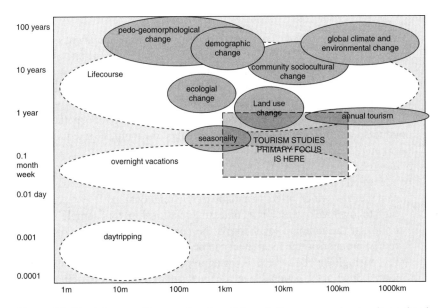

Figure 2.5 The influence of temporal and spatial resolution on assessing tourism-related phenomenon (adapted from Hall 2004a, b).

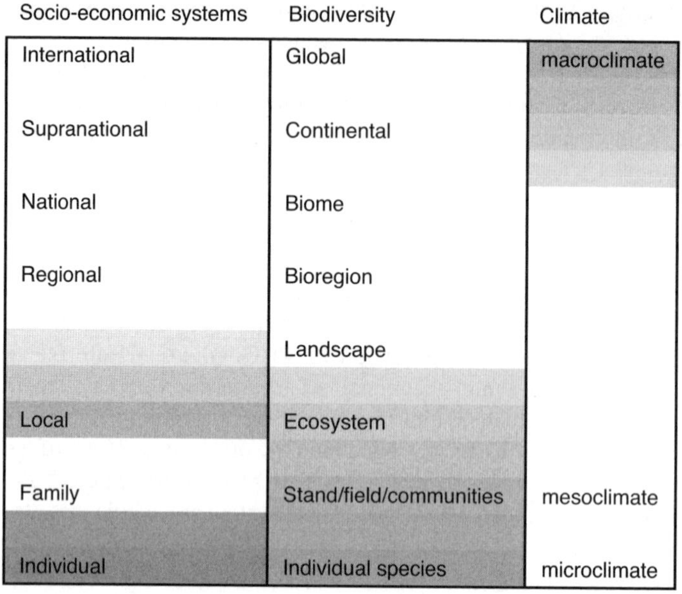

Socio-economic systems	Biodiversity	Climate
International	Global	macroclimate
Supranational	Continental	
National	Biome	
Regional	Bioregion	
	Landscape	
Local	Ecosystem	
Family	Stand/field/communities	mesoclimate
Individual	Individual species	microclimate

Figure 2.6 Scale in tourism analysis: primary foci in socio-economic systems, biodiversity and climate research in terms of research outputs (Hall 2004b).

for the collection of comparable tourism statistics and the understanding of international tourist flows.

- *Scale standardization*: do we have comparable data from equivalent sampling frames? This issue often arises when comparing the tourism statistics from one country or region to another. Not only do we need to know that the methodologies of collecting tourism statistics are the same, but also the areas being investigated must be equivalent in some way. Similarly, the aggregation of data from a number of case studies often using different methods and conducted in different contexts also creates difficulties of equivalency between the various cases.
- *Scale linkage*: three different connections between the various scale levels can be identified (Harvey 1969):
  - same level – which refers to a comparative relationship;
  - high to low level – which is a contextual relationship, for example, tourism policy at the national level forms the context within which changes in tourist numbers at the local level can be analysed; and
  - low to high level – which is an aggregative relationship, for example, tourist flows at the national level are the result of the activities of individual, local firms.

Issues of scale and the boundary definition of systems are also important for tourism management, planning and policy-making as there is a need to try to ensure that different levels – or scales – are in sync with one another so as to increase planning effectiveness. Environmental issues are particularly problematic with respect to trying to connect jurisdictional or governance scale (Gössling and Hall 2006a; Hall 2008c).

Tourism is stretched over time and space, and therefore it is extremely important to set boundaries for the system being examined, particularly as the smaller the size of the sub-system being examined the more it is open to external forces for change. The selection of the boundary of a destination, or any boundary in analysing impacts, will affect the relative size and degree of system change within that boundary. Figure 2.7 illustrates some of the relativities of scales that affect tourism systems.

This issue has significant implications for understanding the spatial boundaries of destinations because governance jurisdictions may not match with the relative boundaries of a destination as perceived by visitors and tourism firms. To further complicate matters the destination boundary may not match with ecological boundaries, such as a watershed. Figure 2.8 illustrates some of the implications of the relativities of scale by highlighting how tourism policies, actors, climate and weather, and research intersect with the issue of tourism and climate change. Understanding the issue of scale is therefore fundamental to being able to assess the impacts of tourism and effectively manage them.

Networks Microscale networks – the set of interactions between agents – are an essential feature of CAS. Networks allow the system to solve problems using the large numbers of individual agents that have local interactions with other agents.

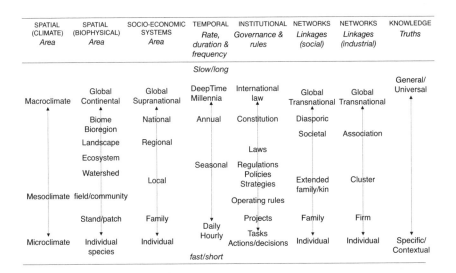

Figure 2.7 Relativities of scale in analysing tourism.

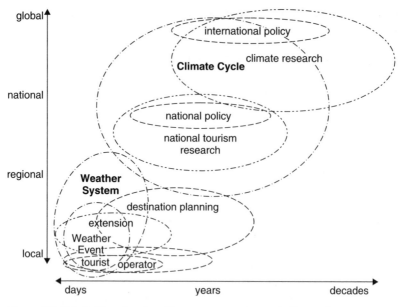

Figure 2.8 Relativities of scale with respect to tourism and climate change.

The agents 'themselves need not be "aware" they are contributing to this endeavor. They are following their own micromotivated rule-sets and interacting with local network neighbors. Such behavior allows the system to process information, and thus to learn' (Rogers et al. 2005: 10). Networks have been a major focus of tourism research since the 1990s, particularly with respect to the role of collaboration, which is regarded as an important feature of learning and innovation systems (Hall and Williams 2008). Advances in computer and telecommunication technology have allowed social and collaborative networks to transcend the traditional limits of geographic space and scale, allowing microscale networks to operate across global distances. This has created expansive marketing opportunities for destinations and highly specialized and personalized touring opportunities for tourists (Lew 2009).

Behaviours Local interactions in networks can lead to the emergence of new structures and behaviours at the next higher level of organization. The collaborative efforts of a chamber of commerce, for example, will offer new opportunities that no single member of the organization can accomplish on their own. In the case of tourism, organizational behaviour is often geared toward joint-marketing schemes (see Case 2.1). However, this is particularly significant with respect to the role of firm networks that may set new rules and regulations with respect to managing tourism impacts, such as developing codes of conduct for tour operators or tourists, and setting green accreditation standards for accommodation and restaurants.

The concepts of tourism systems and CAS clearly has implications for how we understand tourism and how the tourism system responds to change (Farrell and Twining-Ward 2004; Scott and Laws 2005). These concepts have been instrumental in attempts to develop a more integrated approach to understanding tourism impacts and their management (Hall 2008c), as well as analysing tourism competitiveness and innovation (Hall and Williams 2008) (see Case 2.4). Such an approach brings about the integration of different social, economic and environmental systems, within the context of tourism, and with the goal of effective and meaningful action and management.

Integrated tourism management is

> The process of promoting the coordinated development of tourism in relation to biophysical, socio-cultural and related resources in order to maximize the resultant economic and social welfare in an equitable manner without comprising the sustainability of ecosystems or social systems.

Such an approach recognizes the complex interaction between natural and human systems and the need for integration mechanisms within and between these to support sustainable tourism development objectives (see Figure 2.2), and it is closely related to ideas of *adaptive management* (also referred to as adaptive resource management) (Holling 1978). A systems approach, therefore, allows us not only to identify change within and between tourism and other agents, but also to develop mechanisms by which undesirable impacts may be lessened.

Complex adaptive systems, such as tourism, require an appropriate management framework that is self-learning and able to adapt to changing circumstances. Adaptive management is a systematic process for continually improving management policies, strategies and practices by learning from the outcomes of operations, policies and research. These planning and management responses to tourism impacts are dealt with in detail in Chapter 6, but prior to that we provide an overview of tourism's economic impacts (Chapter 3), social and cultural impacts (Chapter 4) and environmental impacts (Chapter 5), before seeing how decision-making responds to tourism impacts.

Case 2.4: Tourism innovation systems

Innovation systems and associated ideas such as clusters and networks are an increasingly important concept in tourism research (e.g. Hjalager 1999; Nordin 2003; Mattsson et al. 2005; Novelli et al. 2006). Innovation systems are constituted by 'interconnected agents' that interact and influence the execution of innovation within a given system, the context of which may be any one or a combination of:

- national
- regional
- spatial, and
- sectoral.

The key characteristics of an innovation system are usually summarized as:

- firms are part of a network of public and private sector institutions whose activities and interactions initiate, import, modify and diffuse knowledge, including new technologies;
- linkages (both formal and informal) exist between institutions;
- flows of intellectual resources occur between institutions;
- learning is a key economic resource; and
- geography and location matter (Holbrook and Wolfe 2000).

Systems of innovation, for example, may be thought of as segmented layers of institutions and production modes that integrate national, regional and local groupings of actors, institutions and resources that pose industry-specific issues of governance and the role of the state (Figure 2.9). The innovation systems framework, therefore, consists of analysing the existence of actors (institutions, universities, industries) in a given territory (ranging from the local to the national), their main competences, and their interactions in innovation-informing networks. Undertaking this analysis provides policy-makers with a tool that allows the construction of more competitive and efficient innovation systems. Traditional examples include the clustering of high-technology industries in California's Silicon Valley and in Ottawa in Canada, the dominance of central Connecticut in the insurance industry, and role of London and New York as the leading global financial centres. Examples of tourism-related business agglomerations that are sometimes described as clusters include the gaming, entertainment and hospitality industries of Las Vegas and the theme park, accommodation and attraction industries of the Gold Coast in Australia.

Tourism is embedded in different types of innovation systems, primarily fulfilling the role of an 'enabler' within a system. It does this by enabling and facilitating business travel, as well as contributing to cultural and social capital. At the same time, the tourism industry can also be regarded as a *specific sectoral innovation and production* (SSIP) system in its own right. A SSIP is defined by Malerba (2001: 4–5) as:

> a set of new and established products for specific uses and the set of agents carrying out market and non-market interactions for the creation, production and sale of those products. A sectoral system has a knowledge base, technologies, inputs and a demand, which may be existing, emerging or potential. The agents composing the sectoral system are organizations and individuals (e.g. consumers, entrepreneurs, scientists). Organizations may be firms (e.g. users, producers and input suppliers) and non-firm organizations (e.g. universities, financial institutions, government agencies, trade-unions, or technical associations), including sub-units of larger organizations (e.g. R&D or production departments) and groups of organizations (e.g. industry associations). Agents are characterized by specific learning processes, competences, beliefs, objectives, organizational structures and behaviours.

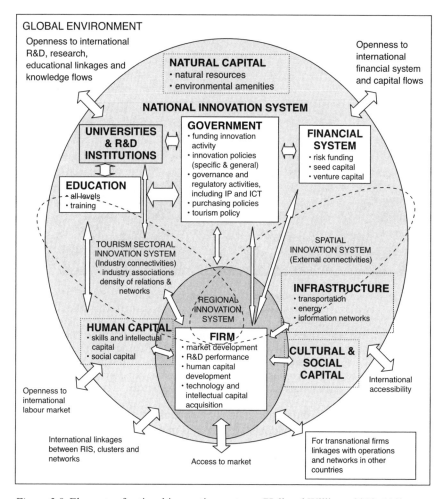

GLOBAL ENVIRONMENT

Openness to international R&D, research, educational linkages and knowledge flows

Openness to international financial system and capital flows

NATURAL CAPITAL
• natural resources
• environmental amenities

NATIONAL INNOVATION SYSTEM

UNIVERSITIES & R&D INSTITUTIONS

GOVERNMENT
• funding innovation activity
• innovation policies (specific & general)
• governance and regulatory activities, including IP and ICT
• purchasing policies
• tourism policy

FINANCIAL SYSTEM
• risk funding
• seed capital
• venture capital

EDUCATION
• all levels
• training

TOURISM SECTORAL INNOVATION SYSTEM (Industry connectivities)
• industry associations density of relations & networks

SPATIAL INNOVATION SYSTEM (External connectivities)

INFRASTRUCTURE
• transportation
• energy
• information networks

REGIONAL INNOVATION SYSTEM

HUMAN CAPITAL
• skills and intellectual capital
• social capital

FIRM
• market development
• R&D performance
• human capital development
• technology and intellectual capital acquisition

CULTURAL & SOCIAL CAPITAL

International accessibility

Openness to international labour market

International linkages between RIS, clusters and networks

Access to market

For transnational firms linkages with operations and networks in other countries

Figure 2.9 Elements of national innovation systems (Hall and Williams 2008: 113).

They interact through processes of communication, exchange, cooperation, competition and command, and their interactions are shaped by institutions (rules and regulations). Over time a sectoral system undergoes processes of change and transformation through the coevolution of its various elements.

As with the analysis of tourism, the aggregation of system components in a SSIP approach may be undertaken in a number of different ways, and may include discussions of sectors and subsectors, agents or functions. Sectoral systems may also be examined in a narrow sense in terms of a small set of product families (e.g. ecotourism), or in a broad way (tourism as a whole). In addition to firms (producers

Photo 2.6 Casino resorts and other tourist attractions on the Las Vegas Strip. The Las
Vegas, Nevada, metropolitan area is one of the fastest growing cities in the
United States, with a population of over 2 million and over 30 million tourist
visits a year. That gives it a very high tourist to resident ratio of 15 to 1.
However, Las Vegas is also a good example of the clustering of tourist
facilities, which both minimizes their impact on the non-tourist parts of the
city and offers opportunities for face-to-face networking and collaboration
among tourism firms and entrepreneurs. (Photo by Alan A. Lew)

and suppliers) and non-firm organizations (state tourism bodies, financial institutions) agents at lower (individuals or firm sub-units) and higher (e.g. public–
private consortia) levels of aggregation may be the key actors in a sectoral system.

Some of the key elements of a SSIP according to Malerba (2001, 2002, 2005a,
b) are:

- *Products* What the system produces.
- *Boundaries and demand* Technologies and demand constitute major
  constraints on the full potential range of diversity of firm behaviour and
  organization in a sectoral system. Such constraints differ from sector to sector
  and consequently affect the nature, boundaries and organizations of sectors,
  as well as being a source of sectoral transformation and growth. A given level
  of firm composition and market demand, along with a specific technological
  environment, defines the nature of the problems that firms have to address in
  their innovation and production activities, along with the resources and
  incentives available for this. However, within these overall constraints, there
  is persistent and substantial diversity in the innovative and productive behaviour and organization of different firms. 'Interdependencies and complementarities define the real boundaries of a sectoral system. They may be at

the input, technology or demand level and may concern innovation, production and scale' (Malerba 2002: 250–1).

- *Knowledge and learning processes* Knowledge is central to innovation and production. Knowledge differs across sectors in terms of particular domains. For example, a scientific or technological field of knowledge can be distinguished from knowledge with respect to marketing and product demand. Furthermore, knowledge also differs in terms of *accessibility* (the extent to which knowledge is internal or external to firms), *opportunity* (the capacity to actually utilize knowledge), *cumulativeness* (the degree to which the generation of new knowledge builds upon previous knowledge) and *appropriation* (the possibility of protecting an innovation from imitation). These vary with respect to the technological and learning regime of each economic sector (Malerba and Orsenigo 1996, 1997, 2000).
- *Agents* Several agents are actors in a SSIP, including firms, their users (including consumers of service products, and intermediate business users) and suppliers. In contrast to an industrial economics approach, in a SSIP approach market demand is not seen as an aggregate set of similar buyers, but is composed of heterogeneous agents with specific attributes, knowledge and competences who interact in various ways with producers (Malerba 2001). Similarly, firms are also regarded as heterogeneous with respect to a wide range of attributes including values, behaviour, experience, organization, learning processes, trajectories, innovative capacities and relations to consumers (demand). Non-firm agents are also elements in a SSIP and may include a full range of governmental, quasi-governmental and non-governmental organizations that support firm innovation, diffusion and production. These also differ greatly across sectors, in the same way that has already been discussed with respect to policy initiatives in national and regional innovation systems. For example, there is substantial variance in national sectoral systems among different countries with respect to the role of government in encouraging knowledge transfer in tourism (Hall and Williams 2008).
- *Networks and mechanisms of interaction* Within SSIPs heterogeneous agents are connected through market and non-market relationships. The types and structures of such interactions and networks among heterogeneous agents will differ between and within sectors due to different knowledge bases, beliefs, values, competences (including learning), technologies, inputs, demand characteristics and behaviours (including the creation of linkages and complementarities) (see Edquist 1997).
- *Institutions* Institutional arrangements, both governmental and non-governmental, are critical to innovation systems and the combination of generic and sector-specific institutions will determine how these intersect with national and regional innovation systems.
- *Processes of variety generation and selection* Processes of variety creation refers to products, technologies, firms and institutions, as well as strategies and behaviour, and are related to entry, research and development (R&D),

innovation and exit mechanisms that interact and contribute to creating variety and agent heterogeneity at different levels. For example, there may be significant differences between sectors and countries as to the entry and survival of new firms in different economic sectors (see Hall and Williams 2008). In contrast, processes of selection reduce heterogeneity and refer not only to the crucial role of the market but also to the effects of non-market selection processes. For example, different governments provide support for some firms, regions and sectors but not others.

The concept of an innovation system is extremely important to being able to understand how firms, public organizations and tourism destinations are able to respond to change via adaptation and mitigation strategies. For example, some destinations clearly appear better able to engage in management and resource capacity building than others, and are therefore more effective in responding to external change, even when other important factors, such as access to financial resources or the strength of the external change, appear equal. Destinations can, therefore, be regarded as a form of *regional innovation system*, 'the set of economic, political and institutional relationships occurring in a given geographical area which generates a collective learning process leading to the rapid diffusion of knowledge and best practice' (Nauwelaers and Reid 1995: 13). Different destinations have different regional innovation systems that will vary with respect to:

- the ability of firms and other relevant non-profit and public organizations to innovate due to their specializations, as well as their functional and organizational characteristics;
- their propensity to interact depending on the existence of clusters and networks, and the attitude of actors towards cooperation; and
- their capacity to construct relevant institutions (for example, in research, education and knowledge transfer) and in their 'governance model', which is dependent on their decision-making powers, financial resources, political legitimacy and their policy orientation.

As a result of these differences, it can be expected that some destinations will have a weak innovation system while others will have systemic interactions to a much higher degree (Tödtling and Kaufmann 1998). This observation is clearly important in understanding how information with respect to specialized knowledge, for example, of sustainable tourism practices and impact minimization, can be effectively disseminated and implemented (Hall 2008c). Unfortunately, much research on tourism planning and sustainability has failed to recognize the role of *blockages* in policy and planning systems, and that the reasons for failure and the non-adoption of sustainable practices lie as much in the realm of the social construction of sustainability at the local level as they do in 'technical' issues. This includes the local cultures of innovation and policy orientation and the role of institutions, including their power relationships, and tendencies toward corruption (Hall and Coles 2008b). The management and mitigation of impacts, therefore, need to be understood within the context of innovation systems at the various scales that respond to them.

Sources: derived from Hall and Williams 2008; Hall 2009a.

Summary and conclusions

The first two chapters of this book have provided a comprehensive framework for understanding tourism impacts by outlining a conceptualization of tourism that relates to issues of contemporary mobility and issues of analysis, particularly with respect to the problem of scale and the nature of the system within which tourism actually occurs. We note that historically tourism impacts have only been studied at a local level and there has often been a failure to appreciate that impacts occur at all stages of the tourist trip rather than at just the eventual destination. The failure to account for impacts at the national and global scale is regarded as particularly important given the significance of economic and cultural global-ization as a process influencing and influenced by tourism.

This chapter has also identified the importance of sustainability as a concept for understanding impacts and change and their management. Sustainability is not easy to define but captures a feeling that the contemporary 'state of the world' is 'somewhat precarious' (Mannion 1991: 309). There are three components to sustainability: economic, socio-cultural and an environmental and natural resource dimension. Environmental sustainability aims to maintain, unimpaired, the sink and source capacities of the biophysical environment. In other words, humankind should learn to live within the physical and biological limitations of the environment – in its role as both a provider of goods and as a sink for wastes (Goodland and Daly 1996). Economic sustainability refers to the maintenance of capital as well as greater equity in the distribution of capital (Goodland 1995). Social sustainability guarantees for both present and future generations an improvement of the capabilities of well-being for all through both the aspirations of equity, as intergenerational distribution of these capabilities, as well as their transmission across generations (Lehtonen 2004). All three dimensions of sustainability are important for tourism (Gössling et al. 2009). The next three chapters discuss the economic, socio-cultural and environmental impacts of tourism. A number of issues will be raised but, where possible, impacts will be discussed in terms of (see Figure 2.10):

- *Scale*: global/supranational, national, community/regional, and individual.
- *Environment*: the type of environment within which impacts occur.
- *Time*: short-term and long-term implications are introduced where possible to illustrate the way in which tourism processes evolve over time.
- *Form*: in some instances particular forms of tourism, such as ecotourism or cultural tourism, will be discussed in relation to their specific impacts.

However, as already noted above with respect to the difficulties in undertaking research on tourism, many of the impacts we discuss are diffuse, non-point impacts

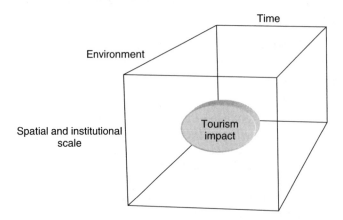

Figure 2.10 Illustrative framework for discussing impacts.

rather than those associated with a particular project or event. Our understanding of the impacts of tourism remains fragmented because of reasons of:

- different scales of analysis
- the limited locations at which research has been undertaken
- the use of different and inconsistent research methods
- a lack of longitudinal studies
- few genuinely comparative studies of different places
- a lack of baseline data, i.e. what existed before tourism commenced
- the paucity of information on the adaptive capacities of ecosystems, communities and economies
- the difficulty of distinguishing between changes induced by tourism and those induced by other activities
- the concentration of researchers upon particular environments or types of tourism
- a lack of an effective ergodic hypothesis for tourism (see Hall 2008c: 34–5).

Ergodic hypothesis A system is ergodic if the long-term observation of a single motion leads to the same frequency of measured values as the observation of many motions with different starting points. In ecology an ergodic hypothesis is an expedient research strategy that links space and time so that different areas in space are taken to represent different stages in time. For example, with respect to ecological succession whereby different locations within an area at one point in time are taken to represent the sequence of changes in species composition that would occur in that area over longer periods of time. In tourism studies possibly the nearest that exists to an ergodic hypothesis is the concept of a tourist area cycle of evolution (e.g. Butler 2006a, b), in which different observations at different points in time (and potentially space) are taken to represent the sequence of changes in a destination (including such factors as market, impacts, community and environmental change) that would occur in that area over a longer period of time. This model is then used as an analog for describing and understanding change in other locations (Hall 2005a, 2007d).

Nevertheless, these limitations to our knowledge of tourism impacts does not mean that impacts do not exist. They do exist and they can have substantial effects on tourism and on the human and physical systems in which tourism occurs.

Self-review questions

1   What are the three basic questions that have arisen with respect to scale?
2   What were the five basic principles of sustainability identified in the Brundtland report and how can they be applied to tourism?
3   What are the key elements of a system?
4   What differences are there, if any, between sustainable tourism and sustainable development?
5   Undertake a review of cities bidding for major events such as the Olympic Games, World Cup soccer or World Fairs. To what extent have social impact assessments been undertaken in comparison with economic or environmental assessments?
6   Why might impact assessments not be conducted or required for hosting major events?

Recommended further reading

Butler, R.W. (ed.) (2006a) *The Tourism Life Cycle. Vol.1, Applications and Modifications*, Clevedon: Channel View Publications.
Butler, R.W. (ed.) (2006b) *The Tourism Life Cycle, Vol.2, Conceptual and Theoretical Issues*, Clevedon: Channel View Publications.
The two volumes on Butler's Tourism Area Life Cycle (TALC) model provide a good account of one of the most influential ideas in studies of change at tourism destinations.

Glasson, J., Therivel, R. and Chadwick, A. (2005) *Introduction to Environmental Impact Assessment: Principles and Procedures, Process, Practice and Prospects*, 3rd edn, London: Routledge.
A good introduction to EIA though with a substantial emphasis on the developed world.

Gössling, S. and Hall, C.M. (eds) (2006) *Tourism and Global Environmental Change*, London: Routledge.
Provides a discussion of issues associated with understanding tourism impacts at a global scale and also includes various thematic chapters.

Gössling, S., Hall, C.M. and Weaver, D. (eds) (2009) *Sustainable Tourism Futures: Perspectives on Systems, Restructuring and Innovations*, New York: Routledge.
Several chapters discuss the changing nature of the tourism system and how it may be improved with respect to sustainability.

Hall, C.M. (2008) *Tourism Planning*, 2nd edn, Harlow: Pearson Education.
Chapter 2 provides a concise account of sustainability issues in relation to tourism, with chapter 4 detailing issues associated with systems analysis in tourism.

Hall, P. (2002) *Urban and Regional Planning*, 4th edn, London: Routledge.
This classic planning textbook provides an excellent introduction to general planning issues and the history of urban and regional planning.

Web resources

EU (2006) Assessment of the environmental impact of projects, Summaries of Legislation: http://europa.eu/scadplus/leg/en/lvb/l28163.htm
This is a very useful summary of legislation that contains direct links to the full text of the relevant EU legislation as well as other EU environmental activities.

International Association for Impact Assessment: http://www.iaia.org/
Provides a number of downloadable pdfs on principles of impact assessment.

Pathways in American Planning History: A thematic chronology: http://www.planning.org/pathways/history.htm

United Nations University: http://eia.unu.edu/
The United Nations University has an EIA open education resource complete with EIA course module, wiki, as well as an instructional guide. A very good resource for learning about general principles of EIA.

Key concepts and definitions

Complex Adaptive System (CAS) A complex, self-organizing, self-similar (it will look like itself at different scales) collection of interacting adaptive agents (also referred to as elements) and can be used to refer to both natural and social systems.

Environmental conservation The management of human use of the biosphere so that it may yield the greatest sustainable benefit to present generations while maintaining its potential to meet the needs and aspirations of future generations (*World Conservation Strategy* definition).

Environmental Impact Assessment (EIA) The process of identifying, predicting, evaluating and mitigating the biophysical, social and other relevant effects of development proposals prior to major decisions being taken and commitments made (International Association for Impact Assessment (IAIA) definition).

Gentrification The reinvestment of capital into a given location, usually urban, which is designed to produce space for a more affluent class of people than those that currently occupy that space.

Impact A change in a given state over time as the result of an external stimulus.

Integrated tourism management The process of promoting the coordinated development of tourism in relation to biophysical, socio-cultural and related resources in order to maximize the resultant economic and social welfare in an equitable manner without compromising the sustainability of ecosystems or social systems.

Regional innovation system The set of economic, political and institutional relationships occurring in a given geographical area which generates a collective learning process leading to the rapid diffusion of knowledge and best practice. If occurring at a national scale this is described as a national innovation system.

Social Impact Assessment (SIA) The process of identifying the future consequences of a current or proposed action which are related to individuals, organizations and social macro-systems.

Sustainable development Development that meets the needs of the present without compromising the ability of future generations to meet their own needs (Brundtland Report definition (WCED 1987: 43)).

System A group of elements organized such that each one is in some way either directly or indirectly interdependent with every other element. A system therefore comprises:

1 A set of elements (also called entities)
2 The set of relationships between the elements
3 The set of relationships between those elements and their larger environment
4 Definition or identification of the system's boundaries
5 For some analysts, identification of the system's function, goal or purpose, even if that only means the ongoing maintenance of the system.

3 Tourism and its economic impacts

Learning objectives

By the end of this chapter, students will:

- Be able to identify the variables and challenges in measuring the economic impacts of tourism on destinations and countries.
- Understand the role of the tourism industry in the economic changes that impact communities today, and especially why communities may want tourism and why they may not.
- Be aware of the potential impacts of tourism development on the economic livelihood of residents in destinations – both negative and positive.
- Understand the contribution of tourism to international trade and national and regional economies.

Introduction

During the mid- to late nineteenth century, tourism's popularity in the newly industrialized countries of North America and Western Europe increased as a result of several important developments (Towner 1996), including:

1 Increased standards of living, especially with respect to education and literacy, as well as levels of disposable income.
2 Improvements in transportation technology, especially railways and steam-ships, which enabled more people to travel at a lower per unit cost.
3 Introduction of annual holidays and a 40-hour week for some workers, especially as a means of ensuring a healthy workforce. This provided more time for travel and recreation over longer distances away from home.
4 Improvements in print communication technology, which meant that there were more books, magazines and newspapers available to the emerging middle classes, at relatively low cost, in which they could read about travel and leisure experiences and opportunities.
5 A growing appreciation of the importance of leisure in society.

As a result of these combined effects, a mass travel culture began to develop as the newly emerged middle classes and upper working classes aspired to travel during their holidays. Governments and the private sector, especially railway companies, quickly recognized and encouraged these trends and thus tourism became another industry for making money (Votolo 2007). It became a tool 'to achieve certain goals of economic growth and restructuring, employment generation, and regional development through the provision of financial incentives, research, marketing and promotional assistance' (Hall 2000: 25). These goals were, and still are, very important for governments at all levels around the world.

Travel and tourism are vital elements of the world economy. They provide employment, tax revenues and development incentives to growing numbers of regional and local economies. In addition the travel and tourism industry serves to enable other industries to develop through its role in supporting business, meetings and convention travel, as well as providing an attractive leisure environment in many locations. This chapter examines the economic importance of tourism, its role in contemporary economies and its contribution to the livelihood of individuals in a destination community. To understand the tourism economy first requires some fundamental understanding of microeconomic and macroeconomic concepts.

Basic and non-basic industries

Tourism is different from traditional industrial sectors in many ways (see also Chapter 1). One difference is that it is considered an export industry, yet the international tourism product never leaves the country where it is produced. Instead of bringing the product to the consumer, the consumer is brought to the product (see Figure 1.7 for a typology of tourism and travel-related international business and different modes of supply of services in GATS). To understand this requires an understanding of the concept of *basic* and *non-basic industries*.

Basic industries are those that bring money directly into a community by exporting all or nearly all of their production. Traditional examples of basic industries are *primary* industries, such as agriculture and mining, and *secondary* manufacturing industries, like automobile and home appliance production. All of these are usually *export* industries – products are exported out of a community or country, and money from the outside is brought back to the community or country to pay salaries and to invest in product development and new ventures. Urban theory suggests that economies cannot survive without basic industries as any substantial change in a basic industry's output or level of earnings will have a substantial impact on the entire economy (Tiebout 1962; Blair 1995).

By contrast, *non-basic industries* (also sometimes referred to as *support industries*) are those that circulate money through a local economy, but do not directly bring significant amounts of new money into a place. These are mostly *tertiary* service industries, such as retail sales. Changes in most non-basic industries will only have a limited impact on an economy's overall condition.

Travel and tourism are also tertiary service industries, but because they bring new, outside tourist money into a community, as well as providing a supporting role for basic industries, they function as basic export industries and help to grow the local economy. Thus, companies and individuals who serve tourists are earning money that originated from sources outside of the destination region. They then spend that money in the destination place and region, supporting additional salaries, production and profits beyond the direct tourist transaction. This economic value of tourism is the single most-cited reason why places with resources of potential interest to tourists are so often keen to develop and promote those attractions. According to the World Trade Organization (WTO), tourism is one of the top five export categories for over 80 per cent of the world's countries, and is one of the principal sources of foreign currency earnings for at least 38 per cent of countries (UNEP 2001a). For example, travel receipts are of major importance in Africa's commercial services exports. It is estimated that more than one-half of its commercial services exports consist of travel services and, in contrast to transportation and other commercial services categories, Africa's travel receipts exceed its travel expenditures (i.e. more people spend money in Africa than Africans spend money outside of the continent) (WTO 2006).

As noted in Chapter 1, another distinction of tourism that makes it different from traditional export industries is that both the producer and the consumer must be present in a shared time and geographical place for a tourism transaction to occur (Daniels and Bryson 2002). This is also referred to as co-creation or co-production. Other economic activities that have this same characteristic are health care and higher education. And just as the introduction of online education is transforming that service industry, so too is tourism being transformed by online travel services (Beldona 2005; Lew 2009), though the tourist experience still occurs almost exclusively in a shared physical space (see also the discussion in Chapter 2 on the relationship between tourism and services).

As an additional benefit, tourism developments are typically considered amenities and improvements to the quality of life in a place (see Chapter 5), and as such their impacts are often perceived to be beneficial. However, like most economic activities, tourism developments take up space, cost money and use resources. Depending on the level of tourism development, its impacts can have broad social, cultural and physical implications. The economy-related social impacts of tourism are discussed in this chapter, while culture-related social impacts are the focus of Chapter 4.

The development of recreation resources are closely related to tourism development. The difference between recreation and tourism is that recreation sites are intended primarily for local residents, while tourism sites are intended for 'outsiders'. Using this dichotomy means that recreation is a non-basic industry, while tourism is a basic industry. Although this distinction is theoretically straightforward, in practice it is very fuzzy, depending on how the difference between resident and tourist is defined (see Chapter 1 for formal definitions of a tourist), and what the market mix is between these two groups at any one point in time at a location where the product is consumed, such as a leisure site.

In addition to site facilities, recreationists and tourists share similar motivations, behaviours and activities – and have similar impacts as well, with the exception of the impact of travel. However, because recreation is more of a local phenomenon than is tourism, recreation impacts are often more locally acceptable than those of tourism. Tourism impacts are attributed to outsiders ('those tourists') and therefore much more has been written about them, especially when the impacts are perceived to be negative in some way.

Supply and demand

A common approach to studying the economics of tourism is to divide it into the broad areas of supply and demand, and then to examine each separately. The *supply-side* of tourism refers to the destination resources that are available for the tourist. These include facilities and attractions of all kinds (such as parks, beaches, shopping and entertainment), as well as supporting infrastructure (such as transportation, hotels and restaurants) and services (such as travel agents, and recreation programmes and activities).

The *demand-side* is the market (or people) who will be drawn to the attractions and facilities. This includes the different types of tourists (or market segments), such as those from different places of origin, of different ages and incomes, and with different motivations and interests. Looking at it from a particular destination's point of view, it is knowing who the clients or tourists are for that place.

Tourism development can be initially stimulated by either the supply-side or the demand-side of tourism, though both are required for ultimate success. In a supply-driven development model, the significance of a destination attraction is significant enough, in terms of what consumers find of value, that it creates its own market demand. Examples of natural features that could do this include the Grand Canyon in Arizona, Milford Sound in New Zealand and Tanzania's Mount Kilimanjaro. Examples of built sites that could probably generate tourism on their own include the Eiffel Tower in Paris, Cambodia's Angkor Wat and the Burj Dubai – the world's tallest building (as of 2008) in Dubai.

The distinctive nature of these prominent sites could generate a small tourism economy with only minimal tourism-specific marketing efforts and infrastructure development. A more significant tourism economy would require improvements to the supply-side resource, including improvements in the transportation system, building new and better accommodations, and hardening portions of the resource so it can handle large numbers of visitors without deteriorating. Transportation connectivity is especially important as without adequate access, tourists will not be able to fill hotels and tours. Media and marketing activities are also extremely important, both to create visitor awareness and to build a brand image for an attraction or destination. Demand-side marketing efforts, therefore, become increasingly important as the tourism market grows.

In a liberal market economy, it is assumed that commodities will be produced to meet the market demand for consumption. When people want to consume something (including tourism), the marketplace will respond by creating more goods and services for them. Economic success comes to those who are best able to meet a market demand. This is essentially the entrepreneurial ability to recognize a potential consumer demand, and then to create a product to meet that demand. Sometimes the entrepreneur can even create a demand that did not previously exist, by the successful marketing of a new product.

Prominent examples of demand-side tourism developments include theme parks, such as Disneyland and Sea World, the cruise ship industry and gaming destinations, such as Las Vegas, Nevada, and China's Macau. Structurally, these are all supply-side products, located in the destination. However, the location, design, amenities and special features of these developments are designed to be attractive to clients and visitors who comprise their principal markets. Market interests are constantly changing, however, in response to changing consumer interests and competition from other destinations, requiring adjustments to meet the new desires and expectations of tourists. It is also important to recognize that entrepreneurial activity does not occur in a vacuum. The governments and

Photo 3.1 The Rio Casino in Macau. Macau consists of a very small peninsula and two
small islands that were once a Portuguese colony on the south coast of China.
Since 2002, when Macau's gaming industry was opened to international
investors and developers, this small backwater city has been transformed into
one of the leading international tourist destinations in Asia (although
approximately 80 per cent of its visitors come from China and Hong Kong).
Most of the hotels and casinos, including the Rio here, are built on land
reclaimed from the sea. (Photo by Alan A. Lew)

agencies of different countries, regions and locations relax regulatory requirements and provide direct financial incentives to encourage highly desired firms and entrepreneurs to establish business in their destination rather than in others. Nevertheless, demand motivations must be understood to successfully design, develop and promote most tourist attractions and products. The tourist demand motivations discussed in this chapter focus on definitions that are used for statistical purposes. Tourist motivations are also discussed in Chapter 4 in the context of identity and self-actualization.

For a tourism economy to exist, *both* demand and supply must be present. The difference is on emphasis. Supply-driven economies generally focus more on the tourist attraction, including preserving its authenticity and educating people about it, while demand-driven economies generally focus more on consumer awareness, image creation and innovative recreational opportunities.

The tourism industry

As a booming transnational and multisectoral industry, there is hardly a place on the planet that is not connected to the tourism economy. In some places, tourists vastly outnumber local residents. Annual tourist arrivals in Hawaii, for example, outnumber residents by over six to one (7.6 million visitors in 2006, with 1.2 million residents; HVCB 2007; see also Photo 2.6). A powerful engine for moving people from one place to another, the tourism industry, as we have discussed in Chapters 1 and 2, is also very complex and difficult to define. Unlike traditional industries that focus on the making, selling and servicing of a single, core product (like automobiles or computers), the tourism industry has many different elements, some of which serve the travelling public exclusively (like passenger airlines), and others of which may serve more non-tourists than tourists (like restaurants and museums). This situation has been termed 'partial industrialisation' (Leiper 1989: 25), which refers to a condition

> in which only certain organisations providing goods and services directly to tourists are in the tourism industry. The proportion of (a) goods and services stemming from that industry to (b) total goods and services used by tourists can be termed the index of industrialisation, theoretically ranging from 100 per cent (wholly industrialised) to zero (tourists present and spending money, but no tourism industry).

In contrast, from a supply-side perspective (see Smith 2004; and Chapter 1), the tourism industry includes any economic activity that contributes to the production of a tourism commodity. In other words, the tourism industry includes those parts of an economy that would greatly diminish in size or even disappear in the absence of tourism. Although 'greatly diminish' is not a statistical constant, we can still identify the major economic sectors that are associated with the travel and tourism industry to include:

- *Passenger transportation*: which includes airlines, airports, trains, buses, taxis, private automobiles, boats and ferries, cruise ships, the service and repair of these transportation modes, and travel agents and online transportation booking services. The transportation sector serves many local and regional residents, day-trippers who do not stay overnight in a destination, and non-leisure business travellers. Day-trippers and business travellers are sometimes considered tourists and sometimes not. Whether or not they are included in a country's or community's tourism arrivals and receipts will have a major impact on the final numbers that they publish. (Note that passenger transportation accounts for about 25 per cent of the total international trade in transportation services. The majority of these services are actually for freight transportation (WTO 2007b).)
- *Accommodation*: which includes hotels, motels, resorts, campgrounds and the homes of friends and relatives, and cruise ships; businesses that service these different accommodations; and travel agents and online hotel booking services. Usually only travellers who stay in hotels, motels and resorts are accounted for in local and national economic statistics because their numbers and expenditures can be estimated from sampling and the sales tax collected by local governments. Tourists staying with friends and relatives are the most difficult to account for statistically.
- *Food and drink*: which includes restaurants and other forms of eating establishments, from street hawkers to cruise ships; coffee shops, bars and other drinking establishments; other providers of food and drink to tourists, including grocery stores; food and drink wholesalers; and businesses that provide other services to restaurants and food and drink providers. In the International Standard Industrial Classification (ISIC), which is widely used for business census statistics, this category is usually referred to as 'cafes and restaurants', and is more narrowly defined (excluding grocery stores and drink wholesalers, for example). Determining the economic impact of tourism on the food and beverage sector is difficult because some restaurants (such as those in hotels) mostly serve tourists, while others mostly serve local residents. Tax revenues and sales figures are the main data source for estimating the importance of food and beverage firms in a local economy, though this usually excludes informal activities in this sector, such as street vendors.
- *Attractions*: almost anything can be an attraction with the right mix of marketing and social context; they vary considerably in scale, form and special interest or mass appeal (Lew 1987; see also Chapter 1). They can be publicly or privately owned, accessed for free or for a fee, and vary in their degree of authenticity. Establishments in all three of the other sectors of the tourism industry (transportation, accommodation and restaurants) can also be significant tourist attractions in and of themselves, especially if they have some heritage or nostalgia character. Most of the more available statistics on tourist attractions are narrowly focused on a few types, such as museums, amusement parks and recreation areas. More serious studies of the tourism economy require creative forms of sampling to understand the economic

impacts of a broader range of tourist attractions (cf. Ksaibati et al. 1994; Stynes 1999). Like food and drink, the balance between tourists and residents at attractions varies considerably from one place to the next.

The accommodation and food and drink sectors together form the 'hospitality industry', though sometimes some forms of transportation are also included in that definition. Some establishments also overlap the four sectors defined above. A cruise ship, for example, is a form of transportation, a type of accommodation, a place for eating and drinking, and an attraction in itself.

As with many service industries there is, in reality, no clear and simple way to meaningfully aggregate their diverse products into a single generic product that can be compared to more traditional manufacturing industries. This is because a large part of what the tourism industry offers is not a material product, but an experience that is purchased and kept more in one's memory than in the living room. The tourism product cannot be stored in a warehouse because it does not exist until the tourist experiences it, and it disappears once that experience is over (see also Chapter 1).

An alternative approach to understanding the economics of tourism is not to think of it as a service or even an industry, but to 'instead think about production chains and information flows within and between companies' (Debbage and Ioannides 2004: 100). Smith (1998) described this process as starting with *primary* inputs and factors, such as land, labour and capital, and then converting them into intermediate facilities, such as airplanes, restaurants and hotels, which finally produce the final output of the tourist experience. While this describes the *supply chain linkages* that exist for the tourism industry overall, the same approach can be applied to any subset of companies to understand the relationships and dependencies that exist in a particular destination.

Tourist expenditures and tourism commodities

The magnitude of tourism as an economic activity in a destination economy (whether a country or a city) is usually measured by determining the total expenditures made by visitors in the course of a trip. This includes expenditures made on behalf of a visitor, such as through contracts with a hotel made by a tour operator in assembling a tour for sale. Certain exceptions made prior to a trip may also be included as trip expenditures, such as the purchase of fuel made immediately before departure. Expenditures such as dry cleaning made immediately after the trip and linked to trip activities may also be counted.

As we mentioned in Chapter 1, the major challenge in measuring the economic magnitude of tourism can be summed up as four dilemmas:

- Visitors consume both tourism and non-tourism commodities
- Residents consume both tourism and non-tourism commodities
- Tourism industries both produce and consume tourism and non-tourism commodities

- Non-tourism industries both produce and consume tourism and non-tourism commodities.

Anything that can be purchased, including both goods and services, is a *commodity*, though some definitions limit commodities to tangible goods. By creating tourist attractions and the promise of tourist experiences, the tourism industry manages to put a price tag on almost all aspects of our contemporary world. This is known as *commodification* (which is the more acceptable term among researchers, though the term 'commoditization' is also widely used). The United Nations World Tourism Organization (UNWTO) has defined a *tourism commodity* as any good or service for which a significant portion of demand comes from persons engaged in tourism as consumers (UNWTO 1994). 'Significant portion' is not defined, but if we arbitrarily assume that this is more than 50 per cent, then a tourism commodity is any product or service for which at least 50 per cent of the buyers are tourists.

Based on this definition, a restaurant with 51 per cent of its customers being tourists is a tourism commodity, while the restaurant next door with only 49 per cent of its customers being tourists would not be. Additional challenges arise when considering that the balance in the proportion of tourist and non-tourist consumption will often vary among communities and neighbourhoods, from one season to another, and between midweek and the weekend. There may also be reasons to use a different cut-off than 50 per cent, such as for a metropolitan city with a large population compared to a small town.

Input–output models

In practice, only a small number of commodities account for the bulk of tourism spending in comparison to commodities oriented toward residents. Identifying the portion of demand for a commodity that is directly attributable to tourism is a task that requires significant data about tourism supply and demand. In the past this was done with extremely complicated and time-consuming microeconomic input–output models (Stynes 1997). These models measure the monetary flows into and out of a community, and between every business and business sector within a local economy (including both tourism and non-tourism businesses). Input–output models can provide a comprehensive snapshot of a local economy. However, they are only applicable to one point in time, when the data were collected, and are usually too difficult to apply to regional and national economies to make their effort worthwhile.

The more common way that countries measure the economic impact of business activities is through macroeconomic surveys and tax data collected from sales made by businesses that are categorized by the ISIC and the Central Product Classification (CPC) (Smith 2004). These are standard systems for assigning a category for every business in an economy for statistical data compilation and analysis.

The categories in these classification systems are used to compare the economies of different cities, states and provinces, and countries, and are aggregated to hide proprietary information about individual companies. They are the basis for the System of National Accounts (SNA). The SNA is the accounting 'framework used by most countries to collect, order, and analyze macroeconomic performance' (Lew et al. 2008: 27). Unfortunately, neither the ISIC nor CPC includes a 'tourism industry' entry, and so tourism has never been included as an industry in any country's SNA. In response the tourism industry, along with tourism-related government agencies and industry lobbying bodies, have developed a separate approach to measure the impact of tourism: the Tourism Satellite Account (TSA) (Smith 2004).

The Tourism Satellite Account system

The national government accountants (statisticians) in France were the first to explore ways to analyse aspects of a nation's economy that are not adequately represented within the SNA. To do this they developed a concept called *comptes satellites* (satellite accounts), which is known as Tourism Satellite Account (TSA) when applied to the tourism industry. The TSA approach identifies the tourism percentage of each ISIC/CPA-listed industrial sector (account) in a country's overall economy (SNA) and identifies the economic contribution of tourism from that calculation (Smith 2004).

For example, tourism may account for 15 per cent of the food and beverage industry's profits in one country, but 50 per cent in another. The tourism portions of each sector can then be added up to determine tourism's overall contribution to a country's economic performance. TSAs are usually constructed for a national economy, so national averages guide the selection of commodities that are included in the analysis. The identification of tourism commodities will, however, vary among nations. Crafts, textiles, agricultural products and fishery products may receive a significant portion of their demand from visitors to some countries, but much less in others. As a result one country may include them in their accounting, while another may totally ignore them. For this reason it is impossible to compare TSA results between two places, unless you are absolutely confident that the same methodology was used in each, from data collection through application to the TSA.

One example of a country that conducts a TSA is Norway (Table 3.1). According to the Norwegian TSA, tourists in Norway increased their consumption from about NOK41 billion (4.97bn euros; US$4.14bn) in 2004 to almost NOK43 billion (5.35bn euros; US$4.46bn) in 2005. Foreign tourists spent NOK26.4 billion in Norway, and business travellers spent close to NOK18 billion (Statistics Norway 2006). According to preliminary figures, the value of the total production of tourism industries was NOK7.4 billion higher in 2005 than in 2004, which in volume is a 5 per cent increase. Hotels and restaurants were the largest sectors and, when combined, accounted for 33 per cent of the total tourism industry production (or total tourism consumption of tourism products). As a result of the TSA, it was

Table 3.1 Total tourism consumption in Norway, 2004 and 2005. NOK million, current prices. (Includes tourism consumption in Norway by resident households, resident industries and non-residents.)

Characteristic tourism products	2004	2005*	As % of total consumption**
Accommodation services	8 418	8 809	10.1
– of this: hotel services	6 942	7 324	8.4
Food and beverage serving services	11 942	12 741	14.6
Passenger transport services	24 710	26 122	30.0
– of this: transport by railway, tramway and suburban transport	1556	1 617	1.9
– of this: transport by scheduled motor bus transportation and taxi operation	3 203	3 572	4.1
– of this: transport by inland water transport	1 322	1 156	1.3
– of this: transport by ocean and coastal water passenger transport abroad	3 419	3 548	4.1
– of this: passenger transportation by air	15 210	16 230	18.6
Package tours and car rental services	9 231	9 844	11.3
Museum, sporting activities, etc.	2 178	2 374	2.7
Total tourism consumption of tourism products	56 479	59 890	*68.8*
Other products			
Food, beverages and tobacco	4 671	4 973	5.7
Clothing and footwear	1 015	1 079	1.2
Souvenirs, maps, etc.	1 286	1 355	1.5
Other transportation costs	7 502	7 506	8.6
Other commodities and services*	11 905	12 286	14.1
Total tourism consumption of other products	26 379	27 199	*31.23*
Total consumption expenditures by tourists	82 858	87 088	*100*

Source: derived from Statistics Norway 2006.
Notes
* Provisional figure.
** Figures are rounded, may not add up to 100 per cent.
Average interbank conversion rate NOK to EUR 31 December 2004: 0.12120.
Average interbank conversion rate NOK to EUR 31 December 2005: 0.12450.

estimated that tourism industries accounted for 3.1 per cent of Norwegian gross domestic product (GDP) in 2005 (Hall et al. 2009).

TSAs can also be used to determine which industrial classifications have the highest proportion of tourism-related sales in order to try and relate tourism economic activity and behaviour to research on standard industrial classifications. For example, in New Zealand the Accommodation, Cafes and Restaurant (ACR) industrial category (the hospitality industry) has the highest proportion of businesses with tourism-related sales (74 per cent) (meaning that 26 per cent of ACR businesses have no tourism-related sales), with the next largest being the transport and storage category at approximately 35 per cent (Statistics New Zealand 2006). TSAs can also be used to calculate labour productivity in tourism compared to other industries. For example, in Australia tourism-related labour productivity was relatively low compared with some other industries (e.g. the 'manufacturing' or 'wholesale trade' category), although higher than others (e.g. the 'agriculture, forestry and fishing' category) (Hall 2007a). Tourism also had a 'significantly lower operating profit margin, 15.2 percent, than the average of all industries, 22.0 percent' (DITR 2002), although it should be noted that service industries overall tended to have a relatively low level of profitability.

One of the most significant problems with the TSA approach is that the selection of sectors to include in the analysis, and the portions of each sector to assign to tourism, is subjective and open to debate. Although the TSA provides a useful basis for understanding the economic effects of visitor-related activity and its inter-industry linkages within an economy, the TSA approach can too easily include items as tourism outputs that are far removed from tourism, at least as it is popularly understood (Productivity Commission 2005). For example, in calculating the global economic size of travel and tourism, the WTTC includes the full budget of the US Federal Railroad Administration, the US National Park Service, and the US Fish and Wildlife Service (Lew 2008). This situation has led some commentators to argue that the use of TSAs, and the promotion of their results, by tourism interest groups has misled people in many countries because they exaggerate the size of the industry and the employment it provides (Leiper 1999). In addition, because the content of the TSA calculations vary so much from one country or place to another, they cannot be used to compare places or even to compare tourism with other industries. The TSA is best used to understand linkages and compare changes within a single economy at one point in time, or over a specific time period for which data is available.

Global and supranational economic impacts

The travel and tourism industry is the world's largest commercial service sector industry. The World Trade Organization (WTO, not to be confused with the UN World Tourism Organization or UNWTO) is the leading international body that monitors international macroeconomic data. The data follow the value of imports and exports between countries and across major economic sectors that are grouped into either Merchandise or Commercial Services. Travel and tourism is part of the WTO's Commercial Services group, but is not encompassed under a single category. Instead, the WTO includes a Transportation category and a Travel category. The WTO defines these categories as:

> *Transportation* covers all transportation services that are performed by residents of one economy for those of another and that involve the carriage of passengers, the movement of goods (freight), rentals (charters) of carriers with crew, and related supporting and auxiliary services.
>
> (United Nations et al. 2002: 36)

Travel covers primarily the goods and services acquired from an economy by travellers during visits of less than one year to that economy. The goods and services are purchased by, or on behalf of, the traveller or provided, without a quid pro quo (that is, are provided as a gift), for the traveller to use or give away. In addition, a *traveller* is an individual staying for less than one year in an economy of which he or she is not a resident for any purpose other than (a) being stationed on a military base or being an employee (including diplomats and other embassy and consulate personnel) of an agency of his or her government, (b) being an accompanying dependent of an individual mentioned under (a), or (c) undertaking a productive activity directly for an entity that is a resident of that economy.

(United Nations et al. 2002: 38–9)

Although the WTO data include both imports and exports, this discussion focused on exports as, from a destination community perspective, they have a more direct visible impact on a destination's economy (although in macroeconomic terms they are of equal importance). For tourism, *exports* are the receipts that a country received from the money that tourists spent in that country (as discussed previously), while *imports* are the expenditures that residents of a country make when they travel outside of their home country, along with products imported by tourism providers. The difference between the two is called the *tourism balance of trade* or *tourism balance of payments*. In the case of the EU, for example, in 2005, tourism expenditure and receipts were nearly in balance with expenditure of 235.6bn (US$196.3bn), and receipts from tourism of 232.6bn (US$193.8bn). About two-thirds of EU member states were in surplus (received more tourist money than their residents spent elsewhere). The economic significance of tourism is especially pronounced in popular small countries where receipts exceeded expenditure by a factor of two or more, including Greece (4.5), Spain (3.2), Malta (2.8), Portugal (2.6) and Cyprus (2.5) (European Communities 2007).

In 2006 *Transportation Services* (not all of them carrying tourists) accounted for 22.9 per cent of international exports in commercial services, while *Travel Services* (mostly the hospitality industry, and excluding transportation) made up 27.1 per cent (Table 3.2). Compared to other commercial service sectors, the Travel Services sector has steadily declined in its relative economic importance at the global level since 2000 when it accounted for 32.1 per cent of international exports in services. Transportation Services, on the other hand, have remained roughly stable over that time period. All other Commercial Services (including communications, construction, finance, insurance, recreation and computer services) comprised the balance of the international service sector trade (see also Coles and Hall 2008).

There were 704 million air passenger trips taken in 2006, and air and sea passengers together accounted for about 25 per cent of the Transportation Services sector, though regionally it ranged from 20.1 per cent in the EU to 33.3 per cent in the United States (WTO 2007b). The rest of this sector is mostly freight

Table 3.2 World exports of merchandise and commercial services, 2006

Product group	Value 2006 ($bn)	Share product group	Share total export trade
MERCHANDISE*	11,497	100.0%	80.7%
Fuels and Mining Products	2277	19.3	16.0
Office and Telecom Equipment	1451	12.3	10.2
Chemicals	1248	10.6	8.8
Automotive Products	1016	8.6	7.1
Agricultural Products	945	8.0	6.6
Other Manufactures**	904	7.9	6.3
COMMERCIAL SERVICES	2755	100.0	19.3
Travel	745	27.1	5.2
Transportation	630	22.9	4.4
Other Commercial Services+	1380	50.0	9.7
[Tourism Services]++	[916]	[33.2]	[6.4]
TOTAL EXPORT TRADE	14,252		100.0

Source: WTO 2007a, b.
Notes
* Merchandise trade products are categorized as Agricultural, Fuels and Mining, and Manufactures; Manufactures include Iron and Steel, Chemicals, Office and Telecom Equipment, Textiles, and Clothing.
** Other Manufactures include Iron and Steel, Clothing, and Textiles.
+ Other Commercial Services include Communications, Construction, Insurance, Financial, Computer and Information, Royalties and Licence Fees, Personal-Cultural-Recreational, and Other Uncategorized Business Services; they range in size from 6 per cent of all Commercial Service exports to less than 1 per cent.
++ Tourism Services is a total of the Travel category and 25 per cent of the Transportation category and 1 per cent of the Other Commercial Services category, this is not an official World Trade Organization category.

transportation. In addition, a record 842 million international travellers spent a record US$745 billion in 2006 (WTO 2007a, b). International tourist receipts grew 9 per cent in 2006 over 2005, and while not the fastest growing sector in international trade, its steady growth and relatively low entry costs make tourism an attractive option for countries and communities seeking to develop the service sector of their economies to expand job opportunities.

Although not formally part of the WTO structure, a *Tourism Sector* can be estimated from the data for Travel Services, the passenger portion of Transportation Services (excluding freight), and the recreation portion of the Other Commercial Services sector. Together, this Tourism Sector comprises a third of international trade in all commercial services, and about 6.4 per cent of all international exports, including merchandise products. This Tourism Sector would have been the sixth largest sector of the global economy in 2006 based on the WTO's categories (Table 3.2), following trade in fossil fuels, telecommunications and computer equipment, automotive products, and agriculture (WTO 2007a, b).

Regional patterns

In addition to comparisons with other exports sectors, the WTO trade sectors can be compared by region and country. Table 3.3 shows the regional distribution of exports and imports in Travel Services between 2000 and 2006. Comparing exports (a country's receipts from tourists) with imports (expenditures by a country's residents abroad) gives us indications of which regions are benefiting the most from tourism and where tourist money is flowing to and from.

Based on these macroeconomic global comparisons, Europe received the greatest share of international receipts in Travel Services. At the same time, Europeans also spent more outside of their home countries than did travellers from other regions (European Communities 2007). Furthermore, Europe's share of both receipts and expenditures had continued to grow, probably reflecting the growing ease of travel within Europe and the relative strength of the Euro currency during this period.

Like Europe, Asia's international tourism receipts (excluding transportation in Table 3.3) has been growing, but unlike Europe, Asia's tourist expenditures had declined as a proportion of global tourist expenditures. The higher share for Asia reflects the growing popularity of Asian countries as tourist destinations, with both China and Southeast Asia being at the forefront of growth in the region. The stable and slightly declining share of expenditures by Asian tourists may reflect a shift to more intra-regional travel in Asia, which would be less expensive than trips to Europe or North America. As a result, Asian tourists may be spending much less per person per trip in 2006 in real terms than they had in 2000.

In opposition to the European pattern, North America had experienced a significant decline in its relative share of both international receipts and expenditures. The huge decline in receipts is directly related to the September 11, 2001

Table 3.3 Share of Travel Services in total exports and imports of commercial services by world regions, 2000 and 2006

Region	% Exports (receipts) 2000	% Imports (expenditures) 2000	% Exports (receipts) 2006	% Imports (expenditures) 2006
Europe	45.8	47.1	48.2	49.3
Asia	18.1	22.6	20.6	22.2
North America	24.4	19.2	17.9	15.4
South and Central America	4.9	3.5	4.6	2.7
Africa	3.1	1.8	4.3	2.1
Middle East	2.7	3.4	2.7	4.7
CIS*	1.0	2.4	1.7	3.5
Total Value (US$bn)	465	434**	745	695

Source: WTO 2001a, b.
Notes
* CIS = Commonwealth of Independent States, which includes Russia and most of the former Soviet Union countries.
** Year 2000 imports were not reported by the WTO; this number was estimated from total commercial services in 2000 and 2006.

terrorist attack on the New York World Trade Center, and the stringent border controls that were subsequently introduced (see Case 3.2). A somewhat smaller, but still significant decline occurred in American tourist expenditures in other countries. The actual expenditures were approximately the same in both years, but relative to the growth witnessed elsewhere around the world, North American travel had been stagnant, at best.

Most of the international travel that takes place in the world is within the Europe Union (EU) and the European Economic Area (EEA). This is a reflection of the relative ease of border crossing among member countries. The 1985 Schengen Agreement among European states allows for the abolition of systematic border controls between participating countries and includes provisions on a common policy on the temporary entry of persons (including visas), the harmonization of external border controls and cross-border police cooperation. The agreement is between most EU states and Iceland, Norway and Switzerland. Ireland and the UK

Photo 3.2 The Schengen Agreement: on a winter day along the border of Finland and Sweden. This is the former customs and immigration office where everyone crossing into Sweden from Finland used to stop. Today, the only reason to stop is to visit the sports bar, with its restaurant and nightclub. The ease of travel between the Finnish and Swedish communities has functionally created a single city on the border, which supports a new Ikea store on the Swedish site. Ikea is a Swedish company whose large home furnishing stores are a form of 'shoppertainment' (shopping and entertainment) that attract people from a very large market area. Are Ikea stores tourist attractions? (Photo by Alan A. Lew)

only participate in cross-border policy cooperation measures (Hall 2008a). As a result, the EU member states comprised 72 per cent of the total of Europe's exports in the WTO's Travel Services sector, and 41.8 per cent of the global exports in this sector in 2006. Beyond Europe, the majority of international travel takes place between Europe and North America, followed by travel between those regions and Asia. The dominance of Europe and North America in both tourism receipts (exports) and tourist expenditures (imports) is a reflection of the high cost of travel in these two regions – relative to most of the developing world. The high ranking of Asia is due to the very large, and growing, middle-class travel population in that region.

The rest of the world receives far fewer tourists, and their receipts, and sends many fewer tourists abroad. Two patterns, however, are notable. The Central and South America region and Africa both receive considerably greater receipts from tourists than their residents expend as tourists in other countries. This reflects the popularity of these regions among travellers from wealthier economies in Europe and North America, and the high level of poverty found in each region, which prevents their residents from participating in international tourism. It also means that in relative terms tourism is often a much more important part of the economy of the countries in Central and South America and in Africa than it is in other regions, which can make them more sensitive to fluctuations in travel demand.

North America also shows the pattern of receiving a higher proportion of tourism receipts (exports) and a lower proportion of tourist expenditures (import). This is most likely due to it being a popular and a relatively expensive destination (during the time period in Table 3.3) for European and Asian tourists. In addition there is the tendency for many Americans not to travel to other countries, with only about 27 per cent of US citizens holding a passport in 2006 (Leffel 2006).

The Middle East and the Commonwealth of Independent States (CIS, mostly members of the former Soviet Union) show the opposite pattern. Their residents spend more money in other countries as tourists than do tourists from other countries do when coming to the CIS and Middle East. This occurs because these regions are not very popular tourist destinations, mostly due to the difficulty of travelling in large portions of the CIS and the Middle East. At the same time, many of the residents of these regions are wealthy enough to easily travel overseas. Travel Services are currently a less significant sector of the economy in these two regions in comparison to the rest of the world. Asia also shows this pattern, and with its rapidly growing middle class, shares some characteristics with the CIS and the Middle East.

Although their proportions of the global Travel Services economy are relatively low in the less developed regions of the world, their tourism receipts (exports) have been growing at a much faster rate than in the more developed economies. The CIS experienced an average annual growth of 24 per cent in Travel Services from 2004 to 2006 (WTO 2007b). In Asia, Travel Services averaged 18 per cent growth, while Africa averaged 16 per cent and Central and South America averaged 13 per cent. By comparison, Travel Services in the developed regions of

Europe and North America both grew at an average of 9 per cent a year for the same time period.

These macroeconomic data show that tourism is a large and important part of the economy in every region of the world. In the larger and more developed economic regions, tourism is a mature industry whose economic impact is seen in the sheer volume of tourist expenditures (imports) and receipts (exports) that it generates. In the smaller developing economic regions, tourism is a rapidly growing sector of their economies that is often vital in providing a positive balance of trade (i.e. more receipts than expenditures) (Coles and Hall 2008).

National and regional economic impacts

Measuring economic impacts

The economic value of tourism to a destination is often calculated by multiplying the number of visitors by the average expenditure per visitor for each day of their visit (Stynes 1999). In addition, the resulting number is multiplied to reflect the number of times that each dollar, or portion of a dollar, is re-spent within a community before it moves completely out of the community. This is known as the *multiplier effect*.

For example, when a tourist pays one dollar to a hotel, $0.20 might go as profits directly to the international corporation that owns the hotel and $0.30 might go to imported supplies and services from outside the community. These would be considered money that leaves the local community almost instantly, and are known as *leakages*. High-end tourism developments, which are often built by multinational corporations, tend to have very high leakage rates.

Of the remaining $0.50 from that original dollar, $0.10 might go to local taxes and utilities, $0.20 might go to employees, and $0.20 might to go to local obtained supplies and service. In this scenario, the $0.50 left from the original $1.00 remains in the community after the first round of the $1.00 in expenditures. When that $0.50 is re-spent on other needs, a portion of it will leave the community and a portion will remain to be re-spent again. This is repeated until all of the original dollar has left the community. In this case, the economic *multiplier* might be about 1.67 before no further amount of the original money is left in the community.

This example also shows how tourist expenditures have both direct and indirect impacts on a local economy. *Direct multiplier effects* are generally those that are related to the initial spending in the industry because they directly lead to employment and growth in that industry. More tourists staying in a hotel will lead to more hotel jobs, higher hotel incomes and hotel expansion. *Indirect* (or 'induced') *multiplier effect* impacts typically occur in the second and third rounds of spending, including the spending by hotels on local supplies and services (such as laundry services) and spending by employees on their personal needs. These expenditures result in increased sales, income and jobs for the larger community. As mentioned above, for each round of spending per unit of initial tourist expenditure, leakages will occur from the local and regional economy until little

104 Tourism and its economic impacts

or no further re-spending is possible. In sum, 'the tourism multiplier is a measure of the total effects (direct plus secondary) which result, from the additional tourist expenditure' (Archer 1982: 237).

Tourism multipliers are, therefore, concerned with the way in which expenditure on tourism filters throughout a particular economy, stimulating other sectors with each round of expenditure. This is based on the Keynesian economic principles of the recirculation of a proportion of income by a region's recipients into consumption spending, which then encourages further employment and income. Several types of multipliers have been developed, but the tourism multiplier may be seen as 'a coefficient which expresses the amount of income generated in an area by an additional unit of tourist spending' (Archer 1982: 236). As noted in the example above it is expressed as the ratio of direct and secondary changes within an economic region to the direct initial change itself.

The size of the tourist multiplier will vary from region to region and will depend on several factors including:

- the size of the area of analysis
- the proportion of goods and services imported into the region for consumption by tourists
- the rate at which the currency circulates
- the nature and patterns of tourist spending, and
- the availability of suitable local products and services.

The size of the tourist multiplier is a significant measure of the economic benefit of tourism because it will be a reflection of the circulation of tourist expenditure through an economic system (Sinclair et al. 2003). In general, the larger the size of the tourist multiplier the greater the self-sufficiency of that economy in the provision of tourist facilities and services. Therefore, a tourist multiplier will generally be larger at a national level than at a state, provincial or local level, because at a state or provincial level, leakage will occur in the form of taxes to the national government and importation of goods and services from other states. Similarly, at the regional and local level, multipliers will reflect the high importation level of small communities and tax payments to state/provincial and national governments (Hall 2007a).

Within a community, each industry generates different multipliers. In addition, tourism industry multipliers vary from one community to the next. Because of this, to fully understand the economic impact of tourism in a place requires very complex modelling of the entire local economy through the input–output modelling approach that was discussed above. If this is done, it can be possible to further subdivide tourist expenditures into:

- different tourist market segments (usually based on accommodation types) to identify their different spending patterns;
- different spending categories, such as accommodation, restaurants, gasoline, groceries, amusement, museums and souvenirs;

- different economic sectors, such as manufacturing, transportation, service, and retail.

As a measure of economic benefit from tourism, the multiplier technique has been increasingly subject to question, particularly as its use has often produced exaggerated results (Sinclair et al. 2003). Other forms of economic activity may also produce similar or better regional multipliers than tourism. Nevertheless, despite doubts about the accuracy of the multiplier technique, substantial attention is still paid to the results of tourism economic impact studies that use it. In addition, once a multiplier is identified for a place or region, no matter what the methodology, it tends to continue to be used for years to come, and even applied to other places and regions that are assumed to have similar economic patterns. Neither of these applications are a valid use of the tourism multiplier.

Common errors in tourism economic analysis

There are a number of challenges in identifying the economic impacts of tourism. Given the inherent mobility of the tourist, the first issue is how the boundaries of the local economy are defined. States, provinces, municipalities and countries have clearly marked boundaries, many of which have specific economic importance. Many smaller areas and communities do not have such boundaries.

Boundaries are important in determining which businesses to include and which to exclude from data collection. A related geographic issue is determining who is a tourist and who is not. As discussed in Chapter 1, distance is one of the primary determinants in defining and distinguishing a tourist from a resident, but it can also be a very arbitrary one. In the United States, a common, but not universal, standard distance to determine this is 50 to 100 miles. Other distances may also be used to differentiate between different classifications of tourism mobility. For example, the Western Australian Tourism Commission (WATC) (1997) estimated that in 1996, over 10 million pleasure-oriented day trips were undertaken in Western Australia. The definition of 'day trip' used by the WATC (1997: 1) was 'a trip taken mainly for pleasure which lasts for at least four hours and involves a round trip distance of at least 50 km. For trips to national parks, state forests, reserves, museums and other man-made attractions the distance limitation does not apply'. The reasons for the selection of those specific parameters are not known, although the lack of consistency in concept definitions across jurisdictions highlights one of the difficulties in undertaking comparative research on tourism and human mobility (Hall 2005b).

Time is another issue in determining who is and who is not a tourist. Temporary or seasonal visitor expenditures are usually considered a part of tourism. However, at what point does a seasonal visitor become a local resident? In the United States, if someone stays in a community for one day less than six months, then they are normally considered a tourist. On the other hand, the US government considers anyone who stays in a foreign country for 364 days a tourist – at 365 days they are considered a resident of that country. These distinctions have important implications because the longer someone stays in a place, the higher their total expenditures will be, but the lower their daily expenditures will be.

Additional issues arise as to whether or not the expenditures of certain *types of travellers* should be included as tourism. Categories that can be either tourists or not include business travellers, those visiting friends and relatives (VFR), hospital patients, college and boarding school students, and military personnel. (Also in the United States, if a home is rented out for less than one month, then it might be considered a form of visitor accommodation for tax purposes, depending on local ordinances and laws.)

Once the region and tourist are defined, then one has to decide which *expenditures* to include. From the perspective of the destination (supply-side), expenditures typically begin when the visitor arrives at the border of the community or region under study, although pre-purchased accommodation and attractions could also be included. When assessing total expenditures from the demand-side, one would also include expenditure made related to the trip before departure and after return. This demand-side approach focuses more on tourist behaviour, which is useful for marketing purposes, and less on the impact of tourism on the destination's economy.

Another definitional issue relates to which *industries and activities* to include in different sectors of the economy. Local arts and crafts and local agricultural products are most likely to be truly local in origin and production. Transportation fuel and manufactured souvenirs are examples of tourism products that are typically imported from elsewhere. Many of these products will also be purchased by locals, and separating out the rate of local purchases from tourist purchases can be a challenge. Tourism Satellite Accounting (discussed above) attempts to do this at a national level and is increasingly being used to compare the contribution of tourism to different regions within a country, as well. For example, in the Nordic countries, sub-regional comparisons of tourism using TSAs are available in Finland and Norway (Hall et al. 2008), but a separate analysis is often required for those local and regional economies. It is important, however, to ensure that all variables and definitions are kept constant to make such comparisons valid.

'Guesstimating' economic impact

Complex economic modelling is often not possible for many smaller destinations given the time, expense and statistical data required. In countries where such data are recorded and publicly available a quick and easy way to measure the economic impact of tourism is to review the taxes collected in a community or region from hotels and restaurants. From this, one can calculate the total expenditure in a community in these two categories over a period of time, and then estimate how that might be expanded to total tourist expenditures. Hotels are fairly straightforward. Although some local residents do use local hotels, the vast majority of room revenues come from tourists and related travellers.

Restaurants are more difficult and might be approached by estimating the number of tourists over a period of time (from hotel revenues), and the average number of meals that each tourist might have had during an average stay in the community. This could be used to separate the restaurant revenues into tourist and local expenditures. Other tourism-related activities can be estimated using these same methods – tax revenues and estimates of the number of patrons who are tourists and their expenditures. This approach works, but also requires that all of the assumptions used are made very clear when the final results are presented. Unfortunately this last requirement is often not met.

This quick and easy approach to assessing tourism's economic impact assumes that some form of comprehensive sales or value-added tax exists and that details of its collection are available in an appropriate format. This is often not the case in many economies. However, because Accommodation and Restaurants are a standard international industrial classification at least some data, if only at a national aggregate level, will usually be available. An alternative approach is to survey visitors to a place and ask them how long they stayed there, how much they spent and what they spent their money on. This approach requires more time and resources than the sales tax method, but can also be used to gather valuable additional information on tourists, their expenditures and their behaviour (Stynes 1999) (Case 3.1). The surveys can also be used to refine the estimations and extrapolations of sales tax data to make them more accurate.

Case 3.1: American Indian Country tourism

A survey method approach was used to estimate the number of visitors and expenditures to an Indian reservation in the American Southwest (for proprietary reasons the reservation name is withheld). On this reservation, arts and crafts were sold directly from village homes, as well as at shops, and provided a major source of income that was generally untaxed and untracked in any way at the local level. In addition to expenditure information, visitors were asked to indicate where they spent the night while in the region. This provided the percentage of visitors who stayed in specific locations for which monthly accommodation rates were known. These rates were used to expand the survey sample to estimate the total number of visitors to the reservation in a year, based on their percentage of the estimated total visitors to the region, which was based on accommodations data. In this case, a known number was used to estimate an unknown number for the region, and then a known percentage, based on the survey, was used to estimate an unknown number for the community. The total estimated number of annual visitors was multiplied by the reported average expenditures from the survey to determine the total annual tourist expenditures for the Indian community.

A lot of assumptions were required to get to the total visitors and total expenditures derived in this study. A major assumption was that the visitors to the reservation stayed overnight in the region at the same rate as non-reservation visitors. This could be accurate, or it could be considerably different. To compensate for this uncertainty, the study included a range of total visitors and expenditures based on lower and higher assumptions.

Photo 3.3 Roadside American Indian arts and crafts shops on the Navajo Indian Nation
(Reservation) in northern Arizona. Craft sales can provide a significant income
for American Indians in Arizona and New Mexico. Almost all of these sales,
especially from stalls like those here, are part of the region's informal
economy, as the reservation lands are not subject to state regulation and sales
tax jurisdiction. (Photo by Alan A. Lew)

The survey found that the median income level of the reservation visitors was a
third higher than the median US income level at the time of the survey, their mean
age was 45, and over half had a postgraduate education. Over 60 per cent were
travelling primarily to visit the American Southwest's Indian Country, and they
included visits to several different reservations in their itinerary. This was clearly a
wealthy, well-educated and highly desirable niche market that any destination
would appreciate.

For this particular reservation, the study found that 70 per cent of tourist
expenditures on the reservation were for arts and crafts, generating several million
dollars in direct income each year. This did not include sales of arts and crafts to
off-reservation art galleries and gift shops, which would easily double the income
generated by arts and crafts to tribal members. During their stay, the median
expenditure per visitor was $4.00 for each hour the tourist was on the reservation.
Food and beverages was the second largest expenditure after arts and crafts. The
results showed that tourism, while not large in its total number of visitors, had a
huge economic impact on the reservation, enabling artisans, in particular, to
maintain a comfortable lifestyle.

About a year after this report was given to the Indian reservation (it was not
released to the public), another report was published and widely distributed in the
news media which indicated the significant economic impact of Indian reservation

tourism in the American Southwest. The methodology used in that report was not clearly defined in the media reporting. After some investigation it was determined that the public report had been based on the average tourist expenditure numbers from the earlier, single reservation report. A committee had made a guesstimate as to how those numbers of visitors and expenditures might apply to the many other Indian reservations in the Southwest. The totals were then published as facts that made front-page headline news throughout the region.

Source: Lew and Van Otten (1998).

Economic goals of tourism

The economic impacts of tourism development can be both beneficial and detrimental to a destination. Among the positive economic benefits that have already been mentioned are employment generation and export receipts.

Tourism is a labour-intensive industry and, as mentioned above, job generation is one of the major positive economic benefits of tourism. The most common tourism-related jobs are housemaids, waiters and waitresses, gardeners, and kitchen workers, though higher-level skilled maintenance and managerial positions are also generated. Tourism creates jobs for local people, especially those with low skills, and in many locations offers more income opportunities for women and migrant groups than do other economic sectors. The benefits of the highly gendered nature of tourism employment, however, is not without its criticisms (Riley et al. 2002). For example, in the Australian case, Hall (2007a: 227) noted that although tourism is relatively labour-intensive compared with some other industries, 'that labour input is vulnerable to minimisation by tourism plant owners who wish to restrict costs, thereby creating a situation in which employment is often casual, part time, and underpaid – and lower paid jobs dominated by women and migrants'. The economic and employment opportunities generated by tourism development have thus aroused a great deal of controversy regarding their true value to the host community in comparison with other industries.

Other economic benefits that destinations seek through the development of tourism and recreation resources include:

*   *Diversifying and stabilizing the economy* Tourism is often sought as an alternative to economic decline in other sectors of an economy, especially following the closing of extractive industries such as mining and timber. This is because mining and timber industries are often located in scenic mountain areas that have tourism potential. Even in diverse economies, having a strong tourism sector can provide more stability as different parts of an economy tend to rise and decline at different times.
*   *Increasing the tax base that helps fund the government* Travel and tourism were estimated to have contributed over US$800 billion in direct and indirect taxes worldwide in 1998, and was projected to double that amount by 2010

(UNEP 2001a). Most governments are funded by some combination of taxes on property, sales and income. Hospitality and transportation facilities pay property and sales taxes, and their higher-paid employees pay income taxes. Government funds may also support tourism development by providing infrastructure support, such as roads and utilities, as well as sea walls for beach protection in coastal areas.

These infrastructure improvements usually benefit the entire community, not just tourists, improving local transportation, communication, security, and pride (Torres and Momsen 2004). In the long run, a stronger local economy will benefit education and health programmes. A more advanced tourism economy requires better skilled workers, which local educational institutions can provide. Health facilities are often improved with tourism, as tourists from wealthier countries demand higher levels of care, which the tourism industry helps to develop through lobbying and donations. Improved health and education builds human resource capacities to further diversify and stabilize the local economy.

- *Attracting businesses and services to support the tourism industry* These are businesses that benefit from the indirect multiplier of tourist dollars. While governments may be more directly involved in attracting a large tourism-related enterprise, such as an all-inclusive resort or a major theme park, there will also be a need for producers and services that support these major projects. These include, for example, laundry and linen services, appliance and maintenance companies, construction and architectural services, food and beverage wholesalers and retailers (which may include farmers), and marketing and human resource services, among others. A successful tourism economy may require active recruiting of these businesses and the training of residents in needed skill areas.
- *Attracting businesses and services seeking tourism amenity locations* Many tourism destinations have the climate and scenic, cultural, educational and recreation amenities that also make them attractive places to live. This makes them attractive to service and high-technology companies that are not bound to the location of raw materials, as were heavy industries such as steel and automobile manufacturing. Developing and improving the quality of a destination's amenities can attract these new economy companies because it makes for happier, and therefore more productive, employees. Amenity-seeking businesses also tend to be strong supporters of maintaining the quality of the destination place, including its economy, its culture, its social structure and institutions, and its environment (Hall and Williams 2008).

On the other hand, negative economic impacts can arise from tourism development through the costs of building or expanding supporting infrastructure and facilities (transportation, utilities and supporting structures), especially when these require materials and services that must be purchased from outside the community. For example, higher taxes for everyone are often needed to pay for improvements to water supplies or sanitation facilities. For less developed countries and regions, there may also be a significant cost in upgrading services and accommodation to meet international standards. This will often require outside technical assistance, resulting in more money leaving the community and country.

Other potential economic costs resulting from tourism and recreation development include:

- *Demands on public services and facilities increase at a faster rate than tax revenues to support expanded services* This is a common problem as governments are often underfunded. It becomes especially problematic when major developments are given tax reductions or tax holidays as a way of recruiting them away from other destinations. This may be done with the hopes that tax revenues from indirect multipliers will compensate for the tax breaks. This is a risk and the result can be seriously insufficient funds to address the increased pressure on roads, water and electricity, as well as recreation, health and education facilities.

- *Increased cost of living due to inflation in land values and the prices of goods and services* Amenity places have good prospects for tourism development, but are also attractive as second-home and permanent home locations for the wealthy and location independent buyers. As tourism develops, more outsiders are exposed to the destination and more may consider buying property there. In addition, entrepreneurs and workers may migrate to the destination drawn by its business potential. The construction industry may see these trends as a positive impact of tourism, but local residents working in lower-wage hospitality jobs will find that housing and the price of basic commodities are beyond the reach of their salaries.

- *Minimum wage and seasonal salary of lower-end employees in tourism and recreation who require additional public assistance and support* While a tourism economy does provide many jobs for people with limited training and skills, these jobs also pay minimum wages that, as already mentioned, may result in a large working poor population. There is a tendency for work in tourism to be socially segmented by education, socio-economic class, gender, race, ethnicity and age (Williams 2004). Employers exploit these divisions of labour to keep wages low. In addition, the seasonality that most tourist destinations experience results in some employees (often the most poor and least educated) only working part of the year. Such a situation is especially problematic when governments pay more attention to the demands of the rich, while ignoring the needs of the poor. In such a situation, it is common for these working poor to seek additional public welfare assistance from government and non-profit/non-governmental agencies to help them meet their basic needs for food, shelter and health. This can be an additional unanticipated burden on government and taxpayers.

An increase in the cost of living and the relatively low pay of tourism sector jobs may drive some local residents, particularly retirees on limited incomes, out of a community. A decrease in the traditional local population results in a decrease in available workers to fill service positions in the tourism sector. Businesses may need to actively recruit and hire non-locals, causing an influx of temporary workers during peak tourist seasons. This in turn results in a rapid employee turnover, challenging employers with the task of frequently training new employees.

- *Volatility and elasticity of tourism businesses to broader social and environmental crises* Tourism is highly sensitive to political upheavals, environmental changes and disasters, and price fluctuations in transportation and goods. For many destinations, the high *price elasticity* of tourism can be like a bungee cord – fluctuating between a big profit season and year, and a low-profit season and year as currency exchange rates, travel costs and consumer confidence changes. Tourism firms tend to adjust to these fluctuations by rapidly hiring or laying off workers, thereby exacerbating labour and income problems in the broader society (Williams 2004).

Examples of crises that can impact tourism to destinations include: droughts and a lack of snow in alpine destinations, as the European Alps have experienced in the 2000s; strikes, protests and riots by disenfranchised populations or unhappy labourers, who sometimes become violent and dangerous; earthquakes and volcanic eruptions, which often occur in coastal areas that are also popular tourist destinations; high visa exit or entry taxes, which were more common prior to the 1990s than they are today; and high fuel prices and transportation taxes that may become more commonplace in the future.

Case 3.2: Two biggest tourism crises of the decade

Since the turn of the millennium in 2000, there have been two major crises that have impacted tourism worldwide. The September 11, 2001, terrorist attacks against the World Trade Center in New York and in Washington, DC, had a huge impact on travel to the United States, specifically, but also elsewhere around the globe. International visitors initially avoided the United States for fear of additional terrorist activities there. At the same time, US border formalities became increasingly so severe that potential international visitors avoided the country to avoid the reputed hassle of passing through US ports of entry. Globally, similar airport security procedures were enacted, though visa and immigrations services were generally less severe than those in the United States. Terrorism concerns, however, also impacted global travel patterns in terms of where people of different nationalities and cultures went and what places they avoided. Tourists from the Middle East, for example, avoided North America and Europe and shifted their holidays to Southeast Asia.

In the United States there was a dramatic decrease in international arrivals. In the year 2000, prior to the 9/11 attacks, the United States received 26 million overseas visitors (excluding Canada and Mexico) (Garcia 2008). This fell to 18 million in 2003, increasing to only 24 million in 2007, with the Travel Industry

Association of America claiming that cumbersome airport security measures and a poor image of the United States abroad had resulted in a loss of US$26 billion in revenues in 2007 (ETN 2008). They estimated that US airlines lost $9 billion, hotels lost $6 billion, restaurants lost $3 billion and $4 billion was lost in government tax revenues. Although the US travel industry has lobbied to make the United States a more welcoming country, the government instead focused on hardening the country's borders (TIA 2008). A weak US dollar through the summer of 2008 did manage to attract a large number of Europeans and Asians to the United States, but this dropped off precipitously when the global credit crisis occurred in October of that year.

In July 2008, crude oil prices rose dramatically to a new historic peak of US$147 a barrel (Marketwatch 2008). Most airlines around the world operated at a loss throughout the northern-hemisphere summer of that year, as they cut passenger services and added fuel surcharges and checked baggage fees. Over 30 airlines worldwide declared bankruptcy in the first half of 2008 (Robertson 2008). Once the summer season was over, the airlines cut staff and reduced their number of flights, especially to smaller market cities, in an effort to increase demand per flight. These moves successfully allowed them to increase airfares, gradually moving them towards earning a profit. However, their respite was short-lived as the global credit crisis suddenly hit in October 2008 (Sharkey 2008). The credit crisis caused consumer confidence in the United States to reach record lows, and both leisure and retail travellers worldwide dramatically reduced their travel purchases, impacting both airlines and destinations (De Lollis 2008). Hotels in many destinations were already suffering from the reduction in air passengers from fewer flights, and now suffered even more when people stopped travelling. Airlines struggled to keep their fares high, but with fewer travellers this proved challenging, and many more bankruptcies were forecast (Robertson 2008). Economic recovery was not expected until well into 2009–2010, at the earliest.

These examples demonstrate how a political crisis and an economic crisis can have significant impacts on travel and tourism at all levels, from the global to the local. They can change global travel patterns, they have shifted global travel expenditures, they can result in job gains, job losses and bankruptcies in different economic sectors in different parts of the world. And they are often totally unpredictable both in when they will occur and how they will impact travel and tourism (see also Chapter 7 for a further discussion of *wildcard* events and their impacts on tourism).

Political economy of tourism

Political economy refers to the study of the 'production and distribution of wealth within a society, and the role of the state in these' (Williams 2004: 61). It is an interdisciplinary field that seeks to understand an economy by examining the social beliefs and political behaviour that it is based on. Many of the economic impacts of tourism are central issues in the study of political economy by social scientists. These include the globalization of investment, production and consumption processes and behaviour, and the process of industrial and regional economic restructuring (Debbage and Iaonnides 2004).

Tourism is an integral part of economic globalization, the commodification of places and cultures, and the transformation of cities from places of production to places of consumption. Tourism has been especially prominent as a tool of neo-liberalism, with its focus on building export economies to grow a country's economy. The political economy questions are: who controls these processes? who benefits from them? who is hurt by them? and how does the economic structure of capitalism shape these processes?

Capitalism, with its focus on generating investor profit, is inherently unequal in its distribution of labour and wealth. Therefore, tourism development, as a capitalist endeavour, has unequal positive and negative impacts. Some people and some places in a country or community benefit from tourism more than others. Tourism can exacerbate existing inequalities in a community, with local elites benefiting much more than the poor and other marginal populations.

As an international industry, much tourism development is led by transnational corporations (TNCs). The term is often used interchangeably with that of multinational corporations, or MNCs. Although the terms are very closely related, TNC is often used to refer to firms that operate at a global scale while MNC tends to be used for those that operate in two or more countries, but not globally. However, for the purpose of the present discussion, we will use TNC to cover both types of international business strategies.

TNCs in tourism include airlines, global hotels, chain restaurants and entertainment corporations. Some of the largest TNCs in these areas, based on foreign assets, include McDonalds (USA), Coca-Cola (USA), Cadbury Schweppes (UK), Accor hotel group (France), Best Western (USA) and Hilton (UK) (UNCTAD 2002; MKG 2005). TNCs control development and expropriate profits from their centres in the core cities of the developed world, while the destinations where the host–guest transactions occur are in the peripheral and less developed world. Table 3.4 shows the high concentration of global hotel ownership in the United States and Europe.

TNCs are closely associated with local destination elites, who identify more with multinational values than they might with the traditional values of local residents. This model of development is known as *dependency* (the periphery being dependent on the core), *neocolonialism* and *leisure imperialism*, and is criticized for having high economic leakage rates (Crick 1989).

Multinational corporations play a key role in globalizing the world economy. The largest TNCs have budgets that are larger than many of the countries that regulate them. They use their resources to lobby governments for concessions and policies that support their development goals. In much of the developed western world, states (national governments) have responded to such lobbying through deregulation.

Government deregulation in the tourism industry has been most evident in the airline industry since the 1980s, which spawned the growth of low-cost carriers, lower air fares, a huge growth in passengers and increases in tourism to more

Table 3.4 Home country headquarters of the 300 largest hotel chains, based on total rooms, 2005

Country	No. of company headquarters in country	Total rooms 2005	Total hotels 2005
USA	146	4068497	32122
England	11	652573	4770
France	5	573842	5027
Spain	15	306405	1544
Japan	16	159495	653
Germany	9	152709	892
China	12	120955	414
Canada	7	83233	466
Netherlands	1	47661	498
Belgium	1	45000	263
South Africa	5	33341	278
Singapore	8	30937	106
Switzerland	2	22827	106
Cuba	4	22554	131
Mexico	2	21590	112
Italy	4	21498	126
Brazil	3	20951	130
India	3	19316	177
Norway	2	18769	143
Thailand	5	16322	72
Hungary	2	13777	91
Australia	4	13697	87
Malaysia	4	12524	45
Poland	1	12000	64
Cyprus	2	11452	63
Portugal	2	11147	94
Finland	2	11046	63
Jamaica	2	9442	34
Ireland	1	7990	35
Sweden	1	7800	36
United Arab Emirates	2	7788	30
Syria	1	7350	20
Scotland	1	6374	71
Hong Kong	2	6151	27
Malta	1	5787	20
Indonesia	1	4512	20
Bahamas	1	4476	11
Tunisia	1	4148	17
Austria	1	4016	25
Egypt	1	3757	17
Korea	1	3529	5
Israel	1	3238	12
Turkey	1	3162	16
Czech Republic	1	2984	30
Zimbabwe	1	1953	17
Kenya	1	1843	21
Total	300	6 620 418	49 001

Source: derived from 'Hotels' corporate 300 ranking' (2006), retrieved 27April 2007, from: www.hotelmags.com

remote destinations. On the other hand, airplane comfort and service standards have declined, airline safety standards have been questioned, and overbooking and flight delays have increased.

In addition to deregulation, governments may intervene in the economy to support the development goals of the tourism industry. This approach tends to be preferred over deregulation in most of Asia. Government intervention is especially common in land development projects, where a government will provide incentives (previously discussed) or use its eminent domain authority (the authority to condemn and confiscate land for public uses) to support a resort or entertainment project. Governments may also intervene in an industry by nationalizing companies, in which the government buys a majority ownership in a private company. This approach is not favoured by entrepreneurs, and is the ultimate threat that governments hold over the private sector. On the other hand, private companies can choose to completely withdraw from a country.

The travel and tourism industry, like other industrial sectors, favours deregulation, and there has been a general decline in regulatory monitoring and intervention in environmental and social areas worldwide since the 1970s (Gordon and Goodall 2000; Hall 2008c). In its place there has been movement of corporations to establish self-regulatory bodies. The World Travel and Tourism Council (WTTC) is the largest body of this type that serves the travel and tourism industry. Its leadership is made up of top executives from MNCs in all of the major sectors of travel and tourism, as well as executives from destination marketing organizations. The WTTC issues reports on sustainable tourism (see Chapter 7), but also lobbies governments worldwide to support deregulation of the industry to make it easier for both its companies and the travelling public to cross international borders.

The relationships between MNCs and country or local governments vary from one place to the next. Some of the major factors that shape this relationship include the relative size of the local economy and the MNC, the economic goals of the local government and the MNC, the strength of political and financial institutions in the local economy, and the degree of local competition and acceptance of an MNC. The political economy under which tourism develops in a destination will shape the way that tourism is produced or made available to tourists, and how it is consumed by tourists and residents. This is why tourism in one place can be very different from tourism in another place, even within the same destination region.

Case 3.3: China's global hotel industry

The growth of tourism in China over the past 30 years, like the rest of the country's economy, has been nothing short of phenomenal. From almost no tourism during the Cultural Revolution (late 1960s to early 1970s), China in 2006 received 41.8 million international overnight visitors, making it the fourth most visited country in the world (CNTO 2007). Meanwhile, the number of domestic tourist trips in China reached 1.5 billion in 2006, and the China National Tourism Administration (CNTA) expected domestic tourist expenditures in 2007 would account for 70 per cent of the US$128.6 billion in travel industry receipts expected for that year (Rein 2007).

In the 1980s, many of the early international investments in China were to build upscale international hotels to meet the urgent needs of a small but growing number of foreign tourists to the country (Zhang and Lew 1997). These early hotels, which distinctly stood out from the drab grey of older post-Second World War buildings, were the first post-Communist structures to appear on the skyline of Chinese cities. They symbolized the modernization of China, and were among the first steps toward capitalizing and commodifying the Chinese landscape.

Chinese government policies restricted local resident access to the hotels, which were considered sources of deviant moral influences. And most of the early hotels were only permitted as joint ventures with Chinese government entities, with the intention of transferring construction and hospitality skills, and some profits, to China.

The result of that early period of hotel development, which lasted to the early 2000s, was a two-tiered hotel industry, consisting of newer high-end internationally owned and managed hotels and older domestic hotels that were low in quality and low in cost. China's domestic hotel industry was small and fragmented into many diverse and separately owned properties. As is the norm in China, the majority of the domestic hotels were owned by government entities, such as military units and agricultural cooperatives.

Thus, since the beginning of modern tourism in China in the late 1970s, international hotel operators have dominated the high-end market in the country, while older state-run hotels comprised the low-end, budget-priced accommodation. This reflected the political economy of the initial post-Mao Zedong era.

In 2005, China joined the World Trade Organization (WTO), which forced it to open its economy to direct foreign ownership of hotels, as well as in other sectors of its economy (Lew 2007a). This step is diversifying the hotel landscape by spurring rapid growth in both international chain hotels and new domestic chain hotels. Chain operations are more efficient in operations, reservations, management and brand development, especially when compared to China's older state-run hotels.

In 2007, Accor, Starwood, and Marriott each announced that they would quadruple the number of hotels they operate in China (Elegant 2006). Accor planned to reach 180 hotels by 2010, while Starwood and Marriott would go to over 100 hotels each in five to six years. Starwood's chains include Sheraton and Westin, while Marriott's chains include the high-end Ritz-Carlton, the mid-range Courtyard by Marriott, and JW Marriott Hotels and Resorts. Accor, based in France, is the world's largest hotel chain and includes the higher-end Sofitel and the budget Ibis and Motel 6 chains.

The 2008 Beijing Olympics prompted international hotel companies to add 11,000 additional luxury hotel rooms in Beijing (Chen 2007). The luxury hotel market also expanded rapidly, creating an oversupply in Shanghai, Guangzhou and Shenzhen after China's 2005 WTO accession. However, the hottest growth area for new hotels in China has been in mid-price and budget-priced hotels

(US$75 to $150 a night). Domestic demand for these hotels is high, especially along the new limited-access highways that are spanning across the country.

China's domestic hotel companies have recently begun to develop chain operations (Elegant 2006). The largest hotel operator in China is the Shanghai Jinjiang International Company, and is credited with building China's first modern budget hotel in 1996 (Lew 2007a). Its budget franchise chain, Jinjiang Inn, had grown to 250 hotels in 2007 (20 per cent of China's budget hotels). It plans to more than double that number by 2010 and will also start building hotels outside of China. Home Inns, founded in 2002, is the second largest domestic hotel chain, with 171 properties as of 2007.

In response to all this new hotel construction, renovations and in some cases demolitions, were initiated for the oldest state-run hotels. One impact of all this rapid hotel growth has been higher hotel prices as the lowest priced hotels are taken off the market. The rapid growth of all hotel sectors raises concerns that China's entire hotel industry, not just the high-end properties, may become overbuilt, which could bring prices down, along with profits.

The case study of China's hotel landscape shows the relationship between multinational companies and local companies, and how that relationship is affected by the political economy of both the country and larger globalization processes. China's political economy has undergone major changes from the Mao Zedong era to the contemporary period, and these changes directly impact the number and types of tourist accommodation in the country.

Community economic impacts

Commodification

A major issue of MNC growth in a local economy is the commodification of local culture and places for promotion and sale to tourists. Traditional arts, crafts, religion, rituals and festivals can change significantly to make them more available for purchase and consumption by tourists as in the form of souvenirs and photographs. Events may be shortened, offered on a more frequent basis and modified to be made more entertaining (Xie and Lane 2006). Crafts that may have originally been utilitarian, are made smaller so tourists can fit them in their suitcases. Sacred sites may not be fully respected as they become tourist attractions (Lew 1999). On a positive note, tourism can revive traditional arts and dances that may be lost under other forces of modernization and social change.

Policies can be adopted to manage the impacts of commodification, but in today's world, mass tourism and the relentless onslaught of modernization will have their impact (Britton 1991). Cultures change with time and in response to both internal and external influences. Outside of a museum, one cannot expect any people or landscape to remain fixed forever. The question is the degree to which local stakeholders are able to decide their own future within the context of choices available to them.

The commodification process can have four outcomes, reflecting the organization of production and consumption in the local economy (Shaw and Williams 2004):

- *Direct commodification* is charging for access to an attraction, such as a cultural performance.
- *Indirect commodification* consists of payments for services that support access to the attraction. This includes transportation and most of the hospitality industry.
- *Part commodification* is when part of the product is commodified and part of it is not. Volunteer tourism (sometimes called 'voluntourism') is a form of this, where the tourist pays for the trip, but is also expected to contribute work.
- *Non-commodification* consists of tourism experiences that do not involve any purchases, such as a walk-through open-air weekend market.

Any tourist site is likely to have a mix of these different forms of commodification. The exact mix would reflect how the place development is deemed most appropriate, and is influenced by the political decision-making process and the economic system that supports the site and its activities (Williams 2004). Thus, the

Photo 3.4 Souvenirs provide one of the greatest opportunities for cultural commodification, as in this example from a gift shop in the Emirate of Dubai in the United Arab Emirates (UAE). Here the traditional dress of Emirati men and women are made into caricature souvenirs, while traditional water containers are available on the back wall in a variety of sizes to meet the packing requirements of tourists. (Photo by Alan A. Lew)

political economy of one place might have large areas that are non-commodified and part commodified, while the same resources may be directly commodified in another place.

Commodification tends to increase over time (Britton 1991), possibly reflecting changes in the political economy. Decreases in commodification are rare, but possible in periods of economic decline when tourism businesses may close and tourists may try to stay with non-commodified friends and relatives to save money. Thus, different people at different times will benefit from the changes in the production of tourism products and their consumption. (Chapter 4 discusses the cultural impacts of commodification.)

Tourism linkages

Because tourism is such a diverse economic activity, it has the potential to support and interact with a variety of other local economic activities through the development of backward and forward linkages. *Backward linkages* are those connected to activities that enable the development and delivery of intermediate tourism facilities. These include food producers who sell to restaurants and the construction industry that builds hotels. *Forward linkages* are relationships with business activities that would not exist if tourism did not exist. These include tour guides and a significant portion of industrial laundry services and taxi and private bus services. It is through these linkages that tourist expenditures are shared and multiplied to help support and build the larger local economy.

The local tourism economy is less effective when these linkages are not developed. As already mentioned, a major criticism of international tourism in less developed economies is that it is dominated by transnational corporations with limited local linkages, high external leakages, and impacts through imported foods, management services, decor and resource consumption. (See, for example, the pro-poor tourism study Case 3.5 below.) For example, it is estimated that hotel facilities worldwide consume about a 100 TWh of energy and 450–700 million m^3 (25 billion ft^3) of water per annum and generate millions of tons of waste (Bohdanowicz 2009). Being aware of such criticisms, however, has prompted a growing number of international hotels to consciously address these issues through certification programmes such as the Nordic Swan labelling programme. Although it should also be noted that many consumers are themselves demanding that hospitality and tourism industries need to be more sustainable. According to a Green Hotel Association survey almost 43 million American travellers are interested in the environmental issues in the hospitality sector, while 80 per cent of those interviewed by Travelocity claim to be willing to pay extra to visit an ecofriendly destination or business (Hauvner et al. 2008; Bohdanowicz 2009).

Tourism pitfalls

Tourism has the potential to meet the objectives of tourists, residents, businesses and local governments to create higher levels of quality of life in destinations (see

Chapter 5). But, tourism development also has the potential to fail in meeting some or all of these objectives when there is a lack of balance between supply and demand and when there is insufficient focus on the quality of experience for the destination community, as well as for visitors. There are three general scenarios in which this could take place: overcapacity, undercapacity and poor quality (Table 3.5), and it is in the context of these situations that tourism impacts become a concern and receive adverse publicity. If there is an appropriate balance between supply and demand and quality tourism objectives, then the negative impacts of tourism are generally not a significant concern even though tourism continues to affect the destination. Indeed, as the following examples demonstrate, such pitfalls

Table 3.5 Tourism development pitfalls

Pitfall	Causes	Impacts
Overcapacity of infrastructure	• Unanticipated low demand • Over-construction of facilities	• Industry: financial losses from low occupancy rates and competitive price-cuts • Government: insufficient tax revenues to maintain infrastructure; loss of citizen confidence • Tourists: possible lower prices or greater value for their money • Residents: employment layoffs and salary reductions
Undercapacity of infrastructure	• Unanticipated high demand • Lack of resources (human and land) to meet demand	• Industry: income windfall due to the ability to charge high prices; increasing development costs • Government: loss of citizen confidence if competition with tourists results; threat of poor quality facilities due to development pressure • Tourists: lack of supply may lead to disappointments, including not feeling that they are getting value worth the money • Locals: much economic opportunity, but higher costs of living and may feel competition for resources from tourists
Poor quality experiences and facilities	• Poor management, training, maintenance in industry • High crime, civil unrest and economic decline • Pollution and over-exploitation of local resources (natural, culture and labour)	• Industry: lower cost, but shorter visits and fewer repeat visitors • Government: inefficient utilization of resources and economic opportunities • Tourists: disappointing experience, perceived poor value for money • Locals: less opportunity for quality recreation and meeting higher needs; alienation and resentment

can be regarded as temporary stages in the overall pattern of development in a destination, provided that remedial action occurs or that the balance between supply and demand is eventually achieved.

Overcapacity of the tourism infrastructure is a situation in which there are too many hotels, attractions or seats on airplanes than there are tourists to fill them. This is a fairly common situation in all types of real estate property investment, including residential, retail, office and accommodation. Rapid growth in any of these areas tends to attract more investors, despite warnings of an impending overcapacity. Overcapacity in tourism can also occur when political crises or environmental disasters strike a destination and dissuade tourists from going to them. (See Case 3.2 for how airlines have tried to deal with overcapacity issues.) The impact of overcapacity is a slowing of the tourism economy due to lower profits from increased competition, even though the actual number of tourists may not have declined. Tourists may benefit from this situation, but employees and managers may find that they have to work harder for less pay.

Singapore faced a serious hotel overcapacity situation in 1984 when, after a decade of strong and steady growth, there was an unexpected decline in tourist arrivals (Lew 1991). This occurred at a time when the hotel industry was rapidly expanding, with many large, new hotels under construction. These included the Westin Stamford, which at 73 floors stood as the world's tallest hotel for over a decade (until 1997). The precipitous drop in hotel occupancy rates caused some hotels that were under construction to stop building work and 'mothball' the structure until the market situation changed. Other hotels cut their rates to try to boost occupancy levels.

For tourists, there was great value to be found in visiting Singapore at that time, while for the broader tourism industry, and some local residents, there was a loss of income and employment. At the same time, however, tourist attractions were less crowded and, therefore, more attractive for local residents. The Singapore government acted quickly to address the problem, lest it face a loss of confidence by the tourism industry and the citizenry (Yeoh et al 2001). They conducted workshops with the tourism industry to develop a marketing plan for the city-state, and they held contests for residents to recommend new attractions for visitors. Historic preservation projects were also expanded, based on complaints from tourists that the old Singapore had disappeared. Some hotels also turned to human resource training and development as they waited for recovery.

Overcapacity situations are relatively common in destinations that serve markets that have growing populations, growing economies and a growing middle class. In such an environment, overcapacity is probably only a temporary situation – though how long that temporary situation will last is often difficult to determine.

Undercapacity occurs where tourist demand is greater than the ability of a destination to provide supply. This is most commonly associated with special events, such as a sporting event or a music concert. An example of undercapacity was the situation in Hong Kong in 1997, leading up to the former British territory's return to China on 1 July of that year. High demand, particularly for the 1 July handover week, resulted in near-record high hotel prices in an already very expensive city to visit (Chan and Lam 2000). (See the discussion in Chapter 2 on mega-events.)

Undercapacity can also happen when a market suddenly discovers a destination, usually through popular media, such as movies. It can occur when local governments are too incompetent to properly plan and facilitate public and private development projects. And it can be a chronic situation when a popular destination is just not able to expand because it has reached the limits of developable land, which might occur on an island or in an alpine mountain environment.

For the tourism industry, undercapacity can offer a wealth of opportunities. New developments and higher prices can bring profit windfalls, though development and operation costs may also be higher. Government officials may also be faced with a situation in which demand for construction for new developments and infrastructure outstrips local ability to ensure quality development. In this situation, poor quality facilities may be allowed to develop, causing tourist experiences to sour and creating long-term image and development problems for a destination. In addition, if they are not careful, this can lead to an overcapacity situation.

Being a sellers market, undercapacity situations put tourists at a disadvantage. Tourists may feel that they are being taken advantage of and, therefore, are receiving less value for their money. They may not get access to the attractions they want to see and they may pay more for everything. And tourists may compete with residents on streets, in parking lots, in restaurant and at attraction sites. This competition can lead to congestion, higher prices, and increased irritation for both hosts and guests. (See Chapter 4 on host and guest relations.)

With the exception of some high-end resort destinations that purposefully limit development to maintain higher prices, undercapacity situations in tourism are much rarer than are overcapacity situations. Due to seasonal fluctuations, some degree of undercapacity may be desirable during peak seasons, as it will result in a lower level of overcapacity during the shoulder and low seasons.

Poor quality tourism experiences and facilities may arise from a variety of causes and only a summary of these is provided in Table 3.5. Anything that causes the tourist to have an experience below what was expected might fall under this category of tourism pitfalls. While it is impossible to guarantee every tourist that they will have an ideal vacation experience, enough problems will result in a negative image of the destination, and possible overcapacity problems caused by a drop in visitor numbers.

This was the case when crimes against European tourists in Florida made world headlines in the early 1990s (see Rother 1993). Tourism from Europe to Miami declined precipitously, and was very slow to recover. In this situation, the decline in the social fabric of Miami resulted in an increase in poverty and crime, which spilled over and directly impacted the tourism industry. As the economic prospects and social conditions for the poorest of the Miami community declined, so did the visitor's tourism experience.

Poor quality hospitality services and facilities within the industry (mostly hotels and restaurants) may save money for entrepreneurs, but in the long term their businesses will suffer by not achieving their full income potential. In addition, poor quality leisure facilities create resentment from and reduce the experiences for local residents.

The quality of service offered by hospitality businesses is under more direct control by the tourism industry than are the larger community social conditions. Programmes to train hospitality employees to be more courteous to visitors can be very effective in improving the visitor's experience. Similar public information programmes have even been effective in changing local resident attitudes toward tourists and tourism.

More subtle problems can occur when a people or a culture within a community become a tourist attraction (i.e. become commodified). The commercial interests of tourism can be insensitive to cultural privacy, and excessive intrusion can result in a backlash of some sort. Many of the American Indian reservations in the American Southwest (discussed in Case 3.1) have instituted clearly demarcated rules of behaviour for visitors to minimize such intrusion. These rules are intended to maintain the quality of the authenticity of the Indian villages.

Case 3.4: Resident attitudes toward tourism development in British Columbia, Canada

Mountain communities have often turned to tourism as an alternative economic activity following downturns in their traditional extractive industries, such as mining and timber. However, residents of these communities often have mixed feelings toward tourism. They may welcome the income opportunities that it can bring, but they are wary of the lower paying jobs (compared to extractive industries), the visual and aesthetic impacts of large development projects, and the presence of new people and tourists in their community.

This was the situation in Valemount, a remote timber industry-based community of a little over 1,000 people in central British Columbia, when it faced two proposals for large-scale ski resorts (Nepal 2008). By the 1990s, the timber industry had all but stopped in the community and there was a recognition that the town's location high in the mountains near Jasper National Park offered much that the tourism industry would desire. To make itself more attractive to outside investors, the town spent $2 million renovating its retail district, and worked with provincial authorities to find developers. The result was one resort that was built in the late 1990s and the proposals for two additional resorts that together would bring 224 new hotel rooms, 120 condominium units, 400 single family homes and at least three 18-hole golf courses.

While the community's business sector favoured the proposals, community planners wanted to make sure that the development process proceeded in a manner that was acceptable to the larger resident population. A survey of citizen perceptions of tourism and recreation in the Valemount region was undertaken to achieve this. The methodology included a facilitated open town meeting and face-to-face interview surveys. The results are shown in Table 3.6.

Table 3.6 Valemount, British Columbia, tourism development survey responses

1 Highest satisfaction levels (over 50% of respondents satisfied)	
a Number of hotels	67.9
2 Lowest satisfaction levels (below 20% of respondents satisfied)	
a Current economic status of the town	8.3
b Community social events	11.9
3 Concerns: tourism development will increase . . . (cited by over 50%)	
a The cost of living	55.6
b The cost of outdoor recreation	55.1
4 Benefits: tourism development will increase . . . (cited by over 50%)	
a The number of local businesses	88.1
b The quality of outdoor recreation	78.6
c The outdoor recreation opportunities	75.9
d The community events and activities	56.8
e The quality of indoor recreation	53.7
5 Overall support for . . .	
a More diversified tourism development	82.3
b Tourist visits doubling in 5 Years	78.8
c Proposed resort development projects	56.0
d Tourist visits tripling in 5 years	49.4
e Building more hotels in the area	37.8

Source: Nepal 2008.

The survey results in this community show that the residents are concerned with the potential increase in the cost of living that tourism development may bring, but they are more concerned with a potential decline in the local economy if development stops. They also feel that there are plenty of hotels in town, and that there is a need for greater diversification of the local economy, including its tourism economy. Their opinions are mixed on the specifics of the two proposed development projects, but they strongly support economic development efforts and feel that they will improve the overall quality of life in Valemount.

This case study shows the mixed perceptions of the economic impacts of tourism on a community, and how difficult it is to know if a major proposed development will be positive or negative. However, based on the information collected in a survey like this one, public planners can work with developers to adjust their projects to better address the concerns and desires of the local community.

Source: derived from Nepal (2008).

Individual economic impacts

The economics of tourism development and tourism activities have direct impacts on the hosts in a community and on their tourist guests.

Conflicts with traditional land uses and occupations

Tourism development can come into competition with traditional uses of community-based resources, including water, energy, land and labour supply. These resources can become more scarce or degraded in the development process, and community residents may face higher taxes for infrastructure upgrade and maintenance costs. In some instances, land is confiscated from the poor, who may not have documentation of formal ownership, for tourism development. In other instances, places where locals practised their traditional livelihoods have become closed to residents so they can be used by tourists. This has been especially problematic in coastal areas in developing countries where beach resort construction has impacted access to traditional fishing grounds.

Tourism land development can be either concentrated in a single location or dispersed over a broad area. *Concentrated development* is often the result of a planning process that seeks to limit the infrastructure costs and other impacts to a single *sacrifice* location. Examples of concentrated development include the Waikiki area in Honolulu, Hawaii, and Bali's Nusa Dua luxury resort complex. *Dispersed development* is more likely to have developed over a longer time and in an organic manner, though exceptions to each of these generalizations are common.

Sacrifice areas, also known as *tourist enclaves*, are often very controversial for the locations in which they occur, because these tourism enclaves can include major land modifications and settlement relocations. However, they can also be less disruptive to the daily life of the broader community and can limit some of the potential negative social and environmental impacts, if properly planned. But at the same time local vendors and other businesses may have a more difficult time benefiting from the enclave, which has higher leakage rates (importing more goods), and causes fewer tourists to frequent the ex-enclave community. Isolated *all-inclusive resorts* can also have these types of impacts (Shaw and Shaw 1999). For example, in a study of enclave tourism in the Okavango Delta, Botswana, Mbaiwa (2005) reported that foreign domination and ownership of tourism facilities has led to the repatriation of tourism revenue, domination of management positions by expatriates, lower salaries for local workers and an overall failure by tourism to significantly contribute to rural poverty alleviation in the delta region. Such a situation raises significant questions about the socio-economic sustainability of tourism operations in the region, as well as the broader issue of whether tourism can be an effective pro-poor form of economic development (see Hall 2007c).

Employment for local residents in the tourism industry can be very different from traditional employment, especially in an enclave environment. A community's traditional work patterns can be affected, resulting in the abandonment of more traditional occupations in favour of easier, and possibly more lucrative, tourism-related work. This can lead to a transformation of household roles, with women and younger adults earning more money in hospitality jobs than older men, and resulting in significant social conflict in a family and a community (Shaw and Shaw 1999).

Photo 3.5 The Waikiki district of the city of Honolulu, Hawaii, is separated from the rest of the city by the Ala Wai Canal, which can be seen entering the ocean on the left side of this photo. The Waikiki beach was a retreat for the Hawaiian monarchy in the 1800s and some of its hotels date back to the early 1900s. Today it has by far the greatest concentration of hotels and tourists on the Hawaiian Islands, along with a wide variety of shopping and recreation opportunities. (Photo by Alan A. Lew)

The influx of affluent tourists bringing new money into a community not only provides jobs and helps to pay for community improvements, but can also make an economy dependent on tourism for its viability. Smaller and less diversified economies are more likely to come to rely on tourism to meet their basic social needs. This dependence can cause local cultures to abandon traditional work practices, especially agricultural livelihoods, in favour of trying to capitalize on the tourist's willingness to spend money on 'cultural' souvenirs.

Demographics and crime

As tourism grows it fosters changes in the size and the demographic composition of the host population, with increasing numbers of 'outsiders' moving to a destination in search of jobs. This is often accompanied by an increase in crime, which local residents attribute to tourism development. Crime increases with urbanization and population growth in general, and tourism can be one of several sectors in a local economy that encourages economic migration and population increases. Criminal elements are drawn to large numbers of disoriented tourists with money to spend and valuable cameras and jewelry to steal. Robberies, muggings, illegal drug sales, prostitution and confidence schemes target both tourists and local residents in many tourist destinations, especially when poverty is also a significant problem.

According to the US Department of Agriculture, rural tourism and recreational communities tend to have higher crime rates due to low-paid and low-skilled migrant workers entering the community to work in seasonal service positions (Brown and Reeder 2005). This increase in crime weighs on local residents in the form of increased taxes for more community policing and protection. On the other hand, increased employment, possibly through tourism, can reduce crime rates.

Child labour and sex tourism

Child workers (under 18 years of age) are estimated to comprise about 12.5 per cent of all employees in the global tourism industry (UNEP 2001b). They sometimes work long hours, preventing them from attending school, and receive lower salaries than adults. Global competition has pushed prices down throughout the tourism industry, which has made unskilled child labour more attractive. Even more children, and often much younger ones, work in the informal economy hawking souvenirs, snacks and services. Sometimes those services include prostitution. Children are drawn to this because of the money it provides to buy consumer goods that would otherwise be unthinkable for their impoverished families who survive on as little as a dollar a day.

Prostitution tends to increase along with the development of tourism in many parts of the world. In some countries, prostitution provides a way for women to escape from poverty either through the profits earned or by developing a relationship and eventually marrying a foreign tourist. In the Philippines, for example, becoming a prostitute who serves foreign tourists is a much more desired position than one who serves local clients.

The trafficking (sale and slavery) of children and women to brothels for prostitution is an illegal problem worldwide. Tourism is, unfortunately, one of the markets for this illegal trade, and child sex tourism is possibly the worst form of tourism. Legal forms of sex tourism exist in adult entertainment districts in destinations such as Thailand and the Philippines. However, an illegal and underground child sex tourism also exists, and is organized in much the same way as other niche tourism sectors, with established destinations, accommodation and resorts, tour operators and package tours, and local guides (in the form of taxi drivers and brothel staff).

Case 3.5: Pro-poor tourism through agriculture linkages

Pro-poor tourism is an approach to tourism development that advocates specifically for policies and practices that benefit the poorest people and communities in a destination. Torres and Momsen (2004) argued that creating business relationships between the tourism industry and local farmers is one of the best ways that tourism development can directly assist the poor in a less developed economy, where agriculture-based rural poverty is a major challenge.

Tourism development can have a significant negative impact on traditional agricultural areas, which are generally much smaller in scale than agribusiness farming. Tourism and tourism-related urbanization competes with farmers and their agricultural livelihood for land, water and labour. Workers may shift from agriculture to tourism during peak seasons, which may also be harvest seasons, causing farmers to shift to less sustainable, but less labour-intensive, farming practices. Farms may even be abandoned as workers find better paying jobs in tourism and other urban industries. An emphasis on foreign foods for foreign tourists results in increased economic leakages and potential changes in the foods produced by local farmers (Telfer and Wall 2000).

Tourism can also have positive impacts on local agriculture by developing backward linkages in which local farmers sell their produce directly to hotels and restaurants (Torres 2003). This can increase the profits of local farmers, which can lead to investments and improvement in local agriculture, and possibly stem the tied of rural out-migration. Tourism can also promote speciality crops that are unique to a region, both for sale in hotels and restaurants, and possibly for the export market as international tourists become more exposed to them.

In a study in the Mexican State of Quintana Roo, Torres surveyed 60 hotel chefs in the resort city of Cancun to determine where they purchased their food products and what their relationship was to local farmers and their products. She found that government planners expected backward linkages between hotels and farmers to develop naturally as the region's economy and infrastructure developed. Efforts to facilitate business relationships between farmers and the hospitality industry were isolated, poorly organized and ineffective.

The result of the survey was that local agricultural purchases from the State of Quintana Roo were very low compared to more distant locations (Table 3.7). The reasons cited for going outside of their home state to purchase food products included not being able to find the food locally, or not finding it in sufficient supply or quality. This was especially true for fruits and beef. Local vegetables were either too inconvenient or too expensive, while locally produced pork and mutton were considered not suited to the tastes of tourists.

Table 3.7 Percent of food products purchased by Cancun hotels by place of origin

Food category	Quintana Roo State	Yucatan State*	Rest of Mexico	Foreign sources	Unknown
Fruits	4.5	20.1	68.1	0.7	6.6
Vegetables	3.4	22.8	68.1	0.4	5.3
Meats	1.0	20.0	48.0	25.0	6.0
Poultry	9.0	64.0	17.0	5.0	5.0
Seafood	35.3	40.0	17.4	2.8	4.5
Dairy products	0.0	8.0	70.0	7.0	15.0

Source: Torres and Momsen 2004.
Note
* Yucatan State is adjacent to the northeast of Quintana Roo State.

Torres found that the reasons for this disconnect between Quintana Roo farmers and Cancun hotels were multifold. The most productive farmers tended to be farthest from Cancun and were mostly Mayan-speakers who had difficulty communicating with hotel chefs and intermediary buyers. This gave them little negotiating power. In addition, the farmers lived in more impoverished areas, not benefiting from the Cancun development; they lacked technical assistance and the capital and credit to invest in their farms, which resulted in quality and consistency issues. Corruption and kickbacks to food buyers also benefited farmers in the adjacent Yucatan State over those of Quintana Roo.

For the farmers of Quintana Roo to benefit from the significant tourism investment and growth in their state, Torres and Momsen concluded that 'The agricultural sector has to become the focus of intensive investment, training, organization and private sector–farmer joint ventures' (2004: 312).

Source: derived from Torres and Momsen (2004).

Entrepreneurship in tourism

Entrepreneurs are those individuals in the local economy who make the economic activity of tourism work (Harper 1996). They are the ones who take the financial risk to develop and innovate products that they hope tourists will want to spend money on. Entrepreneurs respond to opportunities that changing social and economic conditions offer. At the same time, they can impact the local tourism economy by creating new opportunities and new challenges. Successful innovations can create new and expanded markets (customers), new employment opportunities and can even evolve into new economic sectors, shaping regional and national development (Morrison et al. 1999). An example of this is the rise of microbreweries in the United States and Canada in the 1980s and 1990s, following the brewpub (a restaurant with an on-site brewery) tradition that has existed for hundreds of years in Europe (Flack 1997). These small, independently owned and operated eating and drinking establishments presented a new niche product that quickly spread throughout the country, offering new forms of leisure and consumption and contributing to the revitalization and gentrification of many older retail districts.

Although often portrayed as only a moderately innovative sector (e.g. Shaw and Williams 1998; Hjalager 2002), innovation is rapidly emerging as an important theme in tourism policy and development in the search for increased product, firm and destination competitiveness (e.g. Hall 2007b; Sundbo et al. 2007; Hall and Williams 2008) (see also Case 2.4 on tourism innovation systems). Under guidelines of the Organization for Economic Co-operation and Development (OECD) for collecting innovation data, 'An innovation is the implementation of a new or significantly improved product (good or service), or process, a new marketing method, or a new organisational method in business practices, workplace organisation or external relations' (OECD 2005: 46). Four types of innovations are identified:

Photo 3.6 Small tourism business development in the town of Lawrence in the South
Island of New Zealand. The town is primarily a 'toilet and coffee stop' on the
way between Dunedin and Queenstown and Central Otago. However, it has
used its mid-journey position to expand its business offerings to encourage
longer visits and therefore increased visitor spend. As well as cafés and
restaurants there are also a number of art and crafts shops – all geared towards
the tourist. (Photo by C. Michael Hall)

- *product innovations* – new or significantly improved goods or services;
- *process innovations* – new or significantly improved methods for production
  or delivery (operational processes);
- *organizational innovations* – new or significantly improved methods in a
  firm's business practices, workplace organization or external relations
  (organizational or managerial processes);
- *marketing innovations* – new or significantly improved marketing methods.

An obvious question to ask in relation to innovation and tourism is what is the
rate of innovation in tourism compared to the rate of innovation in other sectors?
Unfortunately, this question cannot be easily answered as, as noted previously,
tourism is not a standard industry classification. Nevertheless, it is possible to
study innovation rates in a sector such as the Accommodation, Cafes and
Restaurant (ACR) industrial category as a surrogate for the wider tourism industry.
This has been done for Australia and New Zealand (Case 3.6).

Case 3.6: Innovation rates in Accommodation, Cafes and Restaurants in Australia and New Zealand

Perhaps refuting the sometimes negative image of tourism businesses in terms of innovation, the 2005 survey of innovation by businesses in New Zealand indicated that the Accommodation, Cafes and Restaurants (ACR) sector's innovation rate of 50 per cent of businesses engaged in innovative activity is just below the overall innovation rate of New Zealand businesses (52 per cent), with the highest innovation rates being identified in the finance and insurance (68 per cent) and manufacturing (65 per cent) industries (Statistics New Zealand 2007) (Table 3.8). Although in terms of the relationship between innovation and firm survival (Hall and Williams 2008), it should also be noted that the sector had the lowest continuation rates for enterprises over the period 2001–6 with only 33.1 per cent of firms surviving to 2006 (Ministry of Economic Development 2007).

Although Australia also uses the Oslo Manual (OECD 2005) for the collection of business innovation data, the Australian definition of sectors and business size included in the national survey is different to that of New Zealand – further reflecting the problems with providing comparative national data, as already noted. However, Australia's data suggest that its ACR sector is slightly above (at 35.7 per cent) the country's overall rate (of 34.9 per cent) of businesses that are innovating (Table 3.9) (ABS 2006). In 2007, the Australian Bureau of Statistics (ABS) undertook a comparison of Australia and New Zealand business innovation using the ABS definition of business (five or more employees; the New Zealand survey uses six or more employees). They identified that overall the proportion of businesses innovating in New Zealand (43 per cent) was higher than that of Australia (35 per cent), and with respect to the ACR sector, 40 per cent of New Zealand businesses in the sector were innovating as compared to 36 per cent in Australia (ABS 2007).

Yet despite the reasonably high level of innovation in the ACR sector and the economic importance of tourism to Australia and New Zealand, Hall (2009b) noted that tourism was barely recognized in the innovation policies of the two countries. It is likely, therefore, that because tourism continues to suffer from its perception as being low skilled, low income and low value, it is consequently regarded as low on innovation, as well.

Source: derived from Hall 2009b.

Tourism-related enterprises range in size and complexity from small, locally owned businesses, to medium-sized enterprises and large, transnational corporations. While many communities actively seek out larger corporations for their large payrolls, it is actually the *small and medium-sized enterprises* (SMEs) that provide the most employment in a regional economy. For example, SMEs are estimated to comprise as much as 99 per cent of all firms in the EU. Although definitions vary, the EU considers small enterprises to be those with fewer than 50 employees (100 is used in the United States), while medium-sized enterprises are those with fewer

Table 3.8 Innovation rate in New Zealand: select industry sectors (last two financial years as at August 2005)

Industry	Total number of businesses	Total innovation rate (%)	Implemented (%)	Ongoing or abandoned (%)	Businesses without innovation activity (%)
Agriculture, forestry & fishing	3,201	42	36	6	58
Mining and quarrying	84	44	44	0	56
Manufacturing	5,604	65	58	7	35
Electricity, gas & water supply	18	52	45	7	48
Construction	3,312	41	39	2	59
Wholesale trade	3,222	61	56	5	39
Retail trade	5,823	46	37	8	54
Accommodation, cafes and restaurants	3,336	50	49	2	50
Transport and storage	1,488	53	49	4	47
Communication services	144	62	57	5	38
Finance and insurance	570	68	64	4	32
Property and business services	4,818	50	46	4	50
Education	567	58	55	4	42
Health and community services	1,950	59	57	2	41
Cultural and recreational services	618	57	51	5	43
Overall (all businesses)	34,761	52	47	5	48

Source: derived from Statistics New Zealand 2007, in Hall 2009b.

Notes

Total figures for all businesses in this industry category.

The target population for the Business Operations Survey 2005 was live enterprise units on Statistics NZ's Business Frame that at the population selection date: had an annual GST turnover figure of greater than NZ$30,000; had six or more employees; had been operating for one year or more; were classified to Australian and New Zealand Standard Industrial Classifications. New Zealand Version. An enterprise is defined as a business or service entity operating in New Zealand, such as a company, partnership, trust, government department or agency, state-owned enterprise, university or self- employed individual.

Table 3.9 Innovation rate in Australia: select industry sectors (2004 and 2005 calendar years)

Industry	Total number of businesses	Businesses innovating (%)	Businesses which started but did not yet complete or abandoned any innovative activity (%)	Businesses which were innovation active (%)
Mining	771	31.4	14.9	34.5
Manufacturing	18,201	41.7	16.5	43.1
Electricity, gas & water supply	187	48.8	26.6	52.1
Construction	13,774	30.8	10.0	31.0
Wholesale trade	13,299	43.4	17.2	46.8
Retail trade	30,644	27.5	7.5	28.2
Accommodation, cafes and restaurants	13,591	35.6	9.6	35.7
Transport and storage	5,477	34.0	10.9	34.3
Communication services	446	35.5	18.2	36.3
Finance and insurance	4 359	37.9	15.1	39.5
Property and business services	36 019	30.3	13.4	32.7
Cultural and recreational services	4,487	32.9	12.6	34.4
Total (all businesses)	34,761	33.5	12.2	34.9

Source: derived from Australian Bureau of Statistics (ABS) 2006, in Hall 2009b.
Notes
The scope of the 2005 Innovation Survey was all businesses in Australia with employment recorded on the Australian Bureau of Statistics Business Register of five or more employees, except those classified to: SISCA 3000 General government; SISCA 6000 Rest of the world; ANZSIC Division A Agriculture, forestry and fishing; ANZSIC Division M Government administration and defence; ANZSIC Division N Education; ANZSIC Division O Health and community services; ANZSIC Division Q Personal and other services.
As at 31 December 2005.

than 250 employees (500 in the United States) (Shaw 2004). Both the EU and United States consider a micro-business to have fewer than ten employees. SMEs are considered to be more innovative and also more likely to develop backward and forward linkages with other local businesses than are large corporations.

Economic migration is also a form of entrepreneurship, with migrant workers seeking new opportunities in new places. These economic entrepreneurs often fill positions that those in the destination society find undesirable (Riley 2004). In tourism these typically include housekeeping and landscape maintenance, though it can also include restaurant employees, taxi and other transportation drivers, front desk receptionists, and even managerial employees. Economic migrants also start their own businesses, either working on contracts with larger companies or serving tourists directly. *Lifestyle entrepreneurs* migrate to start new businesses in high amenity destinations (Ateljevic and Doorne 2000). The lifestyle, climate, scenery and ambience of many popular tourist destinations can compensate for the lower incomes and higher financial risk that lifestyle migrants face. Tourism in many destinations would be considerably more expensive, and possibly less developed, without these sources of lower cost entrepreneurs and workers. For example, in a national survey of bed and breakfast (B&B) accommodation operators in New Zealand, Hall and Rusher (2005) found that the most significant responses with respect to the importance of goals when starting the business were related to issues of lifestyle as well as the desire for social interaction. Earning income is not a significant necessity (slightly more than a third of all respondents). At first glance this might support the idea that such operations are therefore not well managed. However, a series of further questions regarding such perceptions clearly indicate that this is not the case. The vast majority of respondents see profit as being extremely significant and there is also a strong desire to keep the business growing although this is also matched by enthusiasm for lifestyle gains and job satisfaction. As Hall and Rusher (2004) noted such a combination of goals may cause tensions but it does not mean that B&B operations are any less well managed or customer oriented than in the formal tourism sector. Indeed, as Hall and Rusher (2005) argued, the social motivations of running a B&B clearly indicate the potential for stronger customer orientations than in those businesses with staff who are not so interested in making social connections with their consumers. Moreover in terms of attitudes towards government assistance, there was very little support for the notion that such support was essential for business growth.

This national study, along with a previous regional study (Hall and Rusher 2004), found that respondents indicated strongly that the risks and responsibility of operating a B&B business were worth the perceived gains in lifestyle. An examination of the attitudes of owners to the issues of applying strict business principles to a business that is known for its lifestyle benefits demonstrated that there is evidence of a strong business philosophy being balanced against the personal goals of the business owners to enjoy a good lifestyle. These charac-teristics are therefore consistent with those exhibited by small family-owned tourism enterprises (e.g. Getz and Carlsen 2000). Hall and Rusher (2005) argued that there were a number of consequences that emerged from understanding the importance of lifestyle as a strategic business objective of small tourism businesses:

- the understanding of small business performance and entrepreneurial success needs to incorporate quality-of-life measures as an important component of entrepreneurial decision-making (see also Chapter 4 on quality-of-life issues in tourism);
- lifestyle and amenity factors are a significant factor in the locational decision-making of new small tourism business ventures;
- in looking at entrepreneurship and business development issues of small firms it becomes important to look at the stage of the life course of the entrepreneur, or more likely, copreneurs, given that the businesses tend to be run by couples in a relationship who are also partners in a business setting as well;
- the innovative capacity of many small firms to successfully undertake business promotion without participating in collective marketing activities means that there are not necessarily any incentives for joining formal tourism networks;
- lifestyle businesses demonstrate substantial business skills and may actually be more service-oriented than many larger firms in the formal tourism sector.

Not all entrepreneur activities are innovative pioneers (Shaw 2004). Tourism entrepreneurs may copy what they see as successful formats, resulting in the common situation of one shop after another offering the same souvenirs and T-shirts. Furthermore, not all entrepreneur activities are successful, and failures can have significant impacts on destinations. At their worst, entrepreneurs can be seen as exploiting the natural and cultural resources of a destination. Business failures are much higher for smaller businesses due to inexperience, insufficient capital and greater sensitivity to market fluctuations. Failed business ventures, such as hotels and restaurants, can stand as empty blights on the local landscape. This is most likely to be a problem in instances of oversupply than undersupply.

One of the very few tourism specific studies of firm survival and mortality was undertaken by Santarelli (1998) in Italy, with respect to the survival of hotels, restaurants and catering firms between 1989 and 1994. Examining new firm survival rates, defined as the share of new firms starting in 1989 that were still in existence at the end of each subsequent year, it was observed that one year after start up, 68 per cent of firms still operated, dropping to 45 per cent by the sixth year. This was significantly lower than the 59 per cent survival rate identified for Italian manufacturing firms during approximately the same period.

In examining regional variations in survival rates, Santarelli (1998) observed that in those areas where the barriers to entry (measured in terms of such factors as advertising and capital-raising requirements, shortage of bank credit and lack of modern infrastructures in the surrounding area) are lower, the entry process is less selective and a larger share of entry attempts are more likely to fail or have a short life expectancy. In addition, the duration of firm life appeared to be affected by the dynamics of industry evolution within particular regions. It may, therefore, be easier for new firms to survive in those regions in which the tourist industry was growing at higher rates. Since their entry into the industry is less likely to inflict market share losses on their rivals, the likelihood of retaliation by incumbents will be lower (Hall and Williams 2008). In addition, Santarelli (1998) found that firm size is conducive to new firm survival. According to Santarelli (1998: 162), this can be:

> explained by the fact that larger firms survive longer because they are in general more efficient, employ more capital intensive methods, achieve more easily economies of scale, have a larger availability of internal finance besides benefiting from easier access to external finance. Moreover, when the opportunity cost of staying in the market increases, larger firms may decrease in size before they exit, whereas under the same circumstance their smaller counterparts will be the first to leave the market.

Summary and conclusions

To understand the economic impacts of tourism requires an understanding of the challenges of quantitative analysis, the perspective of the entrepreneurial spirit, the social and political context in which tourism development occurs, and the social consciousness to address the fundamental needs of all members of society. Tourism as we know it today is fundamentally a capitalist phenomenon. Its goals are to earn money for tourism entrepreneurs by providing service to tourists. Commodification is an inextricable part of this process, as is competition and the unequal distribution of tourism resources and profits.

Tourists, as consumers, are not without some degree of agency in their role in the tourism economy. They hold the ultimate right to say no to a purchase transaction and are often motivated by goals that are not directly related to the price of transportation and accommodation. This demand-side of tourism forces the industry to act as conservators and developers of culture and nature, which is not typical of most other economic sectors. This is because the real product of the tourism industry is experience. The tourism industry is part of the broader *experience economy* and the *culture economy* (Debbage and Ioannides 2004).

The positive and negative impacts of tourism on culture are discussed in the next chapter. Like its economic impacts, tourism has considerable potential to benefit culture, however one might define that, though it may be more likely to change it in some way. And how well the tourism is managed in the culture economy can have significant monetary impacts on a destination.

Self-review questions

1   What are the challenges of assessing the economic impacts of tourism on a destination?
2   What are the major ways that a local economy changes with the introduction or growth of tourism?

3   Beyond the direct economic impacts, in what ways does tourism impact the quality of life in a place?

Recommended further reading

Smith, S. (2004) The measurement of global tourism: Old debates, new consensus, and continuing challenges, in A.A. Lew, C.M. Hall and A. Williams (eds) *A Companion to Tourism*, pp.25–35, Oxford: Blackwell.
Provides a good overview of some of the issues associated with defining tourism as an industry.

Leiper, N. (1990) Partial industrialization of tourism systems. *Annals of Tourism Research*, 17: 600–5.
Details a partial industrial approach to tourism which provides a different perspective to the supply-side approach and is also useful for helping to explain some of the difficulties in achieving coordination in the tourism industry.

Riley, M., Ladkin, A. and Szivas, E. (2002) *Tourism Employment: Analysis and Planning*, Clevedon: Channelview.
Examines tourism employment in both its workplace context and its wider economic and social environment.

Shaw, G. and Williams, A.M. (2004) *Tourism and Tourism Spaces*. London: Sage.
Provides critical cultural and economic perspectives on tourism development.

Coles, T.E. and Hall, C.M. (eds) (2008) *International Business and Tourism: Global Issues, Contemporary Interactions*, London: Routledge.
Provides a comprehensive account of the various ways in which tourism is part of international business and trade in services including its regulation and connection to place marketing and commodity chains.

Hjalager, A.M., Huijbens, E.H., Björk, P., Nordin, S., Flagestad, A. and Knütsson, Ö. (2008) *Innovation Systems in Nordic Tourism*, Oslo: Nordic Innovation Centre.
Provides a series of business case studies on tourism businesses in the Nordic context.

Hall, C.M. and Williams, A.M. (2008) *Tourism and Innovation*, London: Routledge.
Provides an overview of issues in tourism and innovation including the role of entrepreneurship.

Web resources

Members of the World Travel and Tourism Council (WTTC)
http://www.wttc.travel/eng/Members/Membership_List/index.php

Pro-Poor Tourism Partnership (PPT) – Pro-Poor Tourism Working Papers
http://www.propoortourism.org.uk/ppt_pubs_workingpapers.html

Statistics Norway Tourism Satellite Account website
http://www.ssb.no/turismesat_en/main.html

The Economic Impact of a Proposed Mariana Trench Marine National Monument (2008)
http://www.pewtrusts.org/our_work_report_detail.aspx?id=40478

World Trade Organization – International Trade Statistics
http://www.wto.org/english/res_e/statis_e/statis_e.htm

GATS and Tourism Impacts (Impact Assessment Framework for Developing Countries)
http://www.tourism-futures.org/content/view/1212/45/

Travel Economics: World Economic Crises and Travel and Tourism
http://delicious.com/alanalew/traveleconomics

Key concepts and definitions

Basic industries Those that bring money into a community via exporting all or nearly all of their production.

Innovation The implementation of a new or significantly improved product (good or service), or process, a new marketing method, or a new organizational method in business practices, workplace organization or external relations.

Marketing innovations New or significantly improved marketing methods.

Non-basic industries Those that circulate money through a local economy, but do not directly bring significant amounts of new money into a place.

Organizational innovations New or significantly improved methods in a firm's business practices, workplace organization or external relations (organizational or managerial processes).

Partial industrialization A condition in which only certain organizations providing goods and services directly to tourists are regarded as being in the tourism industry.

Political economy The study of the production and distribution of wealth within a society, and the role of the state in these.

Process innovations New or significantly improved methods for production or delivery (operational processes).

Product innovations New or significantly improved goods or services.

Quality of life The degree of well-being felt by an individual or a community.

Supply-side In the case of tourism refers to the destination resources that are available for the tourist. These include facilities and attractions of all kinds (such as parks, beaches, shopping and entertainment), as well as supporting infrastructure (such as transportation, hotels and restaurants) and services (such as travel agents, and recreation programmes and activities).

System of National Accounts (SNA) The accounting framework used by most countries to collect, order and analyse macroeconomic performance.

Tourism balance of trade (also referred to as the *tourism balance of payments*) The difference between tourism exports and imports.

Tourism commodity Any good or service for which a significant portion of demand comes from persons engaged in tourism as consumers (UNWTO 1994).

Tourism exports The receipts that a country receives from the money that international tourists spend in that country, along with products exported by tourism businesses.

Tourism imports The expenditures that residents of a country make when they travel outside of their home country, along with products imported by tourism providers.

Tourism multiplier A measure of the total effects (direct plus secondary) which result, from the additional tourist expenditure.

4 Tourism and its socio-cultural impacts

Learning objectives

By the end of this chapter, students will:

- Be able to define the concept of culture and its different forms in contemporary society.
- Understand the role of international organizations in defining cultural icons and the politics associated with that process.
- Understand the role of tourism in contributing to, and benefiting from, the broader context of community social change.
- Understand how tourists and residents are affected by tourism roles, relationships and experiences.

Introduction

Tourism is a cultural phenomenon. It both impacts cultures and society, and is shaped by cultures and society. It is often difficult to distinguish cause and effect in the relationship between tourism and society. This is further complicated by the fact that cultures change through time, and despite efforts by preservationists and conservationists to *museumize* sites and landscape to maintain their authenticity, this is never really possible.

Defining culture and society

The United Nations Educational, Scientific and Cultural Organization (UNESCO 2002) defined *culture* as 'the set of distinctive spiritual, material, intellectual and emotional features of society or a social group, and that it encompasses, in addition to art and literature, lifestyles, ways of living together, value systems, traditions and beliefs'. This is slightly different from the definition of *society*, which generally refers to the social relationships that exist between people with a common interest. Geertz (1973) defined 'society' as the arrangement of social relationships in a group, and 'culture' as the group's shared beliefs and symbols.

These are the definitions upon which we base our discussion of the cultural impacts of travel and tourism. We do not consider issues of civilization (another definition of culture), or of culture being exclusively within the domain of the fine arts and social or political elites. Travel and tourism have been described as being at the core of what it means to be modern (Rothman 1998), and all forms of tourism are aspects of modern mass culture. This even applies to the luxury tourism niche market, which at one time was associated with cultural elites. Today, tourism, along with increased education, growth in media access and increases in disposable income, has torn down the wall between the *high culture* of elite social classes and the *low culture* of the masses, and especially the lower social and economic classes. Low culture is sometimes referred to as *popular culture*, though that concept straddles the contradictory ideas of *mass produced culture* and the idiosyncrasies of *local vernacular culture*. Another cultural concept related to tourism is *global culture,* which has a similar definitional bifurcation, referring to *cultural relativism* (the equal value of all cultures, as seen in the 'world music' genre), and *cultural homogenization*, in which cultural diversity is subsumed under a single, worldwide culture (which, to many critics, often resembles American culture).

Mass culture in this context of mass production also encompasses most definitions that are associated with '*popular culture*' and '*global culture*'. In fact, a major cultural distinction that exists both explicitly and implicitly in tourism development, marketing and experiences is that between global culture, which is usually associated with the consumer culture of the developed world and the process of cultural and economic globalization, and local (or vernacular) culture and subcultural groups.

Culture is also closely related to political and economic systems and agendas. Like tourism, political and economic processes are cultural phenomena that are simultaneously defined by culture and influence cultural change. Dominant political leaders and leading businesses, through their behaviours and decisions, help shape cultural values. They do this via their influence on production (what is produced and how), consumption (producers are also consumers), fashion and taste (through marketing and public relations), though they are not always successful in their efforts.

In summary, a culture as we know it today is the result of an ongoing dialectic among numerous stakeholder voices, including, and certainly not limited to, political parties, religious organizations, economic interests, global media, local elites, educational institutions and, of course, the tourism industry and visitors, and through the everyday behaviour of people living their lives. This culture-forming process applies as equally to our interpretations of the past (historic sites) and our treatment of nature (wilderness areas), as it does to the cultures of contemporary societies.

In this chapter we first examine the socio-cultural impacts of tourism at the global level. The focus of this discussion is on the globalization of culture, especially through popular culture from the West (Euro-American), which, along with the continuing emphasis on the exotic in tourism promotions, has overtones of culture colonialism and hegemony (Teo and Leong 2006). At the national level, we examine the impacts of tourism-related migration, which is part of the increasing multiculturalism found in previously homogeneous nation states, as well as the role of tourism in defining national cultures and identities through imaging, branding and the promotion of national icons and events.

Most of the socio-cultural impacts of tourism have been observed and examined at the community and individual levels, because this is where the nexus of the host and guest relations resides. At the community level entire cultural landscapes have been transformed by tourism from both conservation and theming perspectives. Material culture has been commodified into souvenirs, though dying arts and performances have also been revitalized.

At the individual level tourism impacts both the host and the guest. Through a wide range of niche experiences, tourists seek opportunities to express and form their personal identities. Locals who participate in the tourism industry are also changed by the opportunities it provides. The greater the social, economic and cultural division between the host and guest, the greater the potential change in both. The *quality of life* concept presents another perspective on the impacts and relationship between tourism and the individual, by holistically incorporating culture, economics and the environment. For both hosts and guests, tourism activities have the potential to either raise or reduce the quality of life experience.

Global and supranational cultural impacts

Socio-cultural impacts of tourism at the global and supranational level occur formally through international organizations that set policies and fund conservation efforts, and through a less formal 'global popular culture', of which tourism is an integral part.

World heritage

Several major international organizations serve as global culture-brokers by defining taste and designating importance to particular types of cultural heritage (Winter 2007). The organizations that do this include:

• United Nations World Tourism Organization (UNWTO) – an intergovernmental supranational organization of country-level government tourism organizations, as well as associate members from the private sector, regional government and other organizations, that provides tourism development consulting services to national and sub-national governments.
• United Nations Educational, Scientific and Cultural Organization (UNESCO), whose World Heritage Centre approves and maintains the official World Heritage List.
• International Council on Monuments and Sites (ICOMOS), a group of non-governmental organizations that reviews sites seeking Cultural Heritage designation on UNESCO's World Heritage List. (The World Conservation Union, or IUCN, provides the same function for Natural Heritage sites.)

In addition to these mostly cultural organizations, there are numerous other government, industry and non-profit/non-governmental organizations whose activities include cultural conservation activities, but which focus more on economic development and the natural environment. Examples of these include the United Nations Environment Program (UNEP), End Child Prostitution/Child Pornography and Trafficking (ECPAT), and the Pacific Asia Travel Association (PATA).

The relationship between heritage conservation and tourism development is complex and not always harmonious. Conservationists typically seek to preserve a site in its physical originality or authenticity, regardless of the economic goals of the tourism industry that seeks to share the site with visitors. Both motivations, however, can result in a *museumization* of the heritage place, separating it from the natural evolution of the larger society and fixing it in a 'preferred' period of time.

For much of the world, however, the designation of a site on the World Heritage List is highly sought after because it helps to ensure that certain conservation

Photo 4.1 Bayon Temple is one of the many temples, and two former cities, in Cambodia's Angkor Wat World Heritage Site and Cambodian National Park. First built in the twelfth century, the temples contain a mix of Hindu and Buddhist symbolism, with the faces shown here being assigned to both Buddha and the Khmer King who built the temple. Meaning 'City Temple', Angkor Wat was once the capital of the Khmer empire and one of the world's largest cities, but fell into disrepair by the sixteenth century when the first Europeans visited the site. Today, the national park receives about 700,000 visitors (in 2005), about half domestic and half international. (Photo by Alan A. Lew)

standards will be met, and that the site will likely become a major international tourist attraction. Designation of places on the World Heritage List may also provide management assistance and funding through the World Bank's Cultural and Heritage Development Network, as well as through other international foundations and non-governmental organizations.

UNESCO also designated a *World Heritage in Danger* listing of heritage sites where development pressures and lack of funding threaten their maintenance and long-term existence. Some of the threats that heritage sites face include air pollution that erodes older structures and art works, wildlife poaching, the theft of cultural artefacts for sale on the black market, degradation from too many visitors, and impacts from nearby developments, especially roads, visitor accommodation and mining (see Case 1.2 for a discussion of the Galápagos Islands and the *World Heritage in Danger* list).

Despite the positive role that World Heritage listing is seen as having with respect to tourism (e.g. Shackley 1998), UNESCO has been charged with a cultural bias in its designation of World Heritage List sites (BBC News 2004). As of 2007, just under half of all the cultural and natural sites on the World Heritage List were in Europe (Table 4.1). Historically, European standards in defining and managing heritage have formed the basis for reviewing World Heritage applications. Because of this, important cultural sites in Japan and China have been denied designation because their structures were made of new material, even though the architecture and craftsmanship were centuries old. This different concept of heritage prompted the governments of Malaysia, China and New Zealand to call for a separate Asian World Heritage Committee to assess the different traditions and values in that region (Sulaiman 2007).

The designation of a World Heritage Site is a highly selective form of cultural conservation with distinct cultural and social impacts. It provides funding to renovate and reconstruct historical structures, if not larger landscapes. Arts, crafts, music, costumes and performances that are associated with the historical period represented in the cultural heritage site are likely to be revived for the tourist market. These material and performance cultures, however, may be modified to make them more readily consumable by tourists. Listing may also preclude other forms of economic development or human activity (Aa et al. 2004). Arts and crafts are made smaller so they can be packed more easily in a suitcase, while

Table 4.1 World Heritage List sites by status, 1998–2007

Category	Number of sites				Percentage of sites (%)			
	1998	*2001*	*2006*	*2007*	*1998*	*2001*	*2006*	*2007*
Cultural	445	554	644	660	76.5	76.8	77.6	77.6
Natural	117	144	162	166	20.1	20.0	19.5	19.5
Mixed	20	23	24	25	3.4	3.2	2.9	2.9
Total	592	721	830	851	100.0	100.0	100.0	100.0

Note: Percentages may not total to 100 per cent due to rounding.
For the most recent tally of World Heritage Sites see UNESCO: http://whc.unesco.org/

performances may be shortened and performed on a regular schedule so tickets may be sold.

The traditional arts and crafts of the South Pacific have been commodified since at least the late 1800s when the natives of Marshal Island sold tablecloths made from the mats that they originally wore for clothing, prior to the arrival of missionaries (Wilpert 1985). Van der Veen (1995) identified several ways in which traditional South Pacific crafts are changed through tourism commercialization and commodification:

- *Aesthetic change*: designs are modified to meet the artistic tastes of tourists. This may include efforts 'in the name of reviving traditional cultures, to fool the tourist with a diluted version of the real thing' (Tausie 1981: 55), or creating new forms of 'ethno-kitsch' (Graburn 1976) that may have little authentic connection with the local culture, but which meet tourist expectations.
- *Practical change*: introducing new designs in which tourists will find practical value. Examples include the Marshal Island tablecloths of over a century ago, to walking sticks, salad bowls and salt and pepper shakers of today. The traditional shields of the Asmat people of Papua New Guinea are smaller, more rectangular and flatter today so they can more easily fit into standard suitcases. New materials may also be imported to produce these products in forms that are more durable than traditional materials may have been.
- *Uniformity change*: in the South Pacific, increased artistic uniformity is attributed to the influence of Christianity, which replaced traditional iconography based on a pantheon of spirits with more limited church-based images. On the other hand, traditional arts often provided only limited variation in design because of their ritualistic functions (Graburn 1976), which has been changed by tourism.
- *Quality change*: while quality often suffers at the mass tourist and souvenir end of the market, at the high end of local arts, both traditional and modern art is often of very good quality. Art galleries sell exquisite replicas of traditional arts and crafts, along with the work of the best local modern artists. These arts are often supported by Pacific Island government programmes and museums that view them as important for cultural preservation and social identity (Hanson and Hanson 1990).

In some heritage and natural sites, stadium seating and laser light shows provide a modern context to a traditional art form. A section of the ancient Hindu temple grounds, and World Heritage Site, of Prambanan in Yogyakarta, Indonesia has adopted these enhancements for its evening entertainment of visiting tour groups. This form of development may be more likely in the Prambanan case where most of the residents of Yogyakarta are Muslim and do not identify culturally with the Prambanan temple complex.

Cultural sites, however, may not need to achieve World Heritage List status to be popular tourist destinations. Guidebooks (both print and web-based), destination promotional literature, popular media (including film, television and books) (Beeton 2005) and word of mouth all have the power to give significance to a place, and thereby of making the travelling public more aware of some places than others. Material artefacts and performances that support the image and significance of these sites will arise to enhance the guest's experience and the host's pocketbook. The authenticity of this material culture may be questionable from a conservation purist perspective, but are often appreciated from the time-limited tourist viewpoint.

Glocalization

The UNESCO World Heritage List process can be viewed as contributing to a globalization and homogenization of cultures through its Euro-centric guidelines and the almost uniform development model that has naturally (and with the support of consultants) followed an official heritage designation (Ashworth and Tunbridge 2004). In addition to preconceived notions of proper heritage develop-ment, there are many other forces pushing global homogenization into almost every corner of the planet. These include mass media (especially Hollywood movies and mostly English-language pop music), fast-food and restaurant chains, hotel chains, and grocery and department stores. From a structuralist perspective, globalization is a direct outcome of the need for market capitalism to sustain ever-growing profits in an environment of product saturation (Harvey 1989; Hughes 2004). To maintain economic profitability, the capitalist society creates and encourages rapid changes in fashion and taste, opportunities for instant (and fleeting) gratification, and the creation of new markets, which includes expansion into new geographic territories. The commodification of cultures and places (discussed in Chapter 3 and below) occurs under the relentless expansion of market economics (with tourists as foot soldiers) from the global core societies into the periphery.

Globalization has expanded as the world has become increasingly smaller in terms of transportation (cheaper flights), the Internet (instant information), telecommunications (easy connectivity) and international migration (hybrid cultures). Heritage attractions, hotels and restaurants, and the airlines are all integral parts of the travel and tourism industry, which carries globalizing values with each tourist that crosses from one country to another. The use of high-technology devices to educate visitors about a destination both before and during their visit is another way that globalization (in this case via telecommunications and the Internet) shapes the experience and behaviour of visitors, and consequently their impact on the local culture. The impact of globalization in general, and tourism at the global level in particular, is that places and people are potentially becoming increasingly alike. The unique characteristics of different places are disappearing, local traditions are weakening and consumerism is becoming the defining value of people and places.

Not everyone is accepting these social changes without resistance (Macleod 2004). Reactions against globalization are widespread, and are often referred to as *localization* or *localism*. These centre on the search for a personal *identity* that is a direct reflection of a local cultural identity. Formally, it is seen in indigenous rights groups; the teaching of, and support for, traditional arts and dress; and the appreciation of local foods and historic architecture. Ashworth and Tunbridge (2004) saw an emphasis on local identity as part of a nationalist reaction against internationalism. As mentioned in the cast of World Heritage Sites, nationalism can also be a homogenizing force, through the co-optation of local distinctiveness in the interests of national political agendas. For example, the national tourism promotion campaigns of both Malaysia and Singapore highlight Chinese New Year, the southern Indian festival of Thaipusam and the Muslim month of Ramadan as symbols of their harmonious multicultural populations for both inter-national marketing and to promote domestic unity. So even as local distinctiveness is celebrated, it can be co-opted by nationalist political agendas, while international tourism and tourists may further dilute a destination's distinctive character, by imposing global values and interests on those of the local.

While the extremes of globalization and localization are clearly evident in the different interest groups that tend to support each perspective (often business interests against non-governmental activist groups), the result is the emergence of a hybrid 'mashup' that reflects both global and local elements. *Glocalization* is the creation of new cultural forms that are neither exclusively local nor dominated by the global, but are a new creation that recognizes the contributions of both.

To succeed in this environment, international companies must modify their offerings to meet the demands of local clientele. An example of this kind of mashup is the McDonald's Corporation's fast-food restaurants selling red adzuki bean desert pies in East Asia, which is one of many modifications that adapts their standard menu to local tastes. Another example is the first three Disney theme parks outside of the United States (in Japan, France and Hong Kong), which all took several years before they developed the right mix of offerings to attract enough local visitors to become economically successful (see d'Hautserre 1999). Whether they are motivated by entrepreneurial or altruistic values, tourism entrepreneurs and industry leaders have come to realize the importance of local culture to the mix of attractions that are offered in a comprehensive tourist destination.

Tourism and world peace

Among the more lofty goals of international travel is to promote global under-standing and international peace. Through travel, people learn about other places and come in direct contact with people from distant parts of the world, and from very divergent cultures, as well. The potential for education and communication between cultures is often superficial and commercial, but can transcend those limitations under the right conditions. Direct and authentic communication can increase mutual empathy, which can potentially influence political behaviour beyond the trip itself.

To foster this perspective on tourism the International Institute for Peace through Tourism (www.iipt.org) has held biennial conferences and summits since 1994, and has recognized travel and tourism as a global peace industry. While tourism's role in bringing about peace may be debated (Hall 2005a), there is no doubt that peace is a necessary element for a successful tourism industry.

National and regional cultural impacts

The previous section discussed global perspective on the role of tourism in cultural conservation and cultural change. This included the role of international organizations in cultural conservation and the impact of cultural globalization on local cultures. We now turn to issues of cultural authenticity, with some focus on the perspective of regional influences and national policies.

Traditions and past landscapes

One of the often-cited benefits of tourism to local culture is in the revitalization of traditional arts (as previously discussed). Tourists to destinations with different cultural traditions from their own yearn to take home concrete symbolic representations of their experience, typically in the form of handicrafts, arts, clothing, photographs and miniature replicas of iconic sites and sights. Tourists also seek to witness or experience traditional practices and ceremonies unique to a local culture.

This desire creates a demand for these objects (or souvenirs) that can often be met by local suppliers, who may then rejuvenate traditional skills that had been abandoned for more profitable employment. In this way, tourist demand for traditional crafts and performances has the potential to revive and strengthen disappearing cultural traditions. The authenticity of these cultural artefacts, however, is open to negotiation. To the tourist souvenir hunter, authenticity can range from an object that was handmade by an indigenous resident to historically accurate standards, to an object that was simply purchased in the destination but whose origin was made in China (Reisinger and Steiner 2006). Authenticity is therefore as much in the eyes of the consumer as it is for the producer or observer.

Authenticity and destination images

Definitions of *authenticity* are among the most contested topics in tourism research. One reason for this is the confusion that arises between the authenticity of an object or product (such as a historical site or a painting) and authentic experiences and ways of life. Object authenticity can normally be measured by objective criteria, such as a scientific fact, a known location or a historical personage. Experiential authenticity is a personal and subjective experience, including physical and psychological feelings (Wang 2000). Both object authenticity and experiential authenticity lie at the interface between people, places and tourism. Experiential authenticity will be discussed in greater detail later in this chapter (in the section 'Individual impacts').

Not all tourists are interested in the authenticity of a tourist attraction or place, but many are. For example, if you were to visit a remote destination such as the city of Lhasa in Tibet, would you settle for seeing a replica of the holy Potala Palace, or would you insist on visiting the real building? To reduce the physical impacts of too many visitors, the Chinese government had proposed in 2007 to build a scale model of the Potala Palace, which would contain relics from the real building, and would be used as an alternative for people who would not be allowed to visit the real palace due to limits designed to protect its deterioration (Xinhua 2007). This proposal has its merits, as it addresses issues of too many visitors in a delicate environment. At the same time it raises issues of object authenticity, which can affect the experiential authenticity of cultural tourists. Hall (2007e) argues that replication is not intrinsically bad, what is important is the different experiential depth between the original and the replication. More problematic is when there is a deliberate attempt to deceive. Authenticity is derived from the property of connectedness of the individual to the perceived, everyday world and environment, the processes that created it and the consequences of one's engagement with it.

The UNESCO World Heritage Committee is one of the more authoritative arbiters of heritage authenticity. Even their site approval process, however, can be politically influenced at the national level where local and regional governments vie to have their sites put forward to UNESCO by their national governments. The overt goals of the local governments are typically two-fold: cultural conservation and economic development. At another level, however, it is common for governments to manage heritage tourism sites and events to support nationalistic and political goals. Because the World Heritage List process is through national governments (proposed sites are submitted by countries), the designation of a site can be a way of nationalizing local or minority heritage that might otherwise be subject to irredentism. This issue arose in 2008 when Cambodia's 900-year-old Preah Vihear temple was granted World Heritage status, prompting public protests and military action in neighbouring Thailand (Winter 2008). Minority areas that are disenfranchised from their national governments may also be subject to World Heritage List controversies.

Wang (2000) has named this *Constructive Authenticity* and defined it as socially constructed authenticity that is based on beliefs and viewpoints that represent those of the politically powerful and publicly influential. This form of authenticity is negotiable and often ideological. Efforts by many countries to deny or erase embarrassing events from the past are commonplace, and seldom successful in the long run. One of the most prominent examples of this are ongoing denials of the Jewish Holocaust by some political groups.

More frequent, however, is the prominence that governments give to some festivals, individuals (alive or historical), historical incidents and historical places, while largely ignoring others. Monumental structures and buildings can become an iconic representation for an entire nation, such as the Great Wall for China, the

Eiffel Tower for France and the Statue of Liberty for the United States. Special events can also play this role for a destination, such as the Mardi Gras for Brazil. Yet the contested histories of such icons is often missing from tourism promotion and interpretation.

The loss of historically rooted places, including the attempt to depoliticize them 'decontextualising them and sucking out of them all political controversy – so as to sell . . . places . . . to outsiders who might otherwise feel alienated or encounter encouragements to political defiance' (Philo and Kearns 1993: 24) is relatively commonplace in tourism (Hall 1997). Heritage centres and historical anniversaries typically serve to flatten and suppress contested views of history (Hall and McArthur 1996). This is a significant issue as the representations of memory at tourist sites are not just commodities for visitor consumption, but are also narratives that interact with and influence the composition of personal and collective memory and the official representations of history. (See Case 4.1.)

Case 4.1: Interpreting heritage on Cannery Row

The presentation of one-dimensional views of the past to the tourist and the community is not just encountered at specific attractions but is also encountered at the destination level. For example, the rich and complex ethnic and class history of Monterey, California, is almost completely absent in the 'official' historic tours and the residences available for public viewing. History in Monterey is often 'simplified or simply erased' (Fotsch 2004: 779). In Monterey, as in many other parts of the world, heritage is presented in the form of the houses of the aristocracy or elite. 'This synopsis of the past into a digestible touristic presentation eliminates any discussion of conflict; it concentrates instead on a sense of resolution. Opposed events and ideologies are collapsed into statements about the forward movement and rightness of history' (Norkunas 1993: 36). Narratives of labour, class and ethnicity, which are a significant part of Monterey's history, are typically replaced by romance and nostalgia. Overt conflict, whether between ethnic groups, classes or, more particularly, in terms of industrial and labour disputes, are either ignored or glossed over in 'official' tourist histories.

The overt conflict of the past has often been reinterpreted by local elites to create a new history in which heritage takes a linear, conflict-free form. More recently, past representations of the collective memory of Monterey have become subject to new narratives, such as environmental protection and ethnic diversity, that are creating new tensions with respect to representing the authentic past of Monterey (Walton 2003). Nevertheless, in the case of Monterey, the past as tourists experience it is usually reinterpreted through the physical transformation of the canneries. 'Reinterpreting the past has allowed the city to effectively erase from the record the industrial era and the working class culture it engendered. Commentary on the industrial era remains only in the form of touristic inter-pretations of the literature of John Steinbeck' (Norkunas 1993: 50–1). Steinbeck and his fictional characters, rather than any real cannery workers, have become the focal point of representation. Although as Walton (2003) noted, this was not

Photo 4.2 John Steinbeck's Cannery Row was once a working class, industrial
neighbourhood. Today, like many other Fisherman's Wharf districts, its
former seafood canning buildings are filled with tourists, gift shops,
restaurants and attractions. The Monterey Bay Aquarium, with its full size,
living kelp forest tank, anchors the district, drawing almost 2 million visitors a
year. Cannery Row is one of several major tourist attractions in the Monterey
Bay area. (Photo by Alan A. Lew)

Steinbeck's doing, for he harshly criticized California's 'pseudo-old/new old'
architectural and redevelopment schemes. 'The new story of Cannery Row was
constructed by the commercial interests that took over the street, their public
relations advisers, the newspaper, the city of Monterey, and a variety of popular
writers all of whom discovered cash value and name recognition in this version of
events' (Walton 2003: 274).

The theme of heritage interpretation resulting in a flattened, one-dimensional
history, is a strong thread in critical discussions of heritage tourism and the
contemporary city. For example, in the case of Britain's museums, Hewison
(1991: 175) argued:

> Yet this pastiched and collaged past, once it has received the high gloss of
> presentation from the new breed of 'heritage managers', succeeds in presenting
> a curiously unified image, where change, conflict, clashes of interest, are
> neutralized within a single seamless and depthless surface, which merely
> reflects our historical anxieties.. . . There seem to be no winners, and especially
> no losers. The open story of history has become the closed book of heritage.

Political and social elites have an interest in controlling heritage traditions to establish or fix cultural identities, legitimize existing social and political institutions, and to socialize people into a political ideology. To be successful, however, the story told through heritage must have a semblance of authenticity, believability or credibility that is acceptable to its audience (Fairclough 1992, cited in Jones and Smith 2005).

The process of adapting authentic cultural expressions for tourists is known as '*staged authenticity*' (Table 4.2). This includes performances that represent real traditions and life, but in some form of staged, or tourist-directed, setting. Because of time constraints, most tourists are satisfied with quick and highly structured re-creations of traditional culture and local life. In this situation, staging is inevitable.

Lowenthal (1985) argued that how we represent the past through heritage sites is a direct reflection of the values of the contemporary society. He says that we cannot know the past except through the values of the present society, because every act of recognizing the past alters it by recognizing some parts and ignoring other aspects. If the past were a foreign country, Lowenthal says that nostalgia has made it 'the foreign country with the healthiest tourist trade of all' (1985: 5).

Acculturation

Acculturation is the cultural corollary to the core–periphery model of economic development (discussed in Chapter 3). The acculturation process assumes that when two nationalities (countries or cultures) come into contact over a period of time, there will be an exchange of products, ideas and values. This exchange produces various forms of blending between the cultures, which become more similar with time. However, the exchange and blending is seldom equal, and one culture usually dominates over the other. This results in an unbalanced relationship, both economically and culturally – subjecting the subordinate society to economic dependency and cultural hegemony. The cultural threat is that local

Table 4.2 Production and consumption in the experience of authenticity

Experience	Consumption	
Production	*Real*	*Staged*
Real	*Authentic* Both consumers and producers understand experience as authentic.	*Denial of authenticity* Consumer suspects a staged experience although producer regards as authentic.
Staged	*Staged authenticity* Staged by producers but experienced as real by consumers – covert tourist space.	*Contrived* Both consumer and producer understand experience as staged. Overt entertainment or performance space.

traditions and values will be replaced by those of the dominant culture, which are in a stronger position of support by the political and economic structures of the relationship.

Case 4.2: Tourism and culture clash in American Indian Country

> When every effort was made to wipe out our culture and religion, we made adjustments to insure that there was an outward showing of compliance. We managed to keep our religion and culture going (underground, as it were) so we were able to survive the Spaniards. So too are we able to survive the tourists and culture they represent. – A member of the Taos Pueblo tribe
>
> (Lujan 1998: 145)

This quote reflects the perception of tourism as a tool of post-colonialism, acculturation, and assimilation. Although tourism is not often considered in these terms, it can clearly have many of the same impacts on traditional societies of all types.

American Indians, along with other Native American groups, have been a tourist attraction since the late 1700s, in part because they represent the antithesis of the many ills associated with the industrialization of Euro-American culture. Many non-Indians make annual pilgrimages to the more traditional reservations of Arizona and New Mexico to experience their dances and festivals. For international tourists, the American Southwest holds a special allure as it has come to symbolize much of what makes the United States different from the Old World.

In the almost 400 years since northern Europeans settled the United States and Canada, American Indians have suffered through efforts at annihilation, assimilation and general neglect by the dominant American society. Even today, Indian history and culture are largely ignored by mainstream American society, except when they play roles allotted to them in movies.

Some argue that the tremendous cultural differences that exist between American Indian culture and the dominant American society are so great that Indian culture simply cannot be understood in Western terms (Martin 1987). These conflicting cultural values, based on the major indigenous cultures of the US Southwest, are generalized in Table 4.3. The acculturation process is pushing the dominant American values to replace the traditional American Indian values. This is the concern that the Taos Pueblo Indian quoted above associates with tourism and tourists.

The image of the American Indian held by the majority of Americans and Europeans in the nineteenth century was that of the *Noble Savage*: dignified, stoic, reserved, honourable, hospitable and truthful. The Noble Savage image, in fact, closely reflected the finer qualities of civilized European high culture. These images also reflected the idealization of nature in eighteenth- and early nineteenth-century Romanticism. The natural world was viewed as being an 'Eden', free of the sins of civilization. To the romantics, American Indians epitomized humankind as it once

Table 4.3 Comparison of traditional and modern values on American Indian lands

Traditional American Indian cultural values	Dominant American cultural values
Cooperation and the group	Competition and individualism
Prestige and authority based on family, age and religion	Prestige and authority based upon political position, education, and economic wealth
Education from Elders	Education in schools
Life organized around ceremonial activities	Life organized around work activities
Communal land tenure and management/development	Fee simple land tenure (ownership) and private property rights
Social support from the an extended family group	Social support from government welfare programmes

Source: Lew and Kennedy 2002.

existed in the pure state. It may be argued that any attempt to divorce Indian culture from the natural environments in which it is situated is doomed to be unsuccessful because it neither fits the larger, popular image of Indians, nor even the image that many American Indians have of themselves (Kennedy and Lew 2000).

Today, American Indians continue to be romanticized by the tourism industry (Lew 1998b) and popular culture in general. However, there are some tourists who, upon visiting a reservation, are disappointed to find that the idealized Noble Savage is less primitive and less noble than they had envisioned. Many more tourists, however, will do their best to make sure that their expectations are actually fulfilled, and that their time and money were not wasted. Either way, unfortunately, individuals on reservations frequently must bear the brunt of tourists' desires to mould them to fit a stereotype.

It should not be surprising that American Indians also hold stereotypes of tourists, as well. These stereotypes of Anglo-Americans tend to be similar to those that other non-Americans hold: outgoing, loud, wealthy, disrespectful, racially prejudiced, always in a hurry and ignorant of local (in this case American Indian) culture (Evans-Pritchard 1989).

With time, the tourist can move to the role of 'white friend'. Pueblo Indians are traditionally very open and cordial to guests and most have developed relationships with non-Indians who first came as tourists. The tourist must make frequent visits to the pueblo (usually during the dance season) in order to become a friend. Friendship is marked by being invited to one of the all-day feasts that accompany most dances. With additional time, the friend may be allowed and expected to take part in the preparation of food for the ritual feast. The transformation from tourist to friend is accompanied by an erosion of stereotypes. The American Indian's stereotype of the white tourist becomes less, and the tourist's stereotype of the Indian decreases. Both become real people.

The cultural impacts of tourism on reservations should not be overestimated, however. Change is and will always occur and the influences of the mass media, federal government programmes and dominant American social values in general have had a far greater influence on the loss of Indian traditions than has tourism. Tourism plays a significant role, nonetheless, because it occurs as a physical invasion of the outside world onto the reservation itself. Tourism can contribute to cultural preservation and economic prosperity, but it can also contribute to the detriment of cultural traditions and permanent loss of a way of life.

Sources: see Lew, A.A. and Van Otten, G. (eds) (1998) *Tourism and Gaming on American Indian Lands*, Elmsford, NY: Cognizant Communications.

National government policies

National governments are possibly the single most important stakeholder in designating and developing heritage places and events. This is because most governments have eminent domain control over all land through their condemnation powers. While details vary from one country to the next, governments can condemn land and confiscate it (usually paying some form of compensation) for the broader public good. It is these powers that also allow governments to impose redevelopment and design controls on heritage and other tourism-oriented districts in a community.

In addition to the eminent domain powers, governments are able to appropriate funds (through taxes) and spend money to renovate buildings and sites, and promote special events. This can also be accomplished by reducing taxes on certain locations and businesses. Through these actions, national governments place value on some histories and cultures (and subcultures), and devalue others. The preferred histories reflect policy decisions to establish and promote a national identity (Ashworth and Kuipers 2001). This is further enabled through legal structures, government institutions and agencies, and management and financing systems that enable heritage and cultural conservation.

The sites, events and cultural artefacts that are selected for preservation and memorialization are further accentuated through their interpretation to tourists and residents. Governments then associate themselves with established national heritage to legitimate their right to rule. Thus a tourist historic city represents a nationalist past and is an important icon of national identity (Evans 2001). This can be seen in heritage tourism cities such as Philadelphia and Washington, DC, in the United States, Berlin in Germany, Beijing in China, and Jerusalem in Israel, among many others.

In some countries and communities, tourism marketing campaigns can be indistinguishable from government social policy. The Singapore Tourism Board, for example, had adopted the marketing theme of 'New Asia' in the 1990s, which it defined as 'progressive and sophisticated, yet still a unique expression of the Asian soul' and 'East meets West, old engages new, and tradition marries technology in harmonious co-existence' (STB, *Singapore Tourism 21 Plan*, cited in

Lew et al. 1999). This theme was not only used as a tourism image for marketing Singapore, but was also promoted to the people of Singapore as a national identity. For most countries, however, the relationship between tourism and government values is more subtle, though it is still present.

Community cultural impacts

The introduction of tourists to any site has impacts. At a minimum it increases the number of persons at the site. In the extreme tourism can completely transform the original site, appropriating it from community connections and ownership. The cultural impacts of tourism are emphasized here; economic and related social impacts are discussed in Chapter 3.

Physical damage to cultural resources (including archaeological sites, historic sites, museums and the arts) may arise from vandalism, littering, theft, and the removal and sale of cultural artefacts. All these acts are illegal in most countries, yet the sale of stolen objects from archeological sites is a significant problem in developing economies, such as Egypt, Mexico and Central America, and Cambodia. Poorly paid guards in these developing economies sometimes supplement their income by selling artefacts to tourists. The degradation of cultural sites can also occur when historic sites and buildings are unprotected and the traditional built environment is replaced over time by other economic interests.

Debates over history, identity and image occur at the local level when decisions need to be made on how to develop an iconic place. Different community stakeholders come into play, including local residents, local business people, non-local business ventures and non-local tourists. Local communities have their personal knowledge and belief about their past and current identity. Tourists bring their images of a destination that are built over time by media sources and word of mouth. They are usually more satisfied with their visit when the destination meets their expectations. Entrepreneurial interest groups tend to favour commercial development with less of an emphasis on object authenticity. They tend to be more associated with globalization images and values, including global brands, such as McDonald's. Heritage conservation groups, on the other hand, may be totally opposed to commercial exploitation of their heritage district. Other non-governmental interests may also exist in a community, such as minority rights groups, small business groups and social service organizations. All these stakeholders operate under, and try to influence, government policies and regulations.

History and geography also play a significant role in shaping the range of possible tourism development paths for a community or place. Influential geographic site characteristics include mountain and coastal environments, along with international border locations. Access and relative location of potential visitors are key geographic situational features. Architectural histories impact most human settlements, and colonial histories have had a more subtle influence on the landscapes and cultures of many developing economies.

Each stakeholder group, including the government, has it own subjective framework in which it defines the best approach to authenticity and tourism development. None necessarily has the best solution for any one community. The existing landscape of a tourist district, whether in a city or a rural countryside, reflects the relative political power of each of these stakeholder groups, expressed through the political process, and within the context of local history and geography. Common cultural landscapes that have emerged under the political economy of local tourism development include (based on Ashworth and Tunbridge 2004 and Lew 1989):

- *Historic preservation districts*: where the museumization and object authenticity of older buildings is of primary importance, though this can vary from one place to another; such districts are most common in Europe and North America.
- *Waterfront districts*: where fishing and ocean themes predominate, though commercial fishing may no longer be actively pursued; the Fisherman's Wharf found in many North American cities is an example.
- *Ethnic cultural districts*: where a distinct ethnic group maintains architectural or event tradition to represent and maintain their culture; Chinatown, Little Italy and Little India are some common forms of this theme.
- *Beach recreation districts*: where leisure recreation associated with sun, sand, sea and sometimes sex predominate, usually under warm weather conditions; this form is found with only minor variations throughout the world.
- *Historic or cultural replication districts*: where false façades and architectural ornaments are used to create the past or another culture, often in a Disney-like fashion; examples include fake cowboy towns with the 1880s-style building façades of the American West and imitation European villages in the United States and China.
- *Pedestrian shopping streets*: commercial shopping streets that are closed to automobile traffic either all of the time or on a regular schedule, often associated with heritage or cultural themes; found throughout the world, these streets are most common in Europe and China.
- *Artist colonies or arts districts*: places where artists congregate to make and sell their wares; often associated with historic sites and buildings, but not always; former mining towns in scenic mountain regions provide a common site for artist colonies, as do coastal communities and former warehouse districts in large metropolitan areas.
- *Mediterranean fishing harbour*: a more historical and usually less urban waterfront district, where Mediterranean architecture prevails and fishing as a livelihood is more likely to still be practised; found mostly in the Mediterranean region, but also in other parts of Europe.
- *Landmark business districts*: iconic buildings and other structures that serve as an anchor for retail district tourism and economic vitality; examples include the Sydney Opera House in Australia, the Monterey Bay Aquarium in California, and major sports facilities in cities around the world.
- *Generic heritage districts*: older retail districts that adopt historic preservation themes, but implement them in a generic manner, using common street

Photo 4.3 The Prinzipalmarkt (central marketplace) in the historic district of Münster, Germany, is a mix of truly historic structures, such as St Lambert's church (completed in 1375), as seen here, and newer buildings designed to fit architecturally into the existing fabric of the city. Note the three cages above the church's clock, which displayed the corpses of the Münster Rebellion leaders in 1535, who had promoted polygamy and the rejection of all private property. This is an example of a European *pedestrian shopping street* and *heritage preservation district*. (Photo by Alan A. Lew)

furniture catalogues, for example, rather than designs, fixtures and objects that are representative of the geography, history or culture of the area; elements of generic vernacularism are found in most of the other cultural landscape forms listed here, as well.

- *Tourist accommodation and souvenir districts*: landscapes dominated by hotels, motels and restaurants serving visiting tourist populations; in more dense urban areas these will also include souvenir shops, while in more dispersed rural areas gasoline service stations are also common; these are normally placeless landscapes that are devoid of heritage elements; Las Vegas represents one extreme of this landscape form.

Case 4.3: China's recreation retail districts: from heritage to Disneyfication

A pedestrian shopping street is a type of recreational retail street that is closed to most vehicular traffic either all of the time or part of the time. These types of streets are common in Europe, and increasingly in China where they are very popular and serve as major centres of leisure, shopping and recreation for both residents and tourists (Lew 2007c). The diversity of forms, approaches and architecture that have been adopted in creating these streets are remarkable, generally ranging from historic preservation efforts to postmodern Disneyfication.

Macau, which from 1557 to 1999 was a colony of Portugal, may be the site of the earliest historic preservation pedestrian retail district in China. The historic core area of Macau is famed for its 'European' charm and in the mid-1980s several streets leading from the central plaza of the Largo do Senado to St Paul's Basilica (church ruins) were closed to create a pedestrian-only precinct in which historic preservation efforts were concentrated. Today this area is the heart of non-gaming tourist visitation to Macau.

Only 65 km (40 miles) to the north of Macau is the city of Zhongshan, the capital of the county in which Dr Sun Yat-sen (the founder of modern China) was born. Taking a queue from Macau, the city of Zhongshan replicated central Macau's pedestrian streets in its old downtown area in the early 1990s by creating the Sun Wen Xi Road Tourism Zone. Although it has more neon lights than Macau's historic zone, large numbers of shoppers are drawn to *New* Wen Xi Road on a daily basis, especially after work in the evenings.

Guangzhou, the largest city in southern China, created the pedestrian shopping street of Xiangxiajiu Road in 2000 by closing off a busy street in the evenings and weekends. Xiangxiajiu Road differs from the Macau and Zhongshan examples because of the degree to which its older architecture was embellished in a 'Disneyesque' manner. Its buildings, which are pre-Second World War but without special historic significance, were brightly painted in a variety of colours and decorated with new cornices and other ornaments that do not adhere to historic authenticity. In addition all the windows were replaced with stained glass, and lights were installed directed upwards to highlight the building façades in the evening.

Xiangxiajiu Road was not an especially significant shopping or historic district prior to its renovation, and one end of the road is anchored by a large, modern, multi-storey shopping centre with residential apartment towers above. However, the incredible success of Xiangxiajiu Road led to its becoming a permanent shopping street; the gradual addition of neon signs has actually transformed the European character of the original design into something that is more like Chinatown in San Francisco. This might be considered a type of global–local mashup (glocalization), integrating European and Chinese design values.

Shanghai's Xin Tian Di (New Heaven and Earth) shopping complex opened in 2003 as a more self-conscious effort at mixing Western and Chinese sensibilities. The project transformed two blocks of an old shikumen district (a type of tenement) into boutique shops, stylish night clubs, classy restaurants and an upscale department store, all with a distinctly upscale and Western orientation. The site has some historic significance in that Mao Zedong met in one of the buildings here to form the Chinese Communist Party. Most of the other shikumen buildings in this area have been razed, and its former residential uses have vanished. Xin Tian Di today might be considered a Chinese version of Boston's Faneuil Hall or San Fancisco's Ghiradelli Square. It was the brainchild of American architect Benjamin Wood, and is viewed as one of the most influential heritage retail project in China. The city of Hangzhou has created its own version of Xian Tian Di, and other cities plan to follow suit in coming years.

In the city of Kunming, below the Tibetan plateau in southeast China, a historic-themed shopping complex was built completely from scratch. Kunming's Jin Bi Guang Chang shopping centre is a kind of theme park that replicates a traditional Chinese village. An urban renewal project, the pedestrian streets in this 'village' were never designed to serve anyone but recreational pedestrian shoppers. The architecture is clearly Chinese-themed, but taller and denser than any real village would be. Although it houses some of the city's more Western-oriented night clubs and restaurants, compared to the other districts described here, it has only achieved moderate success.

Other retail districts in China have adopted modern and hyper-modern designs that do not include significant heritage elements. These include Wangfujing Street in Beijing and Beijing Road in Guangzhou. The city of Guilin has placed an underground shopping mall below a large central plaza with two large video screens for viewing advertisements and major national events. Adjacent to the plaza is a 13-storey hotel, down the face of which each evening water cascades, pulsating to the sound of music. Smaller pedestrian streets lead to the city's lakes and river, which are now lined with thematic parks and bridges that replicate famous sites in Europe (Arc de Triumph), North America (Golden Gate Bridge) and Asia (Korean and Japanese gardens).

The transformation of China's urban environment reflects the evolution of Chinese consumers from a pre-consumer (in the Mao Zedong era), through a consumer-based, to a post/hyper-modern consumer society. It is also fostered by competition among Chinese cities to image themselves as modern and attractive environments for economic investment and tourism (Morgan 2004; Wu 2004) by moving factories out of the city, improving the transportation infrastructure, encouraging foreign investment in office and retail sectors, and creating international-class *shoppertainment* and *eatertainment* venues (Lafferty and van Fossen 2001).

These districts can have varying degrees of object authenticity in the architectural design. The most natural and organic is a district that has evolved spontaneously and unselfconsciously over a long period, comprised of many small landscape decisions made by many different local actors. Retail districts that formed in this way are rare in today's world where global corporations and franchises minimize local decision-making. Historic preservation, as a reaction against globalization, is one form of renovation programme that attempts to maintain some level of traditional authenticity.

Tourism revenues, or their potential, provide a major political justification for government funding of infrastructure and marketing in these districts. Tourism may also be used to justify architectural design controls (signage, colours, building ornaments and street furniture), either for historic preservation or for implementing cultural themes. While efforts to accentuate local culture are often stated goals of such developments, Ashworth and Tunbridge (2004) argue that such a generic heritage approach, as just described, can have a positive impact by reducing local political conflict over whose heritage is represented, while also making place images more universally and cross-culturally accessible. Most elements of uniqueness are lost, though not all tourists actually seek that, and the geographic context may in itself be sufficient for marketing purposes.

Positive impacts of tourism on cultural heritage

Not all of tourism's cultural impacts are negative. The development of tourism is often associated with the conservation of traditional arts, including handicrafts and music. Many communities across the world have been able to preserve and manage their heritage and historic sites through the revenues that they have obtained from tourism. This requires both effective management policies and a strong cultural identity. Heritage has become an important segment of the tourism industry. Ideally, heritage resources help keep tourism development authentic to local traditions, while also connecting it to the life of a community. The successful development of heritage resources reflects the distinctiveness of the people and the material culture and events that are important to the lifestyle and history of a community.

Heritage tourism also strengthens communities by providing employment related to heritage sites and events. Tourism development may further contribute to the construction of new or improved facilities for health, transportation, electricity and gas utilities, sports and recreation, restaurants, and the beautification of public spaces. These facilities are used by tourists, but are used more by local residents who enjoy higher living standards as a result.

Tourism can also have a positive impact by encouraging civic involvement and community pride. By recognizing the potential financial benefits of tourism and focusing on local amenities for development, the tourism industry raises awareness among local residents of the social and financial value of their natural and cultural heritage – something that residents are often not aware of. Conserving heritage resources can then become an activity of honour.

The potential positive impacts of tourism are often not realized due to competing interests and political decision-making that is less than transparent. The benefits of tourism are also not evenly distributed across most communities. Tourism must be practised in a sustainable manner that includes public participation and support. An involved citizenship will have a greater sense of ownership and more positive attitudes toward tourism.

Commodification and souvenirs

As mentioned above, the desire of tourists to bring back souvenirs, arts and crafts to remember their travels often results in artisans and craftmakers modifying the styles of their products to better suit the travelling needs and taste of tourists. Critics argue that by modifying the original product to suit the tourist, the product loses its authenticity. This is illustrated in the case of Kuna women of Colombia and the molas that they craft (UNEP 2001b). Molas are the blouses worn by Kuna women, which are pieces of art made with designs that reflect their beliefs about both the physical world and the spiritual life of the Kuna Nation. Tourism has turned the molas, which were previously a Kuna form of religious storytelling, into a commercial business, which has resulted in both the loss of the spiritual nature of the designs, which have changed to meet tourist purchasing interests, and a decline in their quality. Along with this, younger Kuna women are losing the traditional knowledge of the original designs and their spiritual interpretations. In this example, tourism encouraged a sense of pride in the Kuna mola tradition. However, tourism also encourages individuals to trade authenticity and tradition for economic well-being. Cultures change over time in the normal course of events, and perhaps these choices made by the Kuna artisans would have occurred even without tourism; ultimately, however, we cannot really know that.

What we can know is that at least part of the transformation of indigenous arts and crafts is done for profit motives. This is the commodification of culture, which was discussed as an economic activity in Chapter 3. Souvenirs in general, and the Kuna molas in particular, are among the more obvious forms of cultural commodification. The economic value of souvenirs comes less from their production costs, such as labour and raw materials, and more from their 'sign value' (MacCannell 1976). Commodified culture becomes a souvenir or an experience that has a fetish, or even sacred, quality for the tourist. The tourist collects these fetishes as symbols of their class status (Williams 2004). The more expensive and exclusive the tourism commodity (be it a rare work of art or dinner at a posh restaurant), the more memorable, and boastful, the trip.

A less obvious form of commodification is what Ashworth and Tunbridge (2004: 212) call *heritagization*, which is 'the process through which heritage is created from the attributes of the past, whether these are relics, artefacts, memories or recorded histories'. This is part of a larger competitive strategy to enhance products and places to generate profits. Thus, in addition to the desire to express cultural meaning based on local history and tradition, the commodity values of tourists and the tourism industry are also considered in the creation of products, performances and places.

In addition to architecture and arts and crafts, commodification can affect a wide range of cultural attributes, including: food and drink preparation and presentation; performing arts and public celebrations, such as Carnival/Mardi Gras and Chinese New Year parades around the world; clothing styles, both traditional and modern; and recreation and leisure practices, including sport preferences and recreation places. Fusion restaurants that put an international twist on traditional foods, or a local accent on a foreign dish, are among the more popular trendsetters around the world, and offer new levels of creative culinary art to the cosmopolitan cook. Entertainment has become a large part of the tourism-related commodification process, leading to what Hannigan (1998: 89) has called *shoppertainment*, *eatertainment* and *edutainment* (for shopping, eating and education, respectively).

These forms of commodification exemplify Robin's (1999) idea of *enterprise culture*, which stands in contrast to the less commodified *heritage culture*. Enterprise culture is more clearly driven by business activities and the realization of capital, whereas heritage culture is motivated more by the sense of nostalgia and personal and group identity. Enterprise culture is also more reflective of global-ization, but can have its own local hybrid forms, as well. The cultural landscape of a destination, including its buildings, people, products and experiences, is a result of a contested dialect that occurs between these two cultures, and is sometimes referred to as the *global–local nexus* (Chang and Huang 2004).

Cultural change is a natural process, driven by innovations by creative indivi-duals, adaptations to changing conditions and assimilation of external influences. According to Coles (2007) the real question in the issue of cultural change through

Photo 4.4 A sign clearly identifies the sculptures here as being made by non-Native Americans at this gift shop at the Grand Canyon National Park in Arizona. It is not easy to tell the difference between works by ethnic American Indians and non-Indians in most of the tourist shops in Indian Country. Art galleries, however, will always identify these differences because articles made by ethnic Indians are considered more *authentic* and command much higher prices. (Photo by Alan A. Lew)

the commodification process is not about whether the results are as authentic as the uncommodified products – which is an external, and often Eurocentric viewpoint. Instead the issue is whether or not the changes are empowered by local communities. The appropriateness of tourism should be measured by the degree to which it enables 'a strategic use of tourism and its power to provide an identity' (Coles 2007: 944). It is only when local communities have a sense of control over their own destiny that cultural change leads to new forms of authenticity (see also Franklin and Crang 2001).

Individual cultural impacts

Tourist–host encounters

For tourism to exist, there must be a tourist. For a tourist to exist, there must be a host. The host is the person who occupies and defines the destination. A host can also be a collective of persons, such as corporation or land management agency. Sometimes the host is the owner and manager of the destination, as is the case with a theme park (owned and managed by a private corporation) or a national park (owned and operated by a government entity). The host is an insider and the tourist is an outsider in the destination, though that distinction can sometimes be blurred.

The relationship between tourists and the host community that they visit tends to be characterized by four limiting features:

- *Transitory* A tourist is never a permanent resident, though the travelling lifestyle can be long term. For statistical purposes, a resident lives in a place for one year or more; a tourist lives there for less than one year (see Chapter 1 for more on this topic). In reality, most tourists stay in a destination for anywhere from a day to a couple of weeks. As a result they typically lack a sense of commitment and ownership of the destination and its culture. This keeps most relationships at a superficial level, though exceptions do occur, such as with second-home owners who are tourists who have literally bought into a destination (Hall and Müller 2004).
- *Unequal and unbalanced* Tourists have a greater feeling than local residents of being free from the normal limitation of work, the norms of their home society and monetary concerns. This gives tourists greater flexibility in their behaviour and an ability to recreate their personal identity through their relationships with local residents. Most tourists do not have to live with the consequences of their behaviour once they return home. The local host population, however, does. This makes the relationship unbalanced and biased toward the tourist. The larger the cultural and economic difference between tourists and local residents, the more obvious and more significant these impacts can be. On the other hand, local residents have local knowledge and may have had more practice in dealing with, and possibly manipulating, tourist relationships.

- *Lack of spontaneity* Much of the tourist–host encounter is scripted, either formally (companies coaching employees on how to respond to tourist questions) or by adopting the social roles of 'tourist' and 'host', or 'shopper' and 'seller' in the most common form of tourist–host interaction. Scripted roles increase predictability and reduce stress for both the host and guest, though memorable experiences are usually the result of spontaneous encounters.
- *Time and space constraints* In addition to the transitory nature of tourist experiences, tourists are also limited in the number of activities they can undertake within the time they have. This limits their ability to know the full range of geographic, social and cultural diversity in a destination.

Tourist–host encounters can occur in one of three settings:

1   Where the tourist is buying a product or service from the host.
2   Where tourists and hosts occupy the same space at the same time, normally in a public setting, such as a sidewalk, market or park.
3   When tourists and hosts meet and share ideas and information, which can occur on a guided tour or other structured experience.

Shopping and dining are two of the major settings for tourist–host encounters (Timothy 2005). These encounters are normally very short, with the host serving the tourist. Communication is superficial and seldom extends beyond that needed to complete the purchase transaction.

The tourist role

As noted above, tourists on tour are not the same people that they are at home because they adopt the role of 'the tourist'. Many tourists come from societies with different consumption patterns and lifestyles to those current at the destination, and these are further accentuated while they are on holiday, seeking pleasure, spending large amounts of money and sometimes behaving in ways that even they would not accept at home. The tourism industry encourages and supports this role adoption because by behaving as tourists they are more predictable and manageable. The media and government further perpetuates the stereotyping of tourists by referring to non-locals (of which tourists are one form) in derogatory terms, associating them with illegal migrants, unemployed 'hippies', illegal prostitutes and dangerous transients.

Chang suggests that the 'British media frequently portray such visitors as plagues and diseases which contaminate landscapes, harm local societies and contribute to the spread of AIDS' (2000: 345). Similarly, Coles (2008) writes of the relationship between tourism and notions of citizenship observing that the enlargement of the European Union has led to some British newspapers (as well as politicians) expressing concern as to the 'welfare tourists' from the new accession states of Central Europe, while simultaneously highlighting the cheap second home opportunities that exist in these new members states of the EU. Such comments encourage residents to see non-locals as intrusive and as a problem, thereby justifying government policies that prohibit some forms of alternative tourism and nomadism.

Urry (2002) associated the rise of the tourist with the growth of the post-industrial service sector, which has given rise to a new 'service worker class' that is bringing about the rapid growth in contemporary tourism.

> The service class consists of that set of places within the social division of labour whose occupants do not own capital or land to any substantial degree, are located within a set of interlocking social institutions which collectively 'service' capital; enjoy superior work and market situations generally resulting from the existence of well-defined careers, either within or between organisations; and have their entry regulated by the differential possession of educational credentials.
>
> (Urry 2002: 80)

Urry argues that it is their interests, tastes and values that define the globalized culture of tourism. Many tourists even see tourism as an opportunity to play in the waters of socio-economic class levels above their own, as part of their collection of experiences, even if it does put them into debt. Ostentatious behaviour by tourists can, however, further divide them from the communities they are visiting.

Individual tourists vary in the degree to which they adopt the standard tourist role, and in the ways in which they interact and impact the places they visit. Recreational tourists, interested more in hedonistic pleasures, are generally less aware of their impacts and more likely to expect destinations to change to meet their interests. Tourists with a strong interest in culture and history are likely to have less of an overt impact on a destination.

Occasionally, hosts and tourists will have deeper encounters that include the sharing of ideas and information (see Case 4.2). To some extent this can occur on a guided tour between the guide and the tourist; sometimes this can take place over a drink in a bar; and sometimes it occurs in a chance meeting. Tourism can enable individuals who otherwise would never have met to come into contact with one another. Special interest tours, such as culinary tourism, art tours and adventure sports, may offer more opportunity for host–guest interactions. Such encounters can be memorable because they transcend the superficiality of the more common tourist–host encounters.

The ideas, views and opinions shared in these deeper encounters can reduce stereotypes and prejudices, and foster a greater sense of tourist attachment to the destination, increasing a personal interest in the culture and society of the place. This personal involvement in the destination is an effective way to educate the visitor on local practices, traditions and cultural nuances. The exchange of ideas between hosts and tourists may also educate the host about people in other parts of the world, introducing new business and social ideas that could boost their personal well-being. These are the goals sought by those who are advocates of tourism as an avenue for world peace.

Tourism impacts on tourists

For the visitor, the tourist experience has been described as a liminal one. Liminal (from the Latin word for 'threshold') experiences occur at the borders between places, situations or social roles. A rite of passage, such as the Jewish bar mitzvah or a wedding ceremony, is a liminal experience as the person who enters the rite is different from the person who finishes the rite. Liminality causes a disorientation in one's identity and an openness to new ideas and possibilities. This can lead to new perspectives on life and self-understanding, and is similar to the 'self-actualization' level in Maslow's 'hierarchy of needs', discussed below and in Table 4.6.

Tourism is a liminal experience because it physically removes the participant (the tourist) from their normal place and social role, forcing them into a new place and an uncertain social role, and then returns them to their normal home place with a new set of past experiences, memories, souvenirs and bragging rights. This is one of the motivations that people have to travel – because it safely puts one in a state of disorientation, during which opportunities for personal, authentic and transformative experiences are more likely to occur than in one's everyday life. Indeed, both the safety of the trip and the degree of disorientation experienced can be adjusted by the tourist to desirable and comfortable levels.

While it is possible to have negative experiences during a holiday trip, tourists try to make the best of the liminal experience that they paid for, and are more likely to remember the positive experiences over the negative ones. This pattern is found in all forms of tourism, from recreational vacations to volunteer and educational tours. Research on the psychological impacts of travel have shown that both the act of travel planning and the trip itself are associated with higher levels of happiness and psychological well-being (Sangpikul 2007).

All travel impacts the tourist in one way or another. Tourists, however, have different goals when they travel, and while all have liminal opportunities, their motivations will shape the decisions they make while travelling and, ultimately, their assessment of a successful trip. We can see this through an examination of four traditional travel motivations: leisure, education, relationships, and status (adapted from Mayo and Jarvis 1981). Most trips include a combination of these four motivations, though often one or two predominate over the others.

- *Leisure* For many, the goal of travel is simply to relax and recuperate from a stressful work or home environment. This may involve participation in sports or other recreation activities, entertainment or health activities (such as a spa experience). A successful holiday for this type of tourist would be one that results in physical and psychological rejuvenation and a more positive attitude toward home and work. The returning tourists would be able to more successfully accomplish the tasks in their life after a good travel experience.
- *Education* Many tourists are motivated by educational goals, with a desire to learn about the world. Heritage sites, the arts, religion and other mani-festations of culture are the major attractions for this type of tourist, as are scientific and contemporary social conditions. A successful trip for this type

of tourist would leave the tourist with a deeper understanding (or even expertise) of a different geographic place, culture or condition. This could even have the effect of changing the personal values and related behaviours of the tourist.

- *Relationships* The desire to meet new people or to explore personal relationships are major motivations for travel. This includes making new friends away from one's home setting, to reconnect with old friends and relatives, and to research one's personal heritage and genealogy. A successful trip for this type of person would result in a meaningful relationship that continues after the trip has ended. Such relationships could provide new opportunities for career and personal development.
- *Status* Travel to some places and to participate in some activities can enhance one's social status and reputation among family and friends, work colleagues and professional associates, and other social relations upon returning home. While there are general destinations that have mass appeal and prestige, such as Paris or the Giza pyramids in Egypt, it is more common that prestigious places are associated with special interests, hobbies and sports, such as automobile collecting, surfing and mountaineering. A successful trip would bring increased prestige and respect for the traveller, and possibly requests to talk formally about the experience.

Numerous other models of tourist motivation exist, many of which centre on the distinction between the amount of security and risk that a tourist is comfortable with (Lew 1987). Security-oriented tourists have been called psychocentric, passive, guided, recreational, marker involved, sunlust and familiarity seeking. Risk-oriented tourists have been described by the terms allocentric, active, unguided, experiential or experimental, sight involved, wanderlust and exotic seeking.

Case 4.4: Existential tourism in Overseas Chinese travel to China

Many different models have been suggested over the years to divide tourists into different types based on their motivations and behaviour. Cohen (1979) proposed five basic types of tourism, based on the degree to which the tourist is psychologically centred in their home culture.

- *Recreational tourist*: someone who comfortably identifies with their home culture and as a tourist mostly seeks leisure and recreation activities for their holiday vacations.
- *Diversionary tourist*: someone who is alienated in their home culture, and sees alienation in other cultures, as well; as a tourist they simply seek diversions from their everyday life.
- *Experiential or observatory tourist*: someone who is alienated in their home culture, but sees (or gazes upon) centredness in other cultures; as a tourist

they visit more traditional cultures that have not succumbed to modernity's alienating values.

- *Experimental or participatory tourist*: similar to experiential tourist, except that the tourist wants to participate in the centredness of other cultures through festivals, religious rites and other activities.
- *Existential tourism*: someone who is strongly centred in another place, different from their home; as tourists they travel to their centre.

There are several forms of existential tourism. For some people, centredness is associated with a place that holds special meaning because it was the location of a life-changing event. For others, the centred destination is based on race and ethnicity. West Africa holds this type of appeal for American Blacks, as does Israel for Jews around the world (Lew et al. 2008). Family genealogy is a third common motivation for existential tourism, which is often referred to as 'finding one's roots'. Genealogy can be combined with ethnicity, as in the case of Mexican-Americans, Chinese-Canadians, Irish-Australians and many other creolized or 'hyphenated' ethnic groups around the world (see Coles and Timothy 2004). (Note that existential tourism is different from the existential tourist experience, which is a deep, personal feeling of authenticity that may occur at almost any time, given the right conditions, during a trip. See Steiner and Reisinger 2006.)

Overseas Chinese consists of ethnic Chinese who live outside of China. Sometimes this includes those living in Hong Kong and Taiwan, but more often it refers to those living beyond East Asia, with the largest populations in Southeast Asia and North America. In both historic and contemporary times, Overseas Chinese have maintained ties to China through well-established paths, or *networks of ethnicity*, that include clan (or extended family), regional (often a county) and special interest organization, which are also referred to as *voluntary associations* (Lew and Wong 2004). These associations are typically located in the Chinatowns of global cities, and they help to facilitate both migration to foreign countries (helping with employment and housing) and existential tourism trips back to China. They are also closely related to governmental Overseas Chinese Affairs offices that are found in even the smallest towns of southwest China, where most overseas Chinese originated in the nineteenth and twentieth centuries (that geographic pattern is less pronounced in the twenty-first century).

For many Overseas Chinese, existential tourism offers the promise of transcending both geographic and social space by keeping them connected with China and overcoming the uncertainties of migration. The ultimate existential destinations are the ancestral village, the ancestral home and the gravesites of one's ancestors. A trip to these sites supports basic fundamentals of Chinese culture and values, including:

- the importance of the extended family and the clan-based village reference group that is common in southwest China;
- filial piety pressures to care for aged parents and for the graves of ancestors; and

- prestige within the family and clan for those who make monetary donations and home village visits.

Local Overseas Chinese Affairs offices in China encourage these values in welcoming Overseas Chinese home, and of course, encouraging their donations to village schools and medical clinics, and for the building of roads and bridges to help improve the quality of life of their extended relatives. Thus, through the work of voluntary associations and Overseas Chinese Affairs offices, the existential tourism activities of the Overseas Chinese diasporas have changed the landscape and culture of both the ancestral homeland in China and, through the creation of Chinese communities, the new homelands overseas.

The Overseas Chinese tourists are somewhere between insideness and outsideness, and between authenticity and commodification. They are Chinese (insiders), but also tourists (outsiders); they are visiting real ancestral villages (authentic), but there is an expectation that they will make a financial donation (commodification). The desire to participate in existential tourism appears to be stronger for Overseas Chinese who were born in China, and for Overseas Chinese

Photo 4.5 Cui Heng Village, here, is the birthplace of Dr Sun Yat-sen (1866–1925), the founder of the Republic of China who is also honoured by the People's Republic of China, and seen by all Chinese as the father of modern China. This village is now part of the Museum of the Former Residence of Dr Sun Yat-sen, which has renovated the village so that it 'reproduces the living conditions of various socioeconomic strata at the time' that Dr Sun lived there. (Photo by Alan A. Lew)

who live in predominantly non-Asian cultures, such as North America, Australia and Europe. A survey of Overseas Chinese travellers passing through Hong Kong found that these groups had stronger ties to their home villages, and had visited and donated to their home villages at much higher rates than had Overseas Chinese living in Southeast Asia and East Asia (Lew and Wong 2005).

Not all existential travel achieves the goals of the tourist and the destination. Some argue that the existential travel experience (in all its forms), while a major motivation for tourism, is a rare and fleeting one that fades with time (Steiner and Reisinger 2006). In *Imaginary Homelands*, Salmon Rushdie (1991) claimed that we can never go 'home' again, no matter how many visits we make, because the homeland and the home village are as much imaginary places as a real place. We construct our ethnicity, and our own authenticity, from what Mathews (2000) has called the postmodern 'cultural supermarket', in which all identities are impermanent and the only real home is the entire world. Once one enters the cultural supermarket, which is both reflexive and fun, there is no turning back, there is no going home again. In this context, the existential tourist is a socially constructed role, though one which has been clearly informed by traditional Chineseness, shaped by the migration and diaspora experience, enabled by global modernization and seduced by the promise of a sense of home in China.

Culture shock

Culture shock refers to a degree of anxiety or unease that is experienced when one encounters a different cultural or social setting and is confused about what is appropriate and inappropriate behaviour. Culture shock always occurs in tourist travel, though sometimes it is very minor and mostly related to navigating a new geography. At other times, however, culture shock can overwhelm the pleasure of travel, causing significant disorientation. As such it is closely related to the liminal experience. Hosts also experience culture shock when they are confused by the behaviour of tourists. Culture shock is normally not a permanent condition – if it were it might be similar to *alienation*.

Culture shock is greatest between tourist and host when their cultures, ethnicity, religion, values, lifestyles and languages are very different. Such cultural difference can be quaint at first, but can soon lead to irritation or worse, especially if too many tourists exceed the cultural carrying capacity of the local community. This can be further exasperated by income and lifestyle inequalities between tourists and hosts.

Case 4.5: Preventing culture shock

Culture shock is a common experience of travellers in foreign countries or where cultures meet that are very different from each other. It is related to acculturation in that it is most strongly felt by those who are a minority in another culture. It is part of the process of assimilation into a new society and typically develops through several stages. The first is the initial *honeymoon phase* when the new

Table 4.4 Phases of culture shock

Culture shock phase	Attitude toward host culture	Predominant activity	Role
Honeymoon	Mostly positive	Exploration, discovery	Tourist, outsider, student
Negotiation	Mostly negative	Value and identity judgements	Role uncertainty, liminal state
Assimilation	Balanced	Daily life routines	Resident, expatriate, insider

culture is romanticized, and the new places and experiences offer opportunities for exploration and discovery – though occasional irritations still arise (Table 4.4). For many tourists whose travel is very time-limited, this is the farthest that culture shock develops.

After a longer time, though, this can develop very quickly for a tourist, the minor irritations grow and the newcomer enters a *negotiation phase* when differences in culture become more prominent, adapting to the new culture takes more intentional effort, and one starts to long for the food, landscapes and habits from home. In the final *assimilation phase*, the outsider more fully adapts to the new living situation, though irritation may still occasionally arise.

Signs that you may be suffering from culture shock include changes in sleeping patterns (beyond the short-term impact of jet lag), an increased concern for hygiene, increasing number of phone calls to home, hostile feelings toward and avoidance of local people, and fear of public places.

The best way to avoid culture shock, either as a tourist or long-term resident of a new place, is to be prepared. This is mostly done by reading and learning about the destinations and cultures that one is planning to visit. In addition, it requires a sense of openness to different ways of living and viewing the world. The *World Citizens Guide* provides a list of additional tips that were specifically designed for American students studying internationally. Among their suggestions are:

- Look. Listen. Learn.
- Live, eat and play local.
- Be patient – many cultures do not move at the same pace as Americans.
- Celebrate our diversity.
- Try the language.
- Refrain from lecturing – dialogue instead of monologue.
- Be proud of your country, but do not be arrogant.
- Keep your religion private.
- Check the atlas – know where you are in the world.
- Talk about something besides politics.
- Dress respectfully – be aware of what is acceptable in how local people dress.
- Learn some global sports trivia – it helps to form a common bond with others.

- Keep your word – if you say you will do something, do it.
- *Be a traveller, not a tourist.*

More information and traveller tips can be found on their website [www.worldcitizensguide.org]. Following these guidelines should not only help tourists adjust to cross-cultural environments, but should also lessen the potential negative impacts of the tourist presence on those destinations.

Tourism hosts

In addition to the broader impact of cultural change that the tourism phenomenon imposes on destinations, tourists themselves can have direct impacts on the material and social culture of a destination through their simple presence in a place. The direct impacts of tourists on a place generally fall into two types: (1) the deterioration of structures and sites through their use and misuse, and (2) the violation of cultural norms of behaviour.

Tourists, typically out of ignorance or carelessness, often fail to respect local norms of behaviour, including customs and moral values. If unaddressed, the continuing abrogation of cultural norms can heighten a sense of annoyance and resentment by locals against tourists. Cultural clashes of this type are more likely to occur in highly traditional communities – a type that is often sought out by tourists for their authenticity.

In many Muslim countries, for example, women are expected to cover themselves or at least maintain a high degree of clothing modesty in public. Tourists, however, often disregard these standards, appearing in public wearing revealing shorts, skirts or even bikinis, sunbathing topless on the beach (a relatively common practice for women in Mediterranean Europe) or consuming large amounts of alcohol (which is forbidden in Islam). The same types of issues can arise in conservative Christian communities in Polynesia, the Caribbean and the Mediterranean, and are further exasperated by a feeling among tourists that they are 'on holiday' and therefore are more free from the limiting social norms of dress and behaviour that apply in their own home place. An extreme of this behaviour is found in the spring break tradition (and similar traditions with other names) of college students who take a post-winter holiday in sun and surf destinations around the world.

Irritation index

As just mentioned, out of ignorance or carelessness, tourists may fail to respect local customs and moral values. In a desire to remember their experience, they may inadvertently intrude into the privacy of residents in the destination, taking a quick snapshot and then disappearing. Through these and many other ways, they can irritate local residents.

Doxey (1975) suggested that the attitudes of local residents toward tourism unfold through a series of stages over time (Figure 4.1). These stages are known

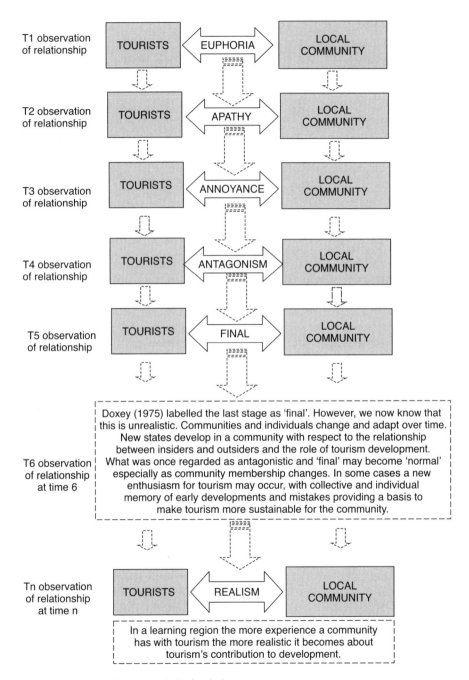

Figure 4.1 Reconsidering the irritation index.

as the *Irritation Index* (also known as the *irridex* model), which suggests that a community's degree and form of irritation toward tourism will vary over time. The process starts with a *euphoria stage*, when locals enjoy contact with visitors and tourism investors, and are very welcoming of them because they bring new experiences and money. Places where local residents actually ask tourists to photograph them, as occurs in many of the more remote islands of Indonesia, are typically in this euphoria stage. Local naivety can lead to a greater potential for outsiders to influence the course of tourism development in communities like this.

The euphoria stage is followed by an *apathy stage* that is characterized by an increasing indifference toward tourists, who have become more commonplace and larger in number. The roles of tourists and hosts become more formalized and less spontaneous. This is usually accompanied by perceptions of tourist rudeness, increasing crime attributed to tourism and its influences, increased traffic congestion, and higher prices for products and property. The tourism industry's role in local political decision-making increases, and tourism planning is mostly focused on marketing the community.

If local residents feel that there are too many tourists, the community may become increasingly irritated and start to question the value of tourism. The community enters the *annoyance stage*, which may indicate that the community's acceptable level of tourism has been reached. Vocal citizens express opposition to the political power of the tourism industry, while community planners focus on infrastructure development to address development growth.

If the growing level of conflict is not adequately addressed, then an anti-tourist *antagonism stage* may emerge in which the local community expresses overt and covert aggression toward tourists and the tourism industry. Non-governmental organizations may form to balance the political and economic power of the tourism industry, while development interests seek to counter negative publicity. Public funds may be spent on promotions to offset negative images, which can have a significant and long-term detrimental impact on the reputation of a destination.

The Irritation Index is not fixed and can be managed. Different cultures have different levels of resilience to tourism. The Hindu culture of Bali, in Indonesia, is famed for its ability to maintain deep religious and cultural traditions while welcoming close to 2 million visitors a year. On the other hand, many of the Pueblo Indian villages of northern New Mexico and Arizona forbid any kind of photography by non-Indians, though it is fully permitted for their tribal members. This restriction was adopted in the early 1900s, soon after the railroad brought the first tourists to these American Indian lands.

The demonstration effect

In addition to creating ill will, the tendency of most tourists toward more lax behavioural norms when they are on holiday can directly influence the behaviour of locals. This occurs because a person on vacation is not the same as someone at home; holidaymakers typically spend more money, behave more freely and demand more services; and tourists usually stay in a destination for a very short time and seldom develop a meaningful relationship with local residents. However, tourists still interact with local residents, introducing behaviours and belief systems that may be different from that of local residents.

The introduction of new behaviours and different values in a destination community as a result of tourism is known as the *demonstration effect* (Doxey 1972; Mcelroy and De Albuquerque 1986). This is where tourists demonstrate new ways of dressing, new norms of interpersonal communication (especially between genders), and different food and religious preferences. Local residents, especially younger people, observe this and adopt similar behaviours. Exposure to modern values can lead to increased migration from rural areas to cities and to more developed countries, resulting in a significant shift in the social structure of the rural place.

Tourism's demonstration effect was probably a more significant force in the acculturation process (see above) in the past than it is today. Recent advances in media (television, music and movies, in particular) and telecommunications (cell phones and the Internet), along with the rapid expansion in global international travel, all contribute to globalization and modernizing values reaching the farthest corners of the planet. This *global acculturation* is often viewed as the *Americanization* of the world, though it can also be argued that it transcends any one dominant society.

Although it is difficult to tease out the impacts of tourism from those emanating from other globalizing forces, such as the media, trade deregulation, information and communications technologies (ICT) and transport technology (Hall 2005a), there is no doubt that the outcome of the relationship between tourists and hosts produces changes in the host community that can affect a place's quality of life, culture, economy and social organization. Tourism contributes to the modernization of society, which has positive and negative impacts, depending on one's perspective. Some cultural traditions may be strengthened in the process, though still changed by it. On the other hand, some local residents may stop respecting their own traditions and religion, leading to tensions within the local community. In this sense the social and cultural impacts of tourism and the resultant change that occurs can be at least as significant as the physical degradation of a place's material culture.

Host perceptions of tourism

Tourism impact studies frequently include surveys of resident opinion on local tourism. Open-ended questions in these surveys can elicit unexpected responses. One survey in Hungary found that the social impacts of tourism included: opportunities to meet interesting people, better restaurant quality, opportunities to improve foreign language skills and an overall increase in employment opportunities (Ratz 2006). The respondents to this survey perceived tourism as contributing positively to their overall quality of life, which they welcomed. Negative impacts that were attributed to tourism included increased crime and lack of security, and a decline in religious values and social morality.

Despite the general similarity between the Hungarian hosts and their mostly European guests, the survey found that 78 per cent of those interviewed perceived significant differences between tourists and themselves. The degree to which these perceptions of impacts and differences are real may be less significant than the fact that they are perceived to exist and that they form the basis for local understanding of the tourism economy and its relationship to social change.

A fundamental difference between resident hosts and tourist guests, which has been briefly mentioned above, is the difference between a sense of *insideness* and a sense of *outsideness* (Relph 1976). Family and social relationships, personal and shared histories, and a common culture and traditions are some of the characteristics of being an insider. These bonds within a community draw clear distinctions between locals and visitors, even when racial and economic difference may be minimal. Chang (2000) has argued that these differences lead to divergent views of tourism and development. 'Feelings of "insideness" and a sense of belonging are pitted against feelings of "outsideness" and not belonging, thereby giving rise to the irreconcilable visions of the way space ought to be used and developed' (Chang 2000: 343).

In many instances, however, the distinction between insider and outsider is blurred. Tourists can sometimes blend in with local residents, particularly when they share language, cultural and physical traits. Local residents participating in recreation activities can be indistinguishable in their behaviour and impacts from outsiders participating in the same activities. Some tourists may even be more familiar with some aspects of the destination than are its residents. (It is common for outside tourists to have visited more different places in a destination than most of its residents.) 'Ignorant' visitors can also be more open to travel and experiential opportunities in a destination than are many residents who may hold fixed stereotypical views on neighbourhoods and sites. The blurring of insideness and outsideness is especially common in large metropolitan destinations, and is one of the reasons why these cities are among the most popular tourist destinations, and why cultural clashes in rural communities can be more prominent than in urban settings.

Leisure-related social and cultural changes are driven by local populations, as well as outside tourists. As people travel more, they become more cosmopolitan and globally experienced, their tastes and desires change, and they tend to become more like tourists. With the rise of online social communities focused on special interests, residents in diverse geographic locations may socialize more with each other than they do with their next-door neighbours. In this way, special interest tourists may also know more about distinctive features in a destination than the majority of locals do. An example of this is the iconic 'Route 66' in the United States. Route 66 was made famous by a popular song written by Bobby Troup in the 1940s and subsequently a hit for the Rolling Stones in 1964, among others, and an early television show in the 1960s. The now-decommissioned *Mother Road* connected Chicago and Los Angeles and was the principal route taken by Depression-era Dust Bowl immigrants seeking a better life in California in the 1930s. Today, some of the most avid Route 66 nostalgia-seekers come from

Europe, especially Germany, and they often know more about the history of the road than those who live on it (Caton and Santos 2007). They trace the old highway's remaining two-lane roads, often through speciality tour companies, as they try to capture the spirit of the American dream.

Case 4.6: Comparing resident and visitor perceptions of place in Simla, India

The image of a place is essential in its marketing as a tourist destination. Destination marketing organizations (DMO, which is often a visitors' and convention bureau) try to define a single and clear image for a place in their marketing efforts. However, most places are far too complex to be collapsed into a single icon. Residents know this through their long place-based biography, which forms their sense of place. Tourists, however, seldom have such a biography. Instead, they know other places mostly through word of mouth and various forms of secondary media exposure, such as movies, daily news sources and books.

Tourists lack detailed knowledge of a destination, especially geography (location and scale) and history (changes over time). Lew (1992) found that this lack of knowledge resulted in tourists having a less clear image of a destination and being more likely than residents to perceive traditional features, authenticity and exoticness in a destination. Their 'outsider perspective did allow tourists to have a more comparative knowledge, gained from touring other cities, which . . . enabled them to see uniqueness in what [residents] would view as commonplace and ordinary' (Lew 1992: 50).

Simla is the capital of the state of Himachal Pradesh. It is located at approximately 2,300 m (7,500 ft) elevation in the Himalayas north of New Delhi, and its cooler climate made it a summer capital for the colonial British rulers prior to Indian independence in 1949. They created an English townscape in Simla, which adds to the dense pine forests and winter snow to create a distinct landscape that draws large numbers of tourists.

A survey was conducted of visitors and residents in Simla to determine their image of this hill station destination (Jutla 2000). As part of this study, each group was asked to describe Simla in one word. The results of that question found two very distinct images, as shown in Table 4.5.

Tourist perceptions focused more on the most prominent visual aspects of the community, especially those that can be observed from public spaces. Resident perceptions were more complex and personal, and included both positive (quiet and safe) and negative (costs and development impacts) attributes that affect their sense of place and quality of life.

Factors that can affect tourist knowledge and perceptions of destinations include motivations (such as recreation, education and heritage interests), number of past visits and the length of the current visit, type of touring (such as independent, walking and speciality tours). Factors that affect the knowledge and perceptions of a place by local residents include their daily work and living space

Table 4.5 Tourist and resident images of Simla, India

Major tourist images	Major resident images
Mountains	Quiet
Trees	Home
Climate	Familiar
Scenery	Tourist congestion
Walking	Peaceful
Zigzag roads	Safe
Fog	Water shortage
Sloping roofs	High cost of living
	Overdevelopment

(where they move or go to the most), length of time in a destination, and degree of civic interest and involvement in local development issues.

Knowledge of how different types of visitors and residents perceive destinations sheds greater light on the complex influence of tourism and the influences of insider and outsider perspectives on human experience and behaviour in the creation of place. Surveys, like this example from India, can help planners to identify key areas that need attention to prevent tourism from becoming a nexus for antagonism between hosts and guests.

Tourism and quality of life

The development of the tourism industry can contribute broadly to changes in the quality of life, social structure and social organization of destination communities. If not carefully managed, however, tourism development can contribute to a decline in the quality of life experienced by residents. Rapid and larger tourism development often results in more impacts, positive and negative, than slower, more organic and smaller-scale development.

Quality of life is both an individual and collective attribute. At the individual level it includes objective and subjective elements. People's objective quality of life requires that their basic needs are met and that they have the material resources necessary to fulfil the social requirements of citizenship and, as such, the concept is closely connected to the social dimensions of sustainability. Quality of life is, therefore, the degree of well-being felt by an individual or a community. It is generally divided into physical well-being and psychological well-being, though it has also been associated with levels of crime, issues of sustainable development, the Human Development Index (HDI), and gross domestic product (GDP), among many other measures – most of which are highly subjective.

Abraham Maslow's theory on the hierarchy of needs (1943) provides a basis for assessing quality of life (Sirgy 1986). The quality of life in a city may be measured by the degree to which the urban environment provides opportunities for individuals to meet their physiological, safety and other needs (Table 4.6).

Maslow's theory was hierarchical, meaning that lower needs (physiological survival and personal safety) must be met before higher needs (self-esteem and

Table 4.6 Abraham Maslow's hierarchy of needs

1	Physiological needs	oxygen, food, water and a relatively constant body temperature – these needs are related to survival and are the most basic and powerful needs
2	Safety needs	protection from elements, disease, fear – these needs are strongest in children and dependents, except in times of emergency
3	Love, affection and belongingness needs	giving (and receiving) love, affection and a sense of belonging, while avoiding feelings of loneliness and alienation
4	Esteem needs	a high level of self-respect and respect from others, including feelings of satisfaction, self-confidence and value (avoidance of feeling inferior, weak, helpless or worthless)
5	Self-actualization needs	pursuing and fulfilling one's 'calling' in life; if not met, the person feels restless, tense and lacking 'something'

Source: Maslow 1943.

self-actualization) can even be seriously considered. Others, however, have argued that these needs are a universal to the human condition, and are equally needed at all times by all people (Max-Neef 1991). Whether hierarchical or not, the needs outlined in Table 4.6 are directly related to issues in tourism development. The relationship between Maslow's list of needs and tourism development focus on employment-related income generation, recreation opportunities related to tourism, cultural conservation and entrepreneurship.

Tourism's contribution to physiological and safety needs. Tourism development offers opportunities for employment and income at many different levels, from those with limited education and training to entrepreneurs and managers. Entry-level positions in hotels and restaurants, in particular, can provide job opportunities to lower-income communities. These jobs provide income to meet basic physiological and safety needs. While some opportunities for advance are available, these positions tend to be low paying and may not meet the safety needs of employees in communities with high costs of living. For tourists, holidays and vacations have direct physiological and psychological health benefits, including reducing heart attacks and relieving chronic stress (Robinson 2003). Workers have been shown to have higher productivity and increased performance following a rejuvenating holiday or sabbatical from a stressful job.

The sheer number of tourists can also directly infringe upon the quality of life for residents. If large numbers of 'outside' visitors cause overcrowding at leisure sites (a situation more common in smaller communities than larger cities), then locals may come to resent the intrusion and feel that their safety and security are threatened. Problems seem to arise either (1) when the number of tourists far exceeds the service infrastructure capacity of a city, and tourists are blamed for increasing congestion, prices and pollution, or (2) when the cultural and socio-economic differences between hosts and guests are major and a feeling of exploitation emerges among local residents.

Tourism's contribution to love, affection, belongingness, esteem and self-actualization needs. Tourism is mostly a social phenomenon. Tourists frequently travel in groups and visit sites with others sharing common interests. Tourists also desire to meet and befriend local residents. And tourists are very interested in learning about the distinctiveness of local cultural history and traditions. Participating in the leisure opportunities associated with tourism can provide a sense of belonging, camaraderie and accomplishment in some settings, such as adventure recreation experiences.

However, because the tourist is a transitory figure, it is more likely that residents may feel a sense of solidarity with each other as insiders protecting their place and culture, in opposition to tourist outsiders. This can be confrontational, but it can also emerge as a sense of education and sharing of local culture with others, and possibly a revival of lost traditions. The social relationship between tourists and hosts also has the potential for local social and cultural change due to increased exposure to capitalist consumption and foreign cultures.

Leisure, in general, is highly associated with opportunities for belonging, enhancing self-esteem, and self actualization. Self-actualization can take place through physical mastery, artistic accomplishments and entrepreneurial success. Tourism development can foster outdoor recreation opportunities, arts and performance classes, and small business ventures. What is most important is that the tourism developments are seen foremost as part and parcel of the local community. Developments that alienate locals from their own community are likely to cause resentment and threaten the quality of life for everyone.

Summary and conclusions

The impacts of tourism on the culture of a place are difficult to unpack from the broader processes of global economic and social change that are affecting every corner of the globe. Since the fall of the Iron Curtain in 1989, new versions of capitalism have come to dominate the world's economy, with dramatically altered patterns of production and consumption. A global sea change in cultural land-scapes, social values and personal identities is accompanying the new, and increasingly universal, economic and political realities.

The fundamental question for tourism is how to balance the competing demands on culture. These demands include: (1) resident desires for a better economic and social quality of life versus their sometimes conflicting desire for an insider culture that reflects their personal identity and community's distinct sense of place, (2) the tourist's desire for comfort and safety versus their conflicting desire for novelty, exploration and authenticity, and (3) the tourism industry's pressure for global-ization, through franchising and copying of successful images and models, versus the conflicting desire to protect local entrepreneurs from competition by multinational corporations.

Can culture survive these debates and tension? Well, yes. In fact, new forms of culture continuously emerge from this ongoing fray. Whose cultures arise from the resulting hybrids and mashup in a country, a community or an individual is one of the aspects that makes travel and tourism interesting for both the tourist and the researcher. For the resident, this is probably only important to the degree that tourism contributes to or detracts from their quality of life, which is, in turn, directly tied to both the economic and environmental effects of tourism.

Self-review questions

1   What is the formal definition of culture and what is its role in the creation of heritage destinations?
2   In what ways are the global and the local merging in contemporary tourism landscapes?
3   How does tourism sometimes change the traditional arts and crafts of a destination?
4   What is the role, experience and relationship of tourists to the destinations they visit?
5   In what ways are the roles of 'tourist' and 'host' both positive and negative for the host–guest relationship experience?
6   How does tourism impact the quality of life of both destination hosts and tourist guests?

Recommended further reading

Rojek, C. and Urry, J. (eds) (1997) *Touring Cultures: Transformations of Travel and Theory*, London: Routledge.
Insight into social theories of culture of travel and cultural meanings of types of places that travel and tourism is creating.

Lowenthal, David (1985) *The Past is a Foreign Country*, Cambridge: Cambridge University Press.
Discusses the role of history and the past, including heritage places, in contemporary society.

Selby, M. (2004) *Understanding Urban Tourism: Image, Culture and Experience*, London: I.B. Tauris.
Cities are the most visited tourist destination and this book discusses that phenomenon from postmodern, place image and urban cultural perspectives.

Coles, T.E. and Timothy, D. (eds) (2004) *Tourism, Diasporas and Space: Travels to Promised Lands*, London: Routledge.
Perspectives on the role of tourism for people who have cultural ancestry in countries far from where they now live, which includes an ever-growing proportion of the world's population.

Beeton, S. (2005) *Film-Induced Tourism*, Clevedon: Channel View Publications.
Discusses some of the impact of the media on communities in terms of socio-cultural impacts as well as commodification for development purposes.

Ryan, C. and Aicken, M. (eds) (2006) *Indigenous Tourism: The Commodification and Management of Culture*, London: Elsevier.
A collection of essays on the impacts and responses of tourism in the most traditional ethnic communities around the world.

Web resources

United Nations Environment Program (UNEP)
Negative Socio-Cultural (and Ethical) Impacts from Tourism: http://www.uneptie.org/pc/tourism/sust-tourism/soc-drawbacks.htm

How Tourism Can Contribute to Socio-Cultural Conservation: http://www.uneptie.org/pc/tourism/sust-tourism/soc-global.htm

United Nations Education, Scientific and Cultural Organization (UNESCO), World Heritage Centre
http://whc.unesco.org/

National Geographic Society, Center for Sustainable Destinations
http://www.nationalgeographic.com/travel/sustainable/

Definitions of Anthropological Terms
http://oregonstate.edu/instruct/anth370/gloss.html

Tourism Concern
http://www.tourismconcern.org.uk/

Conservation International's ResponsibleTravel.com
http://www.responsibletravel.com/

Key concepts and definitions

Acculturation The acculturation process assumes that when two nationalities (countries or cultures) come into contact over time, there will be an exchange of products, ideas and values.

Constructed authenticity Socially constructed authenticity, based on beliefs and viewpoints that represent those of the politically powerful and publicly influential.

Culture The set of distinctive spiritual, material, intellectual and emotional features of society or a social group, and that it encompasses, in addition to art and literature, lifestyles, ways of living together, value systems, traditions and beliefs (UNESCO definition).

Culture shock The degree of anxiety and unease that is experienced when one encounters a different cultural or social setting and is confused about what is appropriate and inappropriate behaviour.

Demonstration effect The introduction of new behaviours and different values in a destination community as a result of tourism.

Experiential authenticity A personal and subjective experience, including physical and psychological feelings.

Glocalization The creation of new cultural forms that are neither exclusively local nor dominated by the global, but a new creation that recognizes the contributions of both.

Heritagization The process through which heritage is created from the attributes of the past, whether these are relics, artefacts, memories or recorded histories.

Object authenticity Authenticity that can normally be measured by objective criteria, such as a scientific fact, a known location or a historical personage.

Society The social relationships that exist between people with a common interest or within a group.

5 Tourism and its physical environmental impacts

Learning objectives

By the end of this chapter, students will be able to:

- Appreciate the scope of the concept of *environment*.
- Define the concept of naturalness.
- Understand how environmental changes can be considered as global in scope.
- Understand global environmental change in relation to tourism with respect to:
  - change of land cover and land use,
  - climate change,
  - the exchange of biota over geographical barriers and the extinction of wild species, and
  - the exchange and dispersion of diseases.

Introduction

Tourism has impacts on the environment at a number of different scales. Until the end of the twentieth century, the predominant focus of tourism's effects were at the destination level. However, we now realize that travelling to and from a destination can also have enormous impacts on the environment, not just in terms of the environment of the transit routes, but at a global scale. As a result, increasing attention is being given to tourism's relationship to global environmental issues, such as climate change. This chapter discusses tourism-related impacts both at the global scale and in relation to particular types of environmental impacts. Cases will be used to illustrate local dimensions of tourism and environmental change. However, before examining these issues it is necessary to discuss what is meant by the concept of the environment.

The environment

Although widely used in everyday speech the concept of the environment is extremely difficult to define with ease, especially as it can be so all-encompassing (see also Chapter 2 on impact assessment). However, with respect to impacts, the term *environment* is usually used in reference to the physical environment of humans. For example, UNEP (2000) noted that for its global environmental information exchange network, INFOTERRA,

> The concept 'Environment and Nature' is very difficult to define in practice. The definition in the INFOTERRA network differs from country to country. How it is defined as dependant on local conditions such as the size of the country, climate, natural resources, industrial activities and so on. For example, some countries have coasts, some have deserts and others have nuclear power plants.

Nevertheless, defining environment is extremely important, especially in the area of the use of legal controls that affect it. Understanding how different countries define the environment, including both popular and legal definitions, may be helpful to identify the key elements of the concept. For example, the US Environmental Protection Agency (EPA) defines environment as 'The sum of all external conditions affecting the life, development and survival of an organism' (EPA 1997: 17), which is a very broad definition that could include almost anything.

In Australia, the *Commonwealth Environmental Protection and Biodiversity Conservation Act 1999* (Section 528), is somewhat more specific, defining the environment as including:

(a) ecosystems and their constituent parts, including people and communities; and
(b) natural and physical resources; and
(c) the qualities and characteristics of locations, places and areas; and
(d) heritage values of places; and
(e) the social, economic and cultural aspects of a thing mentioned in paragraph (a), (b) or (c).

In contrast, in Canadian federal law there is more of a focus on the *natural environment*, in which the term environment means:

• the components of the Earth and includes
(a) air, land and water;
(b) all layers of the atmosphere;
(c) all organic and inorganic matter and living organisms; and
(d) the interacting natural systems that include components referred to in paragraphs (a) to (c).

(*Canadian Environmental Protection Act*, 1999 3(1))

Article 3 of the EU European Council Directive of 27 June 1985 is on the assessment of the effects of certain public and private projects on the environment, which is part of the set of regulations that underlie the EU's environmental impact assessment powers, and includes the following environmental factors:

– human beings, fauna and flora,
– soil, water, air, climate and the landscape,
– the inter-action between the factors mentioned in the first and second indents,
– material assets and the cultural heritage.

(Council of the European Communities 1985, Article 3)

One of the difficulties with assessing the notion of 'the environment' is that ultimately all changes are 'connected with one another through physical and social processes alike' (Meyer and Turner 1995: 304). Nevertheless, for the purposes of management some boundaries, even if artificial, have to be drawn. Therefore, one way in which the physical environment can be understood is by seeing it as a continuum from urban environments to natural physical environments in terms of relative *naturalness* (the absence of environmental disturbance by settled people) and *remoteness* (distance from the presence and influences of settled people) (Hall and Page 2006) (Figure 5.1). Remoteness and naturalness are important elements with respect to the designation of an area as *wilderness*. Using this approach, Grant (1995) provided a relatively simple classification of naturalness:

- *Natural environments*: not disturbed by humans or domesticated animals.
- *Subnatural environments*: some changes, but the structure of vegetation is basically the same (a forest remains as a forest).
- *Semi-natural environments*: the basic vegetation has been altered but without intentional change in the composition of species, which is spontaneous (i.e. overgrazed areas).

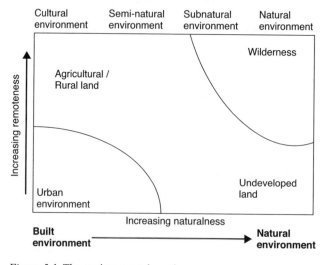

Figure 5.1 The environmental continuum.

Photo 5.1 Tourists climbing Uluru/Ayers Rock, in Uluru-Kata Tjuta National Park of the Northern Territory of Australia. Ayers Rock is both a natural geological feature and a sacred cultural site for the Aboriginal residents of central Australia. For them, the only time anyone would be allowed to climb Uluru was during a rare religious ceremony. Although numerous information signs discourage climbing out of respect to Aboriginal values, it is still allowed by the Australian park service and large numbers of tourists make the climb. (Photo by Alan A. Lew)

• *Cultural environments*: artificial systems such as cultivated land, including forest plantations, and urban environments; the vegetation has been deliberately determined by humans, with loss of the previous habitat.

Naturalness can be determined by both the percentage of endemic biota that would be in a specific location, if it were not for the impact of human beings, and by the *perceived naturalness*. Unfortunately, the former, although being a better objective measure of naturalness, is extremely hard to determine as baseline conditions before human impact occurs are usually unknown. Therefore, the perceived naturalness of a landscape is often used in environmental perception studies. It is, therefore, possible to envisage a situation in which tourists may see a landscape as being highly natural, while in fact it has very low values in terms of retention of its natural biota. Debates over the authenticity of the environment and landscape viewed replicate some of the debates over the authenticity of culture in tourism settings (see Chapter 4).

The relationship between tourism and the environment is bound up in the wider issues over the relationships of people to the environment and the impact of their behaviours and perceptions. Although it has long been recognized that tourism has a wide range of impacts on the physical environment, several significant methodological problems exist in understanding the relationship between tourism and environmental change (see also Chapter 2). These include:

- the difficulty of distinguishing between changes induced by tourism and those induced by other activities;
- the lack of information on conditions prior to the advent of tourism and, hence, the lack of a baseline against which change can be measured;
- the paucity of information on the numbers, types and tolerance levels of different species of flora and fauna; and
- the preference of researchers to concentrate on particular environments, such as beaches, coral reefs and alpine areas. For example, there is extremely limited research in most arid and semi-arid environments.

It is very difficult, for example, to separate the impacts of tourists and residents in specific locations. In fact, it is only in areas where there is no permanent human population that you can identify with absolute certainty the impacts of tourism. Sites of tourism production can have clear environmental effects, while the addition of transient visitor populations to the permanent population can also help indicate the effects of tourism. Nevertheless, the impacts of tourism are certainly not all negative as tourism can provide an economic justification for conserving biodiversity, landscapes and specific species. It is the interest of people in the natural environment that helps provide the base for ecotourism and conservation.

Case 5.1: Tourism and the big cats of Maasai Mara Game Reserve

The Maasai Mara Game Reserve in Kenya is 1,510 km² (583 m²), and is probably the most famous reserve in the country. Set in Africa's Great Rift Valley, the reserve is effectively the northern continuation of the equally famous Serengeti National Park Game Reserve in Tanzania. The Maasai Mara Game Reserve has featured in many wildlife documentaries, most notably *Big Cat Diary* shown on *Animal Planet*. This has only served to reinforce its attractiveness to visitors. Wildlife tourism presently brings in approximately US$200 million every year to Kenya, and is the country's largest earner of foreign currency (Secretariat of the Convention on Biological Diversity (SCBD) 2006).

In early 2008, civil and political unrest in Kenya had led to a dramatic drop in visitor numbers and increasing pressure on leopards and lions in the area known as the Mara Conservancy. The reason for this was that since 2001 a scheme funded via a percentage of park entrance fees paid by tourists had been established that compensated local people for their grazing animals killed by lions or leopards. Since its establishment the number of lions in the reserve has doubled to 80. In late April 2008 the Conservancy was facing a shortfall of $50,000 per month and the fund had to be suspended, with some Maasai threatening to resume hunting any lions or leopards that kill their cows, goats and sheep. According to William Deed, from the Mara Conservancy:

> We have now had several close calls with locals hunting lions and leopards in return for the cattle that have been killed by these predators.. . . Previously, the cattle compensation scheme we had in place would help placate such situations, however with no funding to pay for such a scheme the local communities are no longer seeing the benefits of living so closely with the wildlife.

(BBC News 2008a)

Sources: Mara Conservancy: http://www.maraconservancy.com/

Big Cat Diary: http://www.bbc.co.uk/nature/programmes/tv/bcd/

Mara Triangle news blog: http://www.twitter.com/maratriangle

Tourism has long been a factor in conservation and was the primary reason for the creation of the first national parks, with ecology not becoming a significant influence in park establishment until the 1920s (Hall 1998). Even today tourism remains an important justification for park creation and developments, especially in less developed countries or even in the more peripheral regions of the developed world. However, when national parks were first established the number of visitors was tiny when compared to the present day. The growth of national park systems, as well as the growth in visitors to them, illustrates well the dilemmas of sustainable development discussed in Chapter 2, and the two primary strands of conservation thought: *preservation* and *wise use* (also referred to as *economic conservation*). Preservation implies the setting aside of an area of land or sea that has high degrees of naturalness and then only allowing minimal human impact.

Conservation, as it is usually used with respect to management of the environmental impacts of tourism, tends to have a much more diverse set of strategies, of which setting aside an area as a reserve (preservation) is one of them. This suggests far greater active management by humans, not only of tourism and other activities, but also the manipulation of the environment itself in order to produce desired environmental outcomes. Conservation can also occur at different scales ranging from biomes and ecosystems through to communities and watersheds and individual species. In addition, conservation can also occur at the species level with respect to the maintenance of genetic variety within a species. However, it should be noted that, at times, there may even be management conflicts between the needs of conserving a species versus conservation of a wider ecology with respect to levels of public support and cost of the different strategies that may be required.

Table 5.1 Visitation to the US National Park System

Year	Recreation visits	Recreation hours	Non-recreation visits	Non-recreation hours
1979	205,369,795	1,027,515,046	77,065,306	43,110,345
1980	220,463,211	1,094,294,242	79,860,871	39,292,243
1985	263,441,808	1,303,592,955	82,748,302	39,607,247
1990	255,581,467	1,295,808,178	79,581,270	43,194,844
1995	269,564,307	1,322,447,685	118,239,606	61,462,362
2000	285,891,275	1,271,005,218	143,961,848	78,145,957
2005	273,488,751	1,205,758,305	149,898,447	81,269,826
2006	272,623,980	1,205,394,969	165,768,204	85,556,052
2007	275,581,547	1,224,569,680	163,756,296	84,466,643

Source: obtained from National Park Service, NPS Stats http://www.nature.nps.gov/stats/park.cfm

Many of the early national parks around the world were only made accessible by railway connections and were relatively remote from major urban centres (Frost and Hall 2009). In the case of Yellowstone National Park, Runte (1997: xii) pointed out that:

> The country that invented national parks held just thirty million people [when the park was founded]. As late as World War I, Yellowstone's annual visitation rarely exceeded 50,000. Moreover, the large majority came by train and stagecoach, part of a community of travelers bound to responsibility by limited access, poorer roads, and rustic accommodation. The nation about to carry Yellowstone into another millennium has ten times the population of 1872. Park visitation, both domestic and foreign, now exceeds three million every year.

It was not until the advent of mass car ownership that a significant growth in visitor numbers to national parks occurred, although growth and accessibility has now reached a stage that parks and natural reserves have become integral to many destinations' tourism strategies, although concerns over the environmental stresses of visitation are increasing. For example, in the American situation, Runte (1990: 1) observed that 'Among all of the debates affecting America's national parks, the most enduring – and most intense – is where to draw the line between preservation and use'. Table 5.1 shows the rate of visitation to the US National Park System, which received almost 430 million visits in 2007. The economic impact of such visitation is also substantial (see Case 5.2).

Case 5.2: Economic impacts of Yellowstone National Park visitors

The economic contribution of using tourism to justify conservation activities is perhaps best seen in the role of national parks as tourist attractions. Yellowstone, the world's first national park, was established in 1872 (see Frost and Hall 2009 for a discussion of the role of tourism in the park's creation). In 2006, there were 13.3 million recreation overnight stays in the US National Park System (NPS), representing 3 per cent of all visits. Of park visits, 28 per cent were day trips by local residents, 43 per cent were day trips of 50 miles or more in distance, and 29 per cent involved an overnight stay near the park. Two-thirds of all overnight stays by park visitors were in motels, lodges or B&B's outside the park, another 21 per cent were in campgrounds outside the park and 12 per cent were inside the park in NPS campgrounds, lodges or backcountry sites. A typical group of visitors on a day trip (not overnight) to a national park in 2006 spent US$39 if they were local residents and $69 if non-local, excluding park entry fees.

Spending by visitors on overnight trips varied from $68 for backcountry camping groups to $291 for groups staying in park lodges. Campers spent $95 per night if staying outside the park and $85 if staying inside the park. Based on these averages, park visitors spent a total of $10.73 billion in the local regions within 50 miles surrounding the parks in 2006. Local residents account for 10 per cent of this spending. Visitors staying in motels and lodges outside the park accounted for over half of the total spending, while non-local visitors on day trips contributed about a quarter of all spending. Over half of the spending was for lodging and restaurant meals. Taking multiplier effects into account (see Chapter 3), the $10.7 billion spent by park visitors had a total local economic effect (significance) of $13.0 billion in sales, $4.5 billion in personal income and $7.0 billion in value added, with visitor spending supporting about 213,000 jobs in gateway regions (Stynes 2007).

At the level of individual parks, spending can also be significant. It was estimated that on the 4.125 million recreation visits to Grand Canyon National Park in 2003, $338 million dollars were spent within the local region (Coconino County, Arizona, including gateway communities of Tusayan, Williams, Flagstaff and Cameron on the South Rim and Jacob Lake, Kanab and Fredonia on the North Rim), including $137 million inside the park and $201 million in gateway communities. On average, park visitors spent $201 per party per day in the local area. Park visitor spending generated an estimated $112 million in direct personal income (wages and salaries including payroll benefits) for the region and supported 5,655 jobs in tourism-related industries in the area. Including secondary effects the total economic impacts of visitor spending in 2003 was estimated at $429 million in direct sales, $157 million in personal income, $245 million value added and almost 7,500 jobs (Stynes and Sen 2005). However, at the same time the park is recognized as contributing substantially to the region there are also concerns over the environmental issues associated with having such a high number of visitors and the tourism developments that accompany them (Crumbo and George 2005; Jacques and Ostergen 2006).

Given the potential economic, as well as ecological, benefits of tourism for the environment it is not surprising that the relationship between the two is highly complex (e.g. Buultjens et al. 2005; Naughton-Treves et al. 2005). One of the most influential perspectives on the relationship between tourism and the environment is that of Budowski (1976), who suggested that three basic relationships can occur: conflict, coexistence or symbiosis.

- *Conflict* Tourism and the environment are in conflict when tourism has a detrimental impact on the environment.
- *Coexistence* Tourism and environmental conservation can exist in a situation where the two have relatively little contact, because either both sets of supporters remain in isolation or there is a lack of development. However, this situation 'rarely remains static, particularly as an increase of tourism is apt to induce substantial changes' (Budowski 1976: 27).
- *Symbiosis* Tourism and environmental conservation can be mutually supportive and beneficial when they are organized to ensure that tourists receive benefits and the environment experiences improvements in management practices. This relationship may also have economic advantages and contribute to the quality of life in gateway and host communities.

The following sections will highlight some of the different ways in which these relationships occur at different scales of analysis.

Global environmental impacts

Tourism and global climate change

Carbon dioxide (CO_2) emissions from tourism have grown steadily over the past five decades to the current estimated level of about 5 per cent of annual anthropogenic (human-caused) emissions of CO_2. Although organizations such as the International Air Transport Association (IATA) talk positively of their environmental record (e.g. IATA 2008c) it is the transport sector that contributes the most greenhouse gas (GHG) emissions, including CO_2. As noted in Case 1.3, transport generates 75 per cent of tourism related CO_2 emissions, followed by accommodation (21 per cent) and tourist activities (3 per cent). However, in terms of the relative contribution of *radiative forcing* (defined in Case 1.3) to climate change, the share of transport is significantly larger and ranges from 82 per cent to 90 per cent, with air transport alone accounting for between 54 per cent to 75 per cent of the total (UNWTO et al. 2008).

Knowing the range of potential emissions is important as it highlights the need for further research in the area, particularly as industry groups tend to use values at the lower range estimates when commenting on their emissions. For example, IATA (2008c) states that aviation accounts for 2 per cent of total man-made CO_2 emissions, and that this could reach 3 per cent by 2050. When factoring in all GHG emissions they state that aviation accounts for 3 per cent of the total man-made contribution to climate change and that this could reach 5 per cent by 2050. However, this figure is almost certainly an underestimate as it depends, in part, on research that is increasingly out of date (IPCC 1999).

Significantly, a small number of energy-intense trips is responsible for the majority of emissions. While air-based trips account for 17 per cent of all travel, they cause about 40 per cent of all tourism-related CO_2 emissions, and 54 to 75 per cent of the radiative forcing related to transportation. For example, long-haul travel between the five world regions that the UNWTO use for statistical aggregation purposes (Africa, Americas, Asia-Pacific, Europe and the Middle East) accounts for only 2.7 per cent of all tourist trips, but contributes 17 per cent to global tourism emissions. In contrast, trips by rail and bus/coach account for about 34 per cent of all tourist journeys, but only 13 per cent of all CO_2 emissions. The implications for the development of mitigation strategies are, therefore, substantial as individual trips can cause emissions with up to a factor difference of 1000: from close to zero for a holiday by bicycle and tent to more than 10 tons of CO_2 for a return journey to Antarctica from Europe (UNWTO et al. 2008).

Following the lead of the Kyoto Convention, the European Union has set an ambitious goal of reducing emissions of CO_2 by 20 per cent by 2020 (as compared to the base year 1990). Some countries have even aimed higher. Sweden's government climate policy group, for example, has been working on a climate strategy that it will set an ambitious 30 per cent EU reduction goal by 2020, and a national reduction goal of 75 to 90 per cent by 2050 (Gössling and Hall 2008). In light of this, how then might leisure and tourism be able to reduce its GHG emissions and related impacts on the environment?

Given the relative contribution of transport to leisure-related GHG emissions a key focus will probably be the encouragement of local leisure activities at the expense of long-distance ones. This will likely be done via a mix of marketing and regulatory measures. Such strategies will also need to go hand in hand with efforts to encourage the greater use of green transport systems such as bus and rail, as well as cycling and walking (Gössling et al. 2009). Importantly, for many locations such measures will have substantial economic benefits as monies previously lost from local economic systems will now be retained longer. In addition, it is expected that there will be significant social benefits (such as a greater sense of community) and health benefits (such as lower levels of obesity) as a result of such measures.

Tourism and recreation planners will play a critical role in the design of sustainable communities and destinations as they will likely need to encourage significant localized leisure and tourism resource consumption. Any '100 mile diet' for sustainable consumption must incorporate leisure services in addition to foods and retail goods (see 100milediet.org/). Such changes will likely affect mid- to long-distance leisure mobility (over 100 miles), but here regulatory actions such as environmental taxes or carbon trading will be vital. Many current *carbon offset* schemes, financial instruments that lead to a reduction in greenhouse gas emissions, are likely to be only temporary measures, particularly given the absence of a clear global environmental and social standard for them. Policy settings at various levels of governance will also need to shift if tourism-related climate change is to be mitigated. Unfortunately, the proclamations of organizations such as the UNWTO and the WTTC have little direct impact on what leisure and tourism businesses and consumers actually do. Government policies and regulations with respect to trade, environment and transport will have far more influence on tourism's future GHG emissions than sector-specific policies, such as those that just focus on airlines or accommodation. This means that the future of the environment of tourism probably lies in the integration of leisure and tourism concerns into spatial adaptation and mitigation strategies rather than developing sector-specific approaches in isolation. The tourism sector is so diverse in scope, and the tourism component has been so focused on business as usual, that voluntary behaviours and the operation of the market are not likely to provide the answers we need.

Global environmental change

Global environmental change (GEC) refers to such issues as climate change, modifications of global biogeochemical cycles, land alterations, the loss of non-renewable resources, unsustainable uses of renewable resources, species extinctions and reductions in biodiversity. It is important to note that the environment, whether at a global or a local level, is always changing. However, much of this change occurs so slowly that it has been imperceptible to the human eye. Yet all of these changes have intensified as the size and rate of growth of the world's population, and its accompanying mobility and resource use, has continued to expand – though not uniformly across time and space. In this chapter we are focusing on human-induced environmental change and the role of tourism-related activities, in particular, rather than natural changes.

Human impacts on the environment have a global character in two ways. First, 'global refers to the spatial scale or functioning of a system' (Turner et al. 1990: 15). Here, the climate and the oceans have the characteristic of a global system and both influence, and are influenced by, the tourism production and consumption that takes place anywhere within that system. A second kind of GEC exists if a change 'occurs on a worldwide scale, or represents a significant fraction of the total environmental phenomenon or global resource' (Turner et al. 1990: 15–16). Local environmental change cannot be divorced from global environmental change, as this second form of GEC is the sum of all the environmental changes of a similar type occurring in different locations around the world. Because tourism activities, including transportation, reach almost every corner of the globe and are a large part of the economies of many countries, they are significant for both types of change (Gössling and Hall 2006a).

Gössling (2002) provided the first attempt to overview the global environmental consequences of tourism with respect to five major aspects of the leisure-related alteration of the environment:

- the change of land cover and land use;
- the use of energy and its associated impacts;

- the exchange of biota over geographical barriers and the extinction of wild species;
- the exchange and dispersion of diseases; and
- a psychological consequence of travel, including changes in the perception and the understanding of the environment brought about by travel.

Of these, the psychological aspects are not usually included in the assessment of the impacts of tourism on the physical environment. Gössling developed various estimates for these factors, some of which are included in Table 5.2, along with extrapolations of the number of tourists worldwide, and hence their resource use.

Changes identified at the global scale are the cumulative effects of changes at numerous local levels. Based broadly on Gössling's (2002) approach, the following sections examine the main dimensions of tourism- and leisure-related impacts on the environment by providing examples at different scales, from global to local.

Table 5.2 Dimensions of tourism and global environmental change

Dimension	2001 estimate	2007 estimate
Number of international tourist arrivals	682 million[1]	898 million[1]
Number of domestic tourist arrivals	3,410 million[2]	4,490 million[2]
Total number of tourist arrivals	4,092 million[2]	5,388 million[2]
Change of land cover – alteration of biologically productive lands	0.5% contribution[3]	0.66% contribution[4]
Energy consumption	14,080 PJ[3]	18,585.6 PJ[4]
Emissions	1400 Mt of CO_2-e[3]	1848 Mt of CO_2-e[4] (1461.6 Mt of CO_2)[5]
Biotic exchange	Difficult to assess[3]	Difficult to assess, however rate of exchange is increasing[4]
Extinction of wild species	Difficult to assess[3]	Difficult to assess, particularly because of time between initial tourism effects and extinction events
Health	Difficult to assess[3]	Difficult to assess in host populations, but sickness in tourists assessed at 50% by WHO [4]

Notes
1 UNWTO figures; 2 Author estimates based on UNWTO data; 3 Gössling (2002) estimate; 4 Author's extrapolation based on Gössling's estimates; 5 UNWTO, UNEP and WMO (2007) data for 2005 extrapolated to 2007 by authors.

National and regional environmental impacts

Change of land cover and land use

Land alteration is regarded as the single most important component of global environmental change affecting ecological systems as a result of habitat change and loss. Although some natural land covers, such as those found within national parks and reserves, are relatively unchanged by tourism, much land is specifically converted for tourism use. This includes transportation infrastructure (such as airports), accommodation, convention and meeting centres, and attractions (such as golf courses and theme parks). Gössling (2002) calculated that globally, leisure-related land use might amount to 515,000 km² (200,000 mi²), representing 0.34 per cent of the terrestrial surface of the Earth and 0.5 per cent of its biologically productive area. Less than 1 per cent of these leisure-induced land alterations were due to accommodation establishments, while traffic infrastructure requires 97 per cent of the area. Leisure-induced land alterations, however, are often concentrated in relatively small areas and within ecologically sensitive or significant environments. Furthermore, it should also be stressed that Gössling's calculations are extremely conservative, including only commercial accommodation and not second homes or other secondary dimensions of tourism-related urbanization in destination and resort areas.

Tourism urbanization is not a new process. However, in the late twentieth century what did become different, and what characterizes the *new urban tourism* (Page and Hall 2003), is the scale, complexity and diversity of consumption experiences that now exist in urban and peri-urban landscapes built specifically for tourism and leisure. Several cities and regions around the world are now economically geared towards such consumption, including: the Gold Coast and the Sunshine Coast in Australia; Honolulu and Las Vegas in the United States; Whistler and Niagara Falls in Canada; the Algarve in Portugal; and the Costa Brava in Spain. These cities and regions are only the more obvious examples of the many tourism- and leisure-specialized communities that have developed in the resort regions of the world as a result of increased leisure mobility and amenity migration.

Tourism urbanization tends to be focused in high-value amenity environments and is closely associated with other forms of amenity-related urbanization, such as retirement and lifestyle migration, particularly in coastal areas. Land use change and urban sprawl are major problems along large areas of the coastlines of Europe and North America, bringing urban pollution problems. For example, in destinations with substantial boating recreation, water pollution problems can also exist in marinas, harbours and waterways, where oil and wastes are discharged from boats. Water pollution is a particular problem in sheltered waterways that are not adequately flushed by tidal or wave action.

Over a third of Europe's total population live within 50 km (30 mi) of Europe's coasts and that figure is growing. More than 70 per cent of the Mediterranean coast from Barcelona to Naples had been developed by 2000, and by 2025, the percentage of the population of Spain, France, Greece, Italy and the former

Yugoslavia living in coastal cities is projected to be more than 85 per cent on average, and as high as 96 per cent in Spain (CO-DBP 1999). As the Parliamentary Assembly, Council of Europe (2003, 2.28) stated, 'There is little doubt that human-induced causes, such as population pressures, urbanisation, over-construction, and ill-planned development (as well as protection) of the Mediterranean coasts have led to much of its deterioration or destruction. Many of these human-induced pressures stem from or are closely linked to tourism' (see Case 5.3).

Case 5.3: Land use change and ecosystem stress in the Mediterranean

The World Wide Fund for Nature/World Wildlife Fund (WWF) (2001) estimated that in 2000 the Mediterranean region received approximately 30 per cent of all international tourist arrivals. By 2020 WWF estimates that this will mean approximately 355 million tourists representing about 22 per cent of global international tourist arrivals (WWF 2001; De Stefano 2004). These figures do not include the many domestic tourists in Mediterranean countries, who also utilize the coastal areas. According to the WWF (2001:2), 'the projected growth of tourism development in the region will continue to damage landscapes, cause soil erosion, put pressure on endangered species, further strain available water resources, increase waste and pollution discharges into the sea and lead to cultural disruption'.

Tourism and amenity urbanization have contributed substantially to land use pressures on Mediterranean coastal landscapes (Tanrivermis 2003). Three-quarters of the sand dunes of the Mediterranean coastline from Spain to Sicily have disappeared, and according to the WWF (2001) this is mainly a result of urbanization linked to tourism development. Tourism-related development is also held to be primarily responsible for the urbanization of the Italian coast. In Italy, over 43 per cent of the coastline is completely urbanized, 28 per cent is partly urbanized and less than 29 per cent is still free of construction. There are only six stretches of coast over 20 km long that are free of construction and only 33 stretches between 10 and 20 km long without any construction (WWF 2001).

Tourism urbanization is also impacting the coast of Tunisia. According to De Stefano (2004), the urbanized coastal area of Tunisia extended for 140 km and tourist areas occupied by hotels and second homes occupy a further 80 km of the total urbanized linear space. This combined urban coastline of 220 km represents approximately 18 per cent of the total Tunisian coastline. De Stefano argues that current and planned tourism projects will nearly double tourism's consumption of the coast to about 150 km of the shoreline. She goes on to note that the inappropriate siting of tourist infrastructures on foredunes is accelerating the process of beach erosion and altering the water dynamics of the coastal region.

Similarly, in Cyprus 95 per cent of the tourism industry is located within 2 km (1 mi) of the coast (Loizidou 2003), placing the coastal environment under extreme pressure. Under the Cypriot land use zoning system of 1997–98, some 37 per cent of the coastline (in length) was zoned for tourism, 12 per cent for agriculture, 6 per

Photo 5.2 Tourists on the Mediterranean beaches of the popular Sousse resort area in
Tunis. Tunisia is just south of Italy on the north coast of Africa. Its close
proximity to Europe made it the breadbasket of the Roman Empire two
millennia ago, and makes it a cosmopolitan tourist destination today. The
Sousse beaches are lined with large hotel and timeshare resorts, amusement
parks and shopping centres, together with some historic towns that attract
beach and recreational tourists from throughout Europe and North Africa.
(Photo by Alan A. Lew)

cent for residences and 3 per cent for industry. The remainder (nearly 43 per cent)
was zoned as open space, protected natural or archaeological areas. However, it
is expected that future revisions of land use planning zones will see more of the
coast zoned for tourism use (Loizidou 2003).

The focus of tourism-related land use change on the coast also means that many
of the Mediterranean coastal ecosystems are under severe stress, with over 500
plant species threatened with extinction. Although the most severe environmental
pressure is on the coastal regions, the Mediterranean Sea itself is also suffering
from severe environmental stress with over 10 billion tonnes of industrial waste per
year estimated to be dumped in its waters with little or no purification, and with
consequent impacts on littoral species, including fish populations (Guidetti et al.
2002). No figures are available specifically on the amount of waste generated by
tourists at resorts and on boats. However, there has been increased stress on
marine ecosystems as a direct result of tourist activities such as diving, fishing,
mooring and cruise ships (Badalamenti et al. 2000).

The generation of waste by resident populations and industries is exacerbated
by the seasonal pressures placed on sewerage and water systems by tourism
(German Federal Agency for Nature Conservation 1997). Indeed, one of the most

substantial impacts of tourism urbanization in the Mediterranean is the overall pressure that is being placed on freshwater supplies, particularly through aquifer depletion (CO-DBP 1999). This is not only because of the direct demands of tourism for the immediate water needs of tourists, but also because of the impacts of tourism urbanization on coastal wetlands and lagoons (De Stefano 2004).

The main causes of tourism's impact on freshwater ecosystems are:

- *Higher water consumption due to population increases.* This includes both tourists and the flexible workforce required for tourism production. For example, in Spain's Balearic Islands, water consumption during the peak tourism month in 1999 (July) was equivalent to 20 per cent of that used by the entire local population in the entire year (De Stefano 2004).
- *Higher consumption of water for tourist facilities.* Kent et al. (2002) even argued that in the case of Mallorca, continued over-exploitation of coastal aquifers will be more critical to water quality issues than the direct impacts of any future sea level rise, although the latter would undoubtedly exacerbate these effects. They also note that proposals for the introduction of 'ecotaxes' on tourists to help pay for water and sewerage infrastructure are likely to represent only a partial solution to the long-term water supply problem.
- *Peaks in wastewater volumes and the stress that causes for wastewater treatment facilities.* Scoullos (2003) reports that only 80 per cent of the effluent of residents and tourists in the Mediterranean is collected in sewerage systems with the remainder being discharged directly or indirectly into the sea or into underground septic tanks. Even more alarming, only half of the sewerage systems are actually connected to wastewater treatment facilities, with the rest being discharged into the sea.
- *Inappropriate siting of tourism facilities and infrastructure on foreshores, dune systems and wetlands.* The United Nations Environment Programme – Mediterranean Action Plan – Priority Actions Programme (UNEP/MAP/PAP 2001:14) noted that mass tourism exacerbates issues of urbanization impacts, 'leading to habitat loss for many wildlife species', and estimated there has been a reduction in the wetland area of approximately 30,000 km^2 (11,600 mi^2), or 93 per cent, since Roman times. Of this, 10,000 km^2 (3,900 mi^2) has been lost in the last 50 years (Parliamentary Assembly, Council of Europe 2003: 2.12).

Source: Hall 2006a.

Biotic exchange and species loss

International travel and trade has long been a major transmission vehicle for biotic exchange, including the introduction of invasive species and infectious diseases. Humans are significant vectors for disease, as well as being carriers of pests, which may also host diseases that affect humans, plants and animals. The impacts of invasions can be substantial. Pimental et al. (2001) examined the economic and environmental threats from alien plant, animal and microbe invasions in the United States, UK, Australia, South Africa, India and Brazil, and found that although some introductions, such as grain and farm animals, have been beneficial, many have caused major economic losses in agriculture and forestry, as well as negatively impacting ecological integrity. They estimated that over 120,000 non-native species had invaded the six countries studied and are causing more than US$314 billion per year in damages. 'In terms of its rate and geographical extent, its potential for synergistic disruption and the scope of its evolutionary consequences, the current mass invasion event is without precedent and should be regarded as a unique form of global change' (Ricciardi 2007: 335) (Table 5.3).

In order to combat the introduction of pests and diseases many countries and regions have introduced biosecurity strategies. Biosecurity refers to the protection of a country, region or location's economic, environmental and human health from harmful organisms, and involves preventing the introduction of harmful new organisms, and eradicating or controlling those unwanted organisms that are already present (Biosecurity Strategy Development Team 2001). Although biosecurity and tourism are closely entwined, the tourism industry usually has little overt interest in biosecurity issues, whether at a national or global scale, unless some risk arises that limits tourist mobility or damages the image of a destination. Examples of this include avian influenza and SARS in Asia, and foot and mouth disease in Europe. However, major biosecurity breakdowns should be a significant concern for the tourism industry because one of the most common ways to stop the further spread of a disease or unwanted species is to restrict the movement of human beings. Tourism is, therefore, both a potential source of biosecurity threats, as well as one of the industries that will be most affected by movement controls.

The global size of biotic exchange is such that alien invasions are included as 'Threats to Biodiversity' in the Convention on Biological Diversity's (CBD) framework *2010 target* of a 'commitment to achieve by 2010 a significant reduction in the current rate of biodiversity loss at the global, regional and national level, as a contribution to poverty alleviation and to the benefit of all life on earth' (McGeoch et al. 2006: 1635). The Secretariat of the CBD goes on to state that 'Biodiversity is most often understood as the number of different species of plants, animals and micro-organisms in existence.. . . However, biodiversity also encompasses the specific genetic variations and traits within species as well as the assemblage of these species within ecosystems' (SCBD 2006: 9).

Biodiversity is essential for human survival and economic well-being, and for ecosystem function and stability (Hall 2006c). Biodiversity at the global scale is a balance between the rates of speciation and extinction, and at the ecosystem level it is a balance between the rates of invasion and local extinction. It is unevenly distributed on the Earth, with broad global and regional patterns. The key problem with respect to contemporary biodiversity is that current rates of extinction are 1,000 to 10,000 times higher than the background rate inferred from fossil records (Singh 2002). The SCBD (2006) reported that among 15 indicators used to assess global biodiversity only two, coverage of protected areas and water quality of aquatic ecosystems, were showing trends positive for biodiversity.

Table 5.3 Comparisons between pre-industrial and modern human-assisted biotic invasions

Characteristic	Prehistoric invasions	Contemporary human-assisted invasions
Frequency of long-distance dispersal events	Very low	Very high. Long-distance dispersal of seeds, spores and invertebrates by wind and ocean currents is at the mercy of the vagaries of weather patterns and ocean circulation, but no such constraints apply to human vectors
Number of species transported per event	Low, except during biotic interchange events	High, e.g. transoceanic ships are estimated to be carrying >7,000 species around the planet on any given day
Propagule load per event	Small, except during biotic interchange events	Potentially large, e.g. transoceanic ships carry an enormous number of propagules (estimated up to 10^8 invertebrates per ship entering the Great Lakes)
Effect of geographic barriers	Strong	Nearly insignificant as a result of the speed and density of the global transport network
Variation in mechanisms and routes of dispersal	Low	Extremely large as a result of increased human mobility and number and type of transport vectors
Temporal and spatial scales of mass invasion events	Episodic; limited to adjacent regions	Continuous; affects all regions simultaneously; possibly some variation in relation to seasonality of tourism demand and hence transport vectors
Homogenization effect	Regional	Global
Potential for synergistic interactions with other stressors	Low	Very high – climate change, changes in geochemical cycles, habitat loss and fragmentation, pollution, overharvesting, disruption of community structure in habitats

Sources: Carlton 1999; Duggan et al. 2005; Hall 2006b, c; Ricciardi 2007.

Photo 5.3   Dhow boats with tourists racing with dolphins in Oman's Musandam
Peninsula. The Musandam Peninsula is at the entrance to the Persian Gulf and
the dhow are traditional Omani fishing boats, though many today are used by
tourists from nearby Dubai. This region is sparsely populated and, in addition
to viewing dolphins, is known for turtles, sharks, seabirds, scuba diving,
fishing, camping on isolated beaches, and scattered old towns and villages.
All of this is situated near the busiest oil freighter ship lanes in the world.
(Photo by Alan A. Lew)

Key factors in the loss of biodiversity include: air, water and land pollution;
overharvesting of natural resources; introduction of alien species; transformation
of the physical environment (what is usually termed *climate change* and the
alteration of global biochemical cycles); habitat loss and fragmentation from land
use changes; and disruption of habitat community structures (Novacek and
Cleland 2001). Tourism is directly or indirectly embedded in all of these stress
factors on biodiversity. However, it is also part of the solution. In many locations,
tourism provides an economic justification to establish conservation areas and
programmes as an alternative to other land uses such as logging, clearance for
agriculture, mining or urbanization. Tourism may also be used as one of the
justifications for the reintroduction of species, which may otherwise have been
hunted to extinction in their natural range (Case 5.4). Often such tourism is
described as ecotourism, safari, wildlife or nature-based tourism, or even sustainable tourism. For example, on the Galapagos Islands of Ecuador, which was
designated as the first UNESCO World Heritage Site in 1978, tourism brings in as
much as $60 million annually and provides income for an estimated 80 per cent of

Table 5.4 Tourism in relation to major strategies of biodiversity conservation

Strategy	Element	Role of tourism
In situ conservation (on site)	Establish protected area network, with appropriate management practices and corridors to link fragments; restore degraded habitats within and outside protected areas	Tourism is economic justification for protected area framework given area's role as attraction; volunteer tourism may also assist in protected area management. Tourism also important for development of environmental knowledge
Ex situ conservation (off site)	Establish botanical and zoological gardens, conservation stands; banks of germplasm, pollen, seed, seedlings, tissue culture, gene and DNA	Botanical and zoological gardens are significant tourism attractions; volunteer tourism also significant; significant educational tourism function
Reduction of anthropogenic pressure	Reduce anthropogenic (human) pressure on natural species populations by altering human activities and behaviours. May also include cultivating species elsewhere	Species populations may have role as tourism attractions; some species which can be sustainably harvested may also be used for hunting and fishing
Reduction of biotic pressure	Removal or reduction of invasive exotic species and pests that compete with indigenous species	Ensure good biosecurity practice; interpretation programmes to support eradication of invasive species and pests
Rehabilitation	Identify and rehabilitate threatened species; launch augmentation, reintroduction or introduction programmes	Species may become tourism attractions; volunteers also significant in management

the island group's residents (SCBD 2006) (see Case 1.2). Table 5.4 indicates some of the major strategies of biodiversity conservation.

Case 5.4: Return of the beaver

Once native to Britain, wild beaver were hunted to extinction more than 400 years ago. Although a small number of imported beaver are kept in wildlife reserves and a few have been brought into the country illegally in recent years, there have not been enough for a wild population to exist. In May 2008, the Scottish government announced the European beaver is to be reintroduced legally to Knapdale, Mid Argyll, for a five-year trial period, with the introduction of four families of beavers from Norway. This was undertaken as part of a Scottish Natural Heritage strategy for species management action (Species Action Framework) that included the restoration of the European beaver to Scotland (SWT 2008).

The trial reintroduction of the European beaver was proposed by the Scottish Wildlife Trust (SWT) and the Royal Zoological Society of Scotland (RZSS), who argued that beavers can have a positive effect on both environmental and woodland management, and could also attract tourists to the area. Simon Milne, chief executive of SWT, stated it was a 'chance to be the first country in the UK to bring back this charismatic and useful creature' (BBC News 2007b). Similarly, Iain Valentine, the head of animals, education and conservation at the RZSS, commented,

> Beaver reintroductions have proved to be very successful in over 20 other countries, and we believe that the time is right to bring the beaver back to Scotland.. . . As well as being a keystone species, in terms of the benefits they bring to ecosystems, they will also provide a socio-economic boost by increasing tourism in the local area.
>
> (BBC News 2007b)

A two-month long consultation process undertaken in 2007 by Scottish Natural Heritage revealed that over 73 per cent of respondents from Mid Argyll supported a trial reintroduction of beavers to Knapdale, Mid Argyll. However, opposition to the proposal was voiced by leaders of the Scottish Rural Property and Business Association (SRPBA). According to Doug McAdam, SRPBA chief executive:

> The reintroduction of the European beaver to Scotland would effectively be an introduction of a now alien species. After a gap of 400 years, former habitats have been developed and are now a managed landscape, providing environmental, economic and recreational benefits.
>
> (BBC News 2007b)

Mr McAdam went on to say that without proper impact assessments and controls in place, the reintroduction of beavers 'may in fact prove detrimental to existing species and habitats in the longer term', and also claimed that bringing back beavers would 'inevitably impact' on spawning areas for trout and salmon (BBC News 2007b).

The reintroduction of beaver to Scotland is part of a broader attempt to try to restore environments that have experienced species loss. Known as *wilding*, the return of landscapes to something akin to their prehistoric ecology is gaining substantial interest in Western Europe, including the UK. One of the most significant projects is that of Wild Ennerdale – a remote valley in the western Lake District of northern England covering an area of 4,300 hectares (11,640 acres). Ennerdale is jointly owned by the National Trust, the Forestry Commission and United Utilities. In 2002, after consultation with various stakeholders, the three landowners signed up to a *wild* vision statement for the valley, 'to allow the evolution of Ennerdale as a wild valley for the benefit of people, relying more on natural processes to shape its landscape and ecology' (Wild Ennerdale Partnership 2006: 30). Supporting this vision are eleven key principles:

- The sense of wildness experienced by people will be protected and enhanced;
- The valley's landscape and habitats will be given greater freedom to develop under natural processes, allowing robust and functioning ecosystems to develop on a landscape scale;
- Public support and engagement will remain central to the Wild Ennerdale process;
- Intervention will only occur if complementary to the vision, or where a threat to the vision is posed;
- Opportunities will be sought to develop greater public enjoyment and social benefit;
- The historical and cultural assets of the valley will be considered and respected;
- Management and decision making will be focused more at the holistic landscape scale;
- Wild Ennerdale will be offered as a demonstration to others by sharing results and information;
- Opportunities will be sought for businesses that are sustainable within the vision;
- Monitoring and assessment of change will be carried out on a large scale and over a long period of time;
- An element of set-up and higher level intervention may be required to facilitate natural processes, recognising our starting point is influenced by past activity.

(Wild Ennerdale Partnership 2006: 31)

In practice, this has meant that commercial forestry operations have been reduced. Motorized transport is restricted. Fences are being removed and plantation spruce are being thinned and regenerated with endemic tree species such as juniper, birch, ash, alder, poplar. Almost all commercial extraction will cease in the valley, apart from using the lake for drinking water (Macfarlane 2007).

Ennerdale is designated a 'Quiet Area' within the Lake District National Park and as such has stricter development control policies in terms of tourist attractions, marketing and visitor facilities. Tourism does however play an important part in supporting the local economy and livelihoods in Ennerdale, with an infrastructure which includes B&B's, Inns, holiday homes, self-catering cottages, camping/caravan sites, a Field Studies Centre, YHA's and a bunk barn.

(Wild Ennerdale Partnership 2006: 39)

Such projects are not without tension as different stakeholders have different perceptions of landscape and wild nature in particular (McCorran et al. 2008). For example, Convery and Dutson (2008: 115) suggest that managing the social consequences of a policy to create wild land in an area such as Ennerdale with a farming landscape that 'represents complex interrelationships among people, place, and production', requires careful consideration. In a worst-case scenario,

MacDonald et al. (2000) argue that such policies risk creating a continuing cycle of increasing rural depopulation, deprivation, further land abandonment and loss of traditional land management skills.

Wildlife viewing and authentic natural landscapes are major tourist attractions. While not explicit in the goals of the beaver reintroduction efforts and wilding policies, leisure activities and tourism-related businesses are among the more compatible beneficiaries of these programmes.

Sources: Royal Zoological Society of Scotland: http://www.edinburghzoo.org.uk/

Scottish Rural Property and Business Association: http://www.srpba.com/

Scottish Wildlife Trust: http://www.swt.org.uk/

Wild Ennerdale: http://www.wildennerdale.co.uk

Tourism directly benefits biodiversity maintenance through several means (Brandon 1996; Christ et al. 2003; Hall and Boyd 2005; Hall 2006c):

- Tourism provides an economic justification for biodiversity conservation, including the establishment of national parks and public and private reserves.
- Tourism provides a source of financial support for biodiversity maintenance and conservation.
- Tourism provides an economic alternative to other forms of development that negatively impact biodiversity and to inappropriate exploitation or harvesting of wildlife, such as poaching.
- Tourism can be a mechanism for educating people about the benefits of biodiversity conservation.
- Tourism can potentially involve local people in the maintenance of bio-diversity and incorporate local ecological knowledge into biodiversity management practices.

In 2003 Conservation International, in collaboration with UNEP, focused on the potential role of tourism in biodiversity *hotspots* – 'priority areas for urgent conservation on a global scale' (Christ et al. 2003: vi). Hotspots (Myers 1988) are areas that both support a high diversity of endemic species and have been significantly impacted by human activities. Plant diversity is the biological measure for designating a biodiversity hotspot. To qualify as a hotspot, a place must meet two strict criteria: it must contain at least 1,500 species of vascular plants (> 0.5 per cent of the world's total) that are endemic species, and it has to have undergone modification of at least 70 per cent of its original (pre-modern) habitat.

Of the 31 existing biodiversity hotspots identified by Myers et al. (2000) and Myers (2003), eight are in Africa and over half are located in less developed regions of the world (Table 5.5). The biodiversity hotspots identified by Conservation International 'contain 44 percent of all known endemic plant species and 35 percent of all known endemic species of birds, mammals, reptiles, and amphibians in only 1.4 percent of the planet's land area' (Christ et al. 2003: 3).

Table 5.5 Biodiversity hotspots

Region	Approximate % of original area protected
North and Central America	
California Floristic Province	10
Caribbean Islands	7
Madrean Pine–Oak Woodlands	2
Mesoamerica (a complex mosaic of dry forests, lowland moist forest, and montane forests stretching from Mexico to Panama)	6
South America	
Atlantic Forest	2
Cerrado (Brazilian woodland savannah)	1
Chilean Winter Rainfall-Valdivian Forests	11
Tumbes–Chocó–Magdalena (coastal Columbia, Ecuador and northern Peru)	7
Tropical Andes	8
Europe and Central Asia	
Caucasus	7
Irano-Anatolian (mountains and basins)	3
Mediterranean Basin	1
Mountains of Central Asia	7
Africa	
Cape Floristic Region (evergreen fire-dependent shrubland of South Africa)	13
Coastal Forests of Eastern Africa	4
Eastern Afromontane	6
Guinean Forests of West Africa	3
Horn of Africa	3
Madagascar and the Indian Ocean Islands	2
Maputaland–Pondoland–Albany (East coast of South Africa and southern Mozambique)	7
Succulent Karoo (Eastern coastal South Africa and Namibia)	2
Asia-Pacific	
East Melanesian Islands	0
Himalaya	10
Indo-Burma	6
Japan	6
Mountains of Southwest China	1
New Caledonia	3
New Zealand	22
Philippines	6
Polynesia–Micronesia	4
Southwest Australia	10
Sundaland (Western Indonesia)	5
Wallacea (Eastern Indonesia)	5
Western Ghats and Sri Lanka	11

Source: derived from Christ et al. 2003; Conservation International Biodiversity Hotspots.

Christ et al. (2003) highlighted several key issues with respect to the relationship between tourism and biodiversity:

- Although most biodiversity is concentrated in less developed countries, five tourism destination regions in the developed world were also identified as biodiversity hotspots: the Mediterranean Basin, the California floristic province, the Florida Keys (as part of the Caribbean Islands biodiversity hotspot), Southwest Australia, and New Zealand.
- A significant number of biodiversity hotspot countries in the less developed world are experiencing rapid tourism growth. Twenty-three of them recorded over 100 per cent growth in international tourist arrivals the 10 years up to 2003, and more than half receive over 1 million international tourists per year; 13 per cent receive over 5 million international tourists per year. Argentina, Brazil, Cyprus, the Dominican Republic, India, Indonesia, Macao, Malaysia, Mexico, Morocco, South Africa, Thailand and Vietnam all receive over 2 million foreign visitors per year, while domestic tourism is also of growing significance.
- More than one-half of the world's poorest 15 countries fall within the biodiversity hotspots, and in all of these, tourism has some economic significance, or is forecast to increase in coming years.
- In several biodiversity hotspots in less developed countries (including Madagascar, Costa Rica, Belize, Rwanda and South Africa), biodiversity or elements of biodiversity, such as specific wildlife, is the major international tourist attraction as a result of ecotourism.
- Forecasted increases in international and domestic tourism suggest that pressures from tourism development will become increasingly important in many biodiversity hotspot countries, especially those in South and Southeast Asia.

Case 5.5: The global conservation estate

The global conservation estate is the sum of all the world's protected areas (Table 5.6). The global conservation estate has grown enormously since the first United Nations *List of Protected Areas* was published in 1962 with just over 1,000 protected areas. By 1997 there were over 12,754 sites listed, while the 2003 edition listed 102,102 sites covering 18.8 million km^2 (7.25 million mi^2). 'This figure is equivalent to 12.65 percent of the Earth's land surface, or an area greater than the combined land area of China, South Asia and Southeast Asia' (Chape et al. 2003: 21). Of the total area protected, it is estimated that 17.1 million km^2 (6.6 million mi^2) constitute terrestrial protected areas, or 11.5 per cent of the global land surface. Some biomes, including lake systems and temperate grasslands, remain poorly represented, while marine areas are significantly under-represented. Marine-protected areas occupy an area of approximately 1.64 million km^2 – an estimated 0.5 per cent of the world's oceans, and less than one-tenth of the overall extent of protected areas worldwide.

Table 5.6 Number and area of protected areas under the International Union for the Conservation of Nature Protected Area Management Categories in 2003

Category	Description	Global no. of categories (2003)	Global no. of categories (2003) (%)	Global area of categories (2003) (km²)	Global area of categories (2003) (%)
Ia Strict Nature Reserve: protected area managed mainly for science	Area of land and/or sea possessing some outstanding or representative ecosystems, geological or physiological features and/or species, available primarily for scientific research and/or environmental monitoring.	4 731	4.6	1 033 888	5.5
Ib Wilderness Area: protected area managed mainly for wilderness protection	Large area of unmodified or slightly modified land, and/or sea, retaining its natural character and influence, without permanent or significant habitation, which is protected and managed so as to preserve its natural condition.	1 302	1.3	1 015 512	5.4
II National Park: protected area managed mainly for ecosystem protection and recreation	Natural area of land and/or sea, designated to (a) protect the ecological integrity of one or more ecosystems for present and future generations, (b) exclude exploitation or occupation inimical to the purposes of designation of the area and (c) provide a foundation for spiritual, scientific, educational, recreational and visitor opportunities, all of which must be environmentally and culturally compatible.	3 881	3.8	4 413 142	23.5
III Natural Monument: protected area managed mainly for conservation of specific natural features	Area containing one, or more, specific natural or natural/cultural feature which is of outstanding or unique value because of its inherent rarity, representative or aesthetic qualities or cultural significance.	19 833	19.4	275 432	1.5

Table 5.6 Continued

Category	Description	Global no. of categories (2003)	Global no. of categories (2003) (%)	Global area of categories (2003) (km²)	Global area of categories (2003) (%)
IV Habitat/Species Management Area: protected area managed mainly for conservation through management intervention	Area of land and/or sea subject to active intervention for management purposes so as to ensure the maintenance of habitats and/or to meet the requirements of specific species.	27 641	27.1	3 022 515	16.1
V Protected Landscape/Seascape: protected area managed mainly for landscape/seascape conservation and recreation	Area of land, with coast and sea as appropriate, where the interaction of people and nature over time has produced an area of distinct character with significant aesthetic, ecological and/or cultural value, and often with high biological diversity. Safeguarding the integrity of this traditional interaction is vital to the protection, maintenance and evolution of such an area.	6 555	6.4	1 056 088	5.6
VI Managed Resource Protected Area: protected area managed mainly for the sustainable use of natural ecosystems	Area containing predominantly unmodified natural systems, managed to ensure long-term protection and maintenance of biological diversity, while providing at the same time a sustainable flow of natural products and services to meet community needs.	4 123	4	4 377 091	23.3
No category		34 036	33.4	3 569 820	19
Total		102 102	100	18 763 407	100

Sources: categories identified in IUCN 1994; figures derived from Chape et al. 2003.

Photo 5.4 Porongurup National Park in south Western Australia viewed from the Stirling Range National Park. The Stirling Range is one of the richest areas for flora in the world; 90 families, 384 genera and over 1,500 plant species occur here, 87 of which are found nowhere else. Although not nearly as rich biologically as the more northerly Stirling Range, there are 10 endemic species of plant in the Porongurup Range as well as strands of giant karri and jarrah trees. Perhaps more significantly the Porongurups are substantially wetter than most of the surrounding countryside and even receive occasional heavy snowfalls. The conservation role of national parks is well illustrated by the photo as the parks can be seen as islands of natural areas in a sea of surrounding agricultural land. (Photo by C. Michael Hall)

The size of the global conservation estate raises the question of just how large the global network of protected areas needs to be (Rodrigues and Gaston 2001). The present size of the global conservation estate exceeds the IUCN's earlier target of at least 10 per cent of the Earth's total land area being set aside for conservation purposes, although there are substantial variations among countries and regions in terms of the area set aside and the effectiveness of their management (Chape et al. 2003). Rodrigues and Gaston (2001, 2002) observed that the minimum area needed to represent all species within a region increases with the number of targeted species, the level of endemism and the size of the selection units. They concluded that:

- no global target for the size of a network is appropriate as those regions with higher levels of endemism or higher diversity will correspondingly require larger areas to protect such characteristics;
- a minimum size conservation network sufficient for capturing the diversity of vertebrates will not be sufficient for biodiversity in general, because other groups are known to have higher levels of endemism (Gaston 2003); and

- the original 10 per cent target is likely to be grossly inadequate to meet biodiversity conservation needs.

Rodrigues and Gaston (2001) estimated that 74.3 per cent of the global land area and 92.7 per cent of the global rainforest would need to be protected to conserve every plant species, and 7.7 per cent and 17.8 per cent respectively to protect just the higher vertebrates. However, Gaston (2003) also notes that even reserves of 12,000 km^2 (4,600 mi^2) may not be large enough to maintain the population of many of the larger species. Unfortunately, political considerations often mean that there has been insufficient consideration of the size and shape of conservation reserves when national parks are established and when their dual-management roles of conservation and tourism are developed (Hall 2006b). This situation is further complicated by the fact that many national park agencies manage more than just national parks and different types of protected areas. The IUCN has six different categories of protected areas, with national parks usually being a component of a wider system of protected areas. However, some national jurisdictions have even more categories. For example, in Austria there are 12 different types of protected areas and 11 in Germany (Mose and Weixlbaumer 2007). In addition, 'one cannot deny a certain image-hierachy between the different categories. In contrast to the prestigious and financially lucrative Category II (national park), the Category V (protected landscape) receives only little attention' (Mose and Wixlbaumer 2007: 5). According to Mose and Wixlbaumer (2007: 5) there are several reasons for the dominance of *national parks* over other categories of protected areas in the public and policy imagination:

- The outstanding image of national parks as the 'premium category of the protected areas'
- The stringent legal and spatial planning rules underlying the national parks (for example, legal statutes to regulate management)
- The supra-regional competence of a governmental administrative body
- The differently weighted overriding management objectives, focusing on conservation.

Furthermore, and perhaps not surprisingly, many people often do not recognize the differences that exist between the different types of protected areas or, in some cases, the concept is alien to their world view (Hall and Frost 2009). This situation is significant for the relationship between tourism and conservation, as there may be substantial differences in understanding by different stakeholders as to what impacts are acceptable or unacceptable within a protected area.

Source: United Nations Environment Programme, World Conservation Monitoring Centre: http://www.unep-wcmc.org/

Community environmental impacts

Energy and emissions: landscape and climate change

> If the global tourism industry were represented as a country, it would consume resources at the scale of a northern developed country. International and national tourists use 80 percent of Japan's yearly primary energy supply (5,000 million kWh/year), produce the same amount of solid waste as France (35 million tons per year), and consume three times the amount of fresh water contained in Lake Superior, between Canada and the United States, in a year (10 million cubic meters).
>
> (Christ et al. 2003: 7)

An important, although often taken-for-granted, part of the leisure and tourism environment is climate (Hall and Higham 2005). Along with human influences, climate and weather have an enormous effect on shaping the nature of the environment, for example in terms of vegetation and water availability, as well as affecting many of the demands for leisure. Because climate generally moves in such regular seasonal cycles and is always 'there', its effects on leisure landscapes were never seriously considered by policy-makers, planners, firms and even communities. After all, the attitude has always been that if we have a bad weather season this year we know that things will 'return to normal' next year – don't we? (see Saarinen and Tervo 2006, for an example of this in the Finnish context). However, this certainty appears to be changing.

There are probably few issues affecting tourism and the environment that are now more discussed than climate change. Although the weather is often a focus of everyday conversation in terms of leisure activity, that interest in the weather is being increasingly discussed in terms of long-term patterns and processes that may be affecting not just demand for leisure but also its supply. Arguably, we have all become climate experts now, with the comments of politicians, business people, scientists and those 'who still do not believe' becoming part of the everyday media.

The general features of climate change are well recognized. As of 2007, 11 of the previous 12 years were among the warmest years on record for the planet, with temperature increase being most obvious at high latitudes. At the same time sea levels are increasing millimetre by millimetre each year. The future effects for Europe, for example, while still debated, look challenging for the rest of the century (Alcamo et al. 2007). Effects will include increased risk of inland flash floods, more frequent coastal flooding, and erosion due to stronger storms and sea level rise. Alpine and high latitude areas will face continued glacier retreat, reduced snow cover and extensive species loss. In Southern Europe, an already vulnerable climatic region, climate change is projected to result in more frequent high temperature periods and drought (Amelung and Viner 2006). This will lead to reduced water availability, as well as increases in the frequency of wildfires and heatwave-related health problems (Alcamo et al. 2007). Yet despite our knowledge of climate change at a broad geographical scale, what we know of the effects on specific tourism environments are relatively poor. This gap in knowledge is a function of several factors:

- Research on climate change, especially at micro and meso scales, is a recent phenomenon. The only tourism activity that has been subject to any significant amount of research is winter tourism (UNWTO et al. 2007, 2008; Hall 2008b).
- Effects of climate change on leisure and tourism have traditionally not been regarded as 'serious' as the effects on other economic sectors by researchers (Hall 2008b).
- Leisure and tourism have neither been the recipients of significant research funds nor have governments sought to develop internationally competitive centres of research expertise in the same way that exists for other economic sectors.

However, what we do know is that climate change affects leisure behaviour and resources and that leisure behaviour is also a contributor to climate change (Hall and Higham 2005; Gössling and Hall 2006a; UNWTO et al. 2008). This section will first discuss some of the ways in which climate change has affected and may affect tourism and leisure mobility, while the contribution of tourism to climate change is discussed further below.

The impacts of climate change on tourism

In terms of the impacts of climate change on leisure, several observations can be made. With respect to overall climatic conditions, such as increased average temperatures and changes in average precipitation and sunshine hours, it can be anticipated that shifts in leisure behaviours will include changes in and switching between activities. Obviously, outdoor leisure activities are the most susceptible to behavioural change, yet the capacity to substitute indoor activities in a controlled climatic environment for outdoor experiences may become of increasing importance in the future. Although climate change is global in scope, in tourism terms, it will be most recognized at the destination level, where changes will lead to destinations becoming more or less attractive at certain times of the year, as well as in the tourism-related activities and decision-making of individuals.

Increased daily temperatures will potentially lead to the shifting of leisure activities in both time (such as moving activities and events to cooler parts of the day during summer or new seasonal patterns) and in space (such as relocating activities to places with a more favourable micro-climate) (UNWTO et al. 2008). It is also likely that the timing of holiday seasons will be affected with some destinations becoming regarded as having unfavourable climatic conditions for leisure at different times of the year than under past conditions. However, the difficultly with such predictions is that human behaviour can be extremely adaptive; deterministic approaches to modelling leisure behaviour – for example if temperature increases by *x* amount then *y* will occur – are extremely problematic (Gössling and Hall 2006b). Projected changes in the order of 2°C to possibly 4°C (3.5°F to 7°F) on average will occur gradually in coming decades, allowing time for human acclimatization, along with gradual changes in the perception of leisure destinations.

Potentially far more serious a dimension of climate change for leisure behaviour and destination resources is the impact of greater variability in climate and weather patterns (IPCC 2007). Precipitation, for example, is becoming more variable in Asia, North America and Europe. Mean winter precipitation is increasing in most of northern Europe, while in the eastern Mediterranean, yearly precipitation is decreasing. In the western Mediterranean there has been no significant change in precipitation, but higher temperatures and greater competition for scarce water resources have meant increased conflict between the leisure sector, especially golf clubs and resorts, and the agricultural sector for water rights and access. Nicholls et al.'s (2007: 331) summary of climate-related impacts in relation to recreation and tourism in coastal areas argued that temperature rise (air and seawater), extreme events (storms and waves), erosion (sea level, storms and waves) and biological effects will have strong impacts; floods (sea level and runoff) will have a weak impact; and rising water tables (sea level) and saltwater intrusion will have a negligible impact, or that an impact is not yet established. However, such relative impacts will not be the same in all locations (Uyarra et al. 2005).

By the 2070s, it is expected that approximately 35 per cent of Europe will be suffering from water stress with summer flows being reduced by up to 80 per cent in some parts of southern and central Europe (Alcamo et al. 2007). For example, Lake La Nava in Castile-León, Spain, and the Nestos Lakes in Hrysoupolis, Greece, both important recreational resources, are expected to be subject to a temperature increase of up to 5°C (9°F) and a 35 per cent decline in annual precipitation, resulting in substantial effects on biodiversity, unless there are drastic reductions in the amount of greenhouse gas (GHG) emissions (Hulme et al. 2003). In these locations, like many others in southern and central Europe, the effects of climate change are exacerbated by the lack of effective regulations governing the extraction of water for industrial uses and the maintenance of landscape and natural vegetation. Similarly Lake Larache in Morocco, which is of considerable tourism and ecological significance (it is a Ramsar Convention wetlands site), is facing severe environmental pressure as a result of human impacts, as well as projected decreases in precipitation in both summer and winter as a result of climate change (Hulme et al. 2003: 22): 'The interaction of higher sea level and reduced river flows in summer and winter will affect estuarine processes and habitat composition throughout the wetland system. Saltwater incursion is likely to disrupt freshwater and inland habitats with potentially significant effects on migratory bird populations and also refuge for reptiles and amphibians.'

The effects of climate change on water bodies may mean not just less water for users, but also a decline in water quality. One of the most potentially damaging effects on water-based leisure is an increase in eutrophication of coastal waters, lakes and rivers. This is caused by the combination of increased nutrients and water temperatures leading to algal blooms, which are not only unsightly and smelly, but can also be harmful to humans, fish and wildlife. Reducing the input of human-produced nutrients into river and lake systems is clearly one response to such a problem, but broader measures with respect to climate change mitigation will be needed. The primary sources of such nutrients is usually agriculture and urbanization, rather than leisure activities, although second homes and changes in land use cover from development can be a significant source of nutrients in some locations.

Increased variability in precipitation, including increases in heavy rains, storms and floods, as well as increasing drought conditions, will also substantially affect leisure activities through the potential damage to infrastructure and leisure resources. Such variability will place extreme pressure on ecosystems that have adapted to the relatively stable weather patterns that the planet has experienced over the past century (IPCC 2007). For example, the exposure of peatlands to increased drought conditions will increase the likelihood of severe fires, while a reduction of freshwater run-off into coastal wetlands, combined with increased sea levels and storms, increases the likelihood of groundwater salinization (ACIA 2005). Visitors to peatland areas, which are significant rural recreation resources in a number of countries in northern Europe, will need to be subject to education campaigns as to the risk of starting wildfires in the same way that visitors to forest areas often are today. Such measures are also important because peatlands, along with many forests, have a vital role as carbon-sinks that reduce carbon levels in the atmosphere (Alcamo et al. 2007). Their disruption can have the opposite effect by releasing large amounts of CO_2 and exacerbating global warming.

Unfortunately, the conservation of coastal wetlands requires far more than just education. As discussed above, many of Europe's coastal areas are under severe environmental stress from leisure-oriented urbanization, primarily consisting of tourism, second-home and retirement-related developments, as well as wetland reclamation for golf courses, marinas and intensive agriculture. In such a situation climate change becomes yet one more stress factor. Yet planning solutions do exist to at least minimize the environmental effects of climate change in coastal systems. The most important is to stop 'reclaiming' wetlands and instead recognize them not only for their recreational and ecological values, but also as a natural protection against the effects of sea level rise. Important examples exist in the UK and the Netherlands of previously drained lands being returned back to wetlands and salt marshes (Alcamo et al. 2007). Interestingly, the creation of these 'new' wetlands has also generated new economic opportunities through increased visitation to observe birds and enjoy nature walks (Hulme et al. 2003).

However, the conversion of reclaimed land back to wetland is easier in more developed countries where governments can afford to support such initiatives, including providing compensation for landowners; people in coastal communities in the developing world may not have such an opportunity, even though the threat of sea level rise is just as real. For example, Field et al. (2007: 634) noted that, 'Although coastal zones are among the most important recreation resources in

Photo 5.5 Tourists and zebras at the Hluhluwe-Umfolozi Park in South Africa's KwaZulu-Natal Province. The Hluhluwe-Umfolozi game reserve was Africa's first wild game park, established in 1895. The Big Five (lion, black and white rhino, elephant, buffalo and leopard) live in the park, though they may not be easily seen on any given day. The park is very large (96,000 ha; 370 mi^2) and most tourists are taken through the park on a guided all-wheel drive vehicle (although other touring options are also available). (Photo by Alan A. Lew)

North America, the vulnerability of key tourism areas to sea-level rise has not been comprehensively assessed'. Nevertheless, in much of the developing world the situation is perhaps even bleaker with respect to both potential impacts as well as knowledge and capacity to adapt. In the case of Africa, Boko et al. (2007: 450) stressed that 'very few assessments of projected impacts on tourism and climate change are available' and later noted:

> There is a need to enhance practical research regarding the vulnerability and impacts of climate change on tourism, as tourism is one of the most important and highly promising economic activities in Africa. Large gaps appear to exist in research on the impacts of climate variability and change on tourism and related matters, such as the impacts of climate change on coral reefs and how these impacts might affect ecotourism.
>
> (Boko et al. 2007: 459)

Case 5.6 discusses the problems facing the Senegal city of St Louis.

Case 5.6: Sea level rise and the World Heritage City of St Louis, Senegal

By 2030, three-quarters of the world's population will be urban, and the biggest cities will be found in the developing world. In many cities in developing countries, slum dwellers number more than 50 per cent of the population and have little or no access to shelter, water and sanitation, education or health services and are therefore the most vulnerable to disasters such as high impact weather events; especially because being poor also means being relatively immobile. Africa has the highest rate of urban growth of any region in the world. Despite its low emissions (less than 4 per cent of global emissions as a continent), Africa is likely to be one of the areas worst affected by the effects of climate change (IPCC 2007).

According to the United Nations Human Settlements Programme (UN Habitat), St Louis in Senegal is the African city most threatened by rising sea levels as a result of climate change. Situated on the Atlantic west coast of Africa, the city, which was declared a World Heritage Site in 2000, is built on an island between the mouth of the Senegal River and the ocean. The city was founded as a French colonial settlement in the seventeenth century and was the capital of French West Africa until 1902. It was also the capital of Senegal from 1872 to 1957, when it was replaced by Dakar. The Island of St Louis is listed as a Heritage site on account of the quality of its colonial architecture and its particular relationship to water, and is regarded as an outstanding example of a colonial city as well as exhibiting significant cultural exchange. The city attracts 10,000 foreign tourists every year, as a result of its architecture and culture, however it may not survive rising sea levels in the coming decades (BBC News 2008b).

Sources: St Louis website (in French): http://www.saintlouisdusenegal.com/

St Louis website (in English): http://www.saintlouisdusenegal.com/english/index.php

UN Habitat: http://www.unhabitat.org/

World Heritage Centre: World Heritage List: http://whc.unesco.org/en/list

Increases in extreme weather events also means that heatwaves, such as the one that affected large parts of Europe in 2003, and which was estimated to have contributed to more than 35,000 deaths and 13 billion (US$ 20 billion) in damages, will become commonplace by the end of the century. Apart from the impacts of such heatwaves on individual leisure capacity and behaviour under warm anomalies of up to +7°C (+12.5°F), the heatwave also had an enormous impact on natural resources such as forests. The hot and dry conditions led to many very large wildfires, in particular in Portugal where 390,000 ha (965,000 acres) were affected (Hall and Higham 2005). In 2007, Greece and the western United States were similarly hit by severe wildfires as a result of extreme heatwave conditions that not only damaged tourism resources, but may also have damaged the image of the affected destinations. From a tourism perspective, such events are important not only as one-off events, but also for their potential longer-term impacts with respect to the image of a destination.

A sequence of heatwaves can affect the availability of water resources, as well as the destination image. Yet, if wildfires occur with such rapidity that the natural ecology cannot recover, then the landscape and environment start to change. For example, in parts of Portugal that have been affected by repeated burning, exotic fire-adapted tree species, such as eucalypts from Australia, have started to out-compete native vegetation. These processes may potentially lead to the development of a landscape that, by the end of the century, will look more like the arid outback of Australia than Mediterranean Europe.

One of the clearest effects of climate change on leisure is its impacts on alpine and high latitude ecosystems that are traditionally characterized by significant periods of snow cover (Hall and Higham 2005). The shortening of the winter season in which snow can be guaranteed has already significantly affected low altitude ski resorts to the point that many are now unviable and need government assistance to develop adaptation strategies. Higher altitude resorts have some capacity to adapt through the use of snow-making technologies. However, such technologies also demand access to substantial energy and water, two resources which are also under increasing pressure in many locations (Scott et al. 2003, 2007). As Agrawala (2007: 2) states, 'no amount of grooming can overcome significant declines or the total absence of snow cover'. Therefore, some resorts are beginning to investigate other adaptation strategies for their survival, such as the potential of developing summer tourism, whereby the relative coolness of the mountains in summer may prove attractive for visitors from the low-lying altitudes of Europe that are subject to increasing daily temperatures and heatwaves (Elsasser and Bürki 2002).

Case 5.7: Climate change and the European Alps: the OECD perspective

The European Alps are regarded as being particularly sensitive to climate change with recent warming in these mountains being approximately three times the global average (Alcamo et al. 2007). The years 1994, 2000, 2002 and 2003 were among the warmest on record in the Alps but climate model projections show even greater changes in the future, with less snow predicted at lower elevations and receding glaciers and melting permafrost higher up. From 1850 to 1980, glaciers in the region lost 30 to 40 per cent of their area. Since 1980 a further 20 per cent of the ice has been lost. The summer of 2003 led to the loss of a further 10 per cent. By 2050 about 75 per cent of the glaciers in the Swiss Alps are predicted to have disappeared, rising to 100 per cent in 2100 (Agrawala 2007).

The threat of climate change is obviously a serious situation for the tourism industry as the Alps attract 60 to 80 million tourists and approximately 160 million skier days in Austria, France, Germany and Switzerland each year (Agrawala 2007). 'Insurance . . . can reduce the financial losses from occasional instances of snow-deficient winters, but cannot protect against systematic long-term trends towards warmer winters' (Agrawala 2007: 2). As of the end of 2006, 91 per cent

(609 out of 666) of medium to large Alpine ski areas continued to have their normal adequate snow cover for at least 100 days per year. The remaining 9 per cent, however, were already operating under marginal conditions. According to Agrawala (2007), a 1°C (2°F) increase in temperature would increase the proportion of marginally operating ski areas to 25 per cent, a 2°C (3.5°F) increase would make 33 per cent marginal, and 70 per cent would have difficulty, with many closing, under a 4°C (7°F) warming of the average climate conditions. Germany is the most at risk, with a 1°C (2°F) warming scenario leading to a 60 per cent decrease in the number of naturally snow-reliable ski areas. Practically, none of the ski areas in Germany will be left with naturally sufficient and reliable snow under a 4°C (7°F) warming scenario. Switzerland would suffer the least from climate change though even a 1°C (2°F) increase would reduce natural snow by 10 per cent and 4°C (7°F) warming would halve the number of snow-reliable slopes (Agrawala 2007).

Source: Organization for Economic Co-operation and Development (OECD): http://www.oecd.org/

Such examples suggest that it is possible to adapt to a number of the effects of climate change. However, the capacity of many endemic plant and animal species to adapt to new climatic conditions is far less than that of humans. Yet in the same way that humans may move in response to climate change, so plant and animal species are also extending into new geographical ranges. For example, evergreen broad-life forests are moving higher up the alpine slopes of Europe and North America, while some well-known tree species, such as holly (*Ilex aquifolium*) are moving northward in Scandinavia. Such biological responses may at first be welcomed, as they suggest that some of the familiar recreational landscapes of northern Europe and North America will survive. Yet for the species and ecological communities that do survive many more will not. Many high mountain and alpine species will become increasingly isolated and will not have the opportunity to extend or change their range.

A clear concern in terms of biodiversity conservation is the capacity of national park and nature reserve systems to cope with the impact of climate change (Dockerty et al. 2003), surrounding land use change and human pressure. Given the migration of species as a result of climate change, the location and extent of the present reserve systems may not be suitable for the conservation of target species and ecosystems. However, even given the existence of a substantial national park system, some ecological features recognized as priority conservation habitat, such as the palsa mires of Lapland (wetland complexes characterized by permanently frozen peat hummocks and a rich diversity of bird species), will almost certainly disappear as the permafrost melts as a result of climate change (Alcamo et al. 2007). In such a situation there is no adequate adaptive response. Instead, there is a need for climate change mitigation (UNWTO et al. 2008; UNEP et al. 2008).

Individual environmental impacts

Although all dimensions of environmental change eventually affect individuals some may be regarded as more direct in their impact than others. One of the most immediately apparent dimensions of tourism and change is that of the exchange and dispersal of disease.

Exchange and dispersal of disease

Patterns of human health and disease are the product of interactions between human biology and mobility and the social and physical environments. However, these patterns are constantly changing as a result of new transmissions and outbreaks of disease. As the number of people in the world who have become mobile has increased, so it has meant that the rate at which diseases spread has grown, as has the potential for populations to have contact with pathogens to which they have hitherto not been exposed (World Health Organization 2008).

According to McMichael (2001: 115) something 'unusual' seems to be happening to patterns of infectious diseases, whether of humans or animals. This has also been described as a 'new pathology of public health in the era of emerging and re-emerging infectious diseases' (Fidler 1999: 17). Of great significance to this new historical transition is the role of international trade and travel as a channel for the spread of infectious disease, what Morse (1993) described as *viral traffic*. Since the 1950s, aviation passenger kilometres have increased by a factor of over 50, while air travel has increased total mobility per person 10 per cent in Europe and 30 per cent in the United States between 1950 and 2000 (Ausubel et al. 1998).

The increase in air travel has also increased the potential for diseases to spread for two reasons: first, the increased speed of travel means that people can travel and pass through airports before disease symptoms become evident; and, second, the size of modern aircraft is increasing, with newer planes carrying ever more numbers of passengers. Bradley (1989) postulated a hypothetical situation in which the chance of one person in the travelling population having a given communicable disease in the infectious stage was 1 in 10,000. With a 200-passenger aircraft, the probability of having an infected passenger aboard (x) is 0.02 (200/10,000 = 0.02) and the number of potential contacts (y) is 199. If homogenous mixing is assumed, this means a combined risk factor (xy) of 3.98 (a 3.98 per cent chance of being exposed to an infectious disease). If the aircraft size is doubled to 400 passengers, then the corresponding figures are $x = 0.04$, $y = 399$, and $xy = 15.96$. In the case of the double-decker Airbus A380 that came into service in 2007, the passenger configuration ranges from 550 to almost 800 seats, the latter being if the plane was placed into an all economy class configuration. If we assume a 600-seat aircraft, then the corresponding figures are $x = 0.06$, $y = 599$, and $xy = 35.94$ (Hall 2005a). In other words, tripling the number of seats available increases the risk factor nine-fold, from a 4 per cent to a 36 per cent chance of being exposed to an infectious communicable disease.

Moreover, the larger the aircraft the more likely it will be that it is travelling between major urban transport hubs, thereby increasing the potential of communicable disease contact between those population groups. An additional dimension to the role of aircraft spreading disease as well as other biotic exchanges, is the connectivity of airports to geographically distant airports by high capacity routes (Tatem and Hay 2007)

Case 5.8: SARS

Respiratory infections are one of the leading causes of human mortality. In late 2002 the first reports arrived in the western media of an outbreak of a new respiratory disease in southern China, which came to be known as SARS (Severe Acute Respiratory Syndrome). The first cases of SARS emerged in mid-November 2002 in Guangdong Province, China and the first official report of 305 cases from the area was announced by the World Health Organization on 11 February 2003. Thirty per cent of those cases were healthcare workers. In the six months after its first outbreak in China, the SARS disease had spread to at least 30 countries and regions, including Australia, Brazil, Canada, South Africa, Spain and the United States. By the apparent end of the outbreak on 14 July 2003, the number of probable cases reached 8,437 worldwide. The disease killed approximately 10 per cent of those infected and the countries with the most cases and deaths were China, Hong Kong, Taiwan, Canada and Russia. The global death toll reached 813, including 348 in China and 298 in Hong Kong (Lee and McKibbin 2004).

The disease was noticeable for its rapid global spread, which had substantial repercussions for local economies and human mobility, as well as for health. As Raptopoulou-Gigi commented, 'The rapid worldwide spread of the coronavirus that causes SARS and the fact that by May 25th 2003, 28 countries reported cases of this infectious disease, suggested that as for other infectious diseases, evolution and spread is facilitated by the mobility of the society either through air travel or the densely populated urban areas especially in Asia' (2003: 81).

SARS was regarded as a serious health threat for several reasons:

- the disease had no vaccine and no treatment
- the virus came from a family of viruses that is recognized for its frequent mutations, making it difficult to produce effective vaccines
- available diagnostic tests had substantial limitations
- the epidemiology and pathogenesis for the disease were poorly understood
- the potential impacts on hospital staff presented a major health human resource problem
- a significant proportion of patients required intensive care
- the incubation period of 10 days allowed its undetected spread via air travel between any two cities in the world.

The spread of SARS was checked by a rigorous system of 'old-fashioned public health measures' (Bell and Lewis 2005: 2) on movement at international borders and health isolation and quarantine procedures (see the previous discussion on biosecurity). Preventing the spread of the disease also came at a substantial economic cost due to changes in trade and tourism flows, as well as associated capital investment. SARS travel bans were put in place for the major affected areas in April 2003, initially in China, especially Hong Kong and Guangdong Province, while other cities in Asia and Canada (Toronto) followed. It was estimated that almost half the planned flights to Southeast Asia were cancelled during the month of April and total visitor arrivals declined by about two-thirds over the course of the crisis, with corresponding effects on related parts of the economy, including hotels, restaurants, retail and even shipping (Bell and Lewis 2005).

In Hong Kong tourism declined by 10 to 50 per cent over the period, with ripple effects felt throughout the region's economy, including a 50 per cent drop in retail sales. However, tourism was affected across the entire East and Southeast Asia region, in part because of travel and business connectivities, but also because of inaccurate perceptions in tourism-generating regions of what places were dangerous to travel to. At the height of the epidemic, visitors and tourism declined 80 per cent in Taiwan and almost as much in Singapore (Bell and Lewis 2005). Nevertheless, as Lee and McKibbin (2004) have stressed in their analysis of the economic impacts of SARS, just calculating the number of cancelled tourist trips and declines in retail trade is not sufficient to get a full picture of the economic impact of SARS because there are domestic and international trade and capital linkages across sectors and across economies. Bell and Lewis (2005) estimated that SARS cost the East Asia region nearly US$15 billion, or 0.5 per cent of GDP, while a 'less readily measurable, but arguably more serious, impact was caused by faltering business confidence' (2005: 21). Nevertheless, this estimate is still only on short-term, direct effects. Because of the patterns of contemporary trade, the economic costs from a disease such as SARS goes beyond the direct impacts in the affected sectors in the disease-inflicted countries. As the world becomes more integrated, the economic and human costs of communicable diseases like SARS are expected to rise, and should also include the forgone income and economic activity that result from disease-related morbidity and mortality.

Source: Hall 2006c.

Summary and conclusions

This chapter has examined some of the main dimensions of tourism and environmental change. One of the key observations is that the main factors related to environmental change are strongly interrelated. Tourism has been shown not only to be significant in impacting the environment but, paradoxically in some cases, also to be a major contributor to its conservation and protection, particularly through the creation of national parks. In less developed countries tourism has become one of the most important alternatives for economic development that does not damage the endemic biodiversity. In addition, tourist experiences in the environment clearly become important for developing positive environmental values; what Gössling (2002) also regards as a dimension of tourism and global environmental change. This could be the most important impact, as people become increasingly aware that changes at the global environmental scale, especially with respect to climatic change, affect them at the local scale, and that their individual actions, such as deciding where to holiday and how they get there, has global implications.

The paradoxical roles of tourism in both resource exploitation and environmental management and conservation, as well as the interrelatedness of environmental change factors, clearly point to the need for an integrated approach to planning tourism – which is the subject of the following two chapters. However, such integration is becoming increasingly difficult as the scales and rates of change have risen dramatically because of the types of human behaviours and actions within which tourism is deeply embedded. Indeed, it could be argued that the cultural decision to set aside an area as a national park in order to conserve it is no longer enough. Instead, it is time to recognize that to maximize the benefits of tourism for the environment we must decide to change the very industrial culture of tourism itself.

Self-review questions

1 How might the naturalness of an environment be defined?
2 At what levels is biodiversity defined?
3 What are the ways in which tourism contributes to biodiversity conservation and loss?
4 How does environmental change at the local scale contribute to global environmental change?
5 What role does travel and tourism play in the spread of pests and infectious diseases?
6 How could tourism provide the economic justification for the reintroduction of species to areas where they have been hunted to extinction? Investigate this question with respect to debates over reintroduction of species in your own country or a particular species, i.e. the reintroduction of wolves.

Recommended further reading

UNEP, UNWTO and WMO (2008). *Climate Change Adaptation and Mitigation in the Tourism Sector: Frameworks, Tools and Practice* (M. Simpson, S. Gössling, D. Scott, C.M. Hall and E. Gladin), Paris: UNEP, University of Oxford, UNWTO, WMO.
Details a range of issues associated with adaptation and mitigation mainly from a developing country context. Available from the UNEP website.

Agrawala, S. (2007) *Climate Change in the European Alps: Adapting Winter Tourism and Natural Hazard Management*, Paris: OECD.
Provides a broad assessment of the threats of climate change to the European Alps.

Frost, W. and Hall, C.M. (eds) (2009) *Tourism and National Parks*, London: Routledge.
Edited book that provides an introduction to the historical and contemporary relationships between tourism and national parks as well as examples of the relationship in a range of different countries and environments.

Gössling, S. and Hall, C.M. (eds.) (2006) *Tourism and Global Environmental Change. Ecological, Social, Economic and Political Interrelationships*, London: Routledge.
Provides a broad overview of a range of different types of environmental change as well as the impacts in certain types of environments.

Newsome, D., Moore, S.A. and Dowling, R.K. (2002) *Natural Area Tourism: Ecology, Impacts and Management*, Clevedon: Channelview Publications.
An excellent overview of some of the environmental impacts of tourism in natural areas.

Runte, A. (1997) *National Parks: The American Experience*, 3rd edn, Lincoln: University of Nebraska Press.
Provides an excellent account of the stresses and strains between conservation and use within the US park system over time.

UNWTO and UNEP (2008) *Climate Change and Tourism – Responding to Global Challenges*, Madrid: UNWTO.
Publication contains both declarations from the WTO on tourism and climate change as well as a technical report. Available from the UNEP and UNWTO websites.

World Health Organization (2008) *International Travel and Health*, Geneva: WHO.
Provides information on potential dispersal mechanisms for diseases as well as the health risks of international travel. Available from the WHO website.

Web resources

Biodiversity hotspots (Conservation International): http://www.biodiversityhotspots.org/
Conservation International: http://www.conservation.org/
International Herald Tribune, interesting article on difficulties in maintaining the World Heritage site of Angkor Wat ('Microbes eating away at pieces of history', June 24, 2008): http://www.iht.com/articles/2008/06/24/healthscience/24micr.php
Millennium Ecosystem Assessment: http://www.millenniumassessment.org/en/Index.aspx
The United States Environmental Protection Agency has a searchable terms of environment database: http://www.epa.gov/OCEPAterms/
United Nations Environmental Programme: http://www.unep.org/
World Health Organization: http://www.who.int/
World Resources Institute: http://www.wri.org/

Key concepts and definitions

Biodiversity hotspots Priority areas for urgent conservation on a global scale that both support a high diversity of endemic species and have been significantly impacted by human activities. Plant diversity is the biological basis for designation as a biodiversity hotspot.

Biosecurity The protection of a country, region or location's economic, environmental and human health from harmful organisms; involves preventing the introduction of harmful new organisms, and eradicating or controlling those unwanted organisms that are already present.

Cultural environments Artificial systems such as cultivated land, including forest plantations, and urban environments; the vegetation has been deliberately determined by humans, with loss of the previous habitat.

Environment The sum of all external physical and non-physical conditions affecting the life and development of an individual.

Environmental change Environment in this context is used to refer to change in the physical environment rather than the socio-cultural environment.

Global environmental change Global in this context has two related meanings: (1) the spatial scale or functioning of a system (e.g. climate and the oceans); (2) when an environmental change occurs on a worldwide scale, or represents a significant fraction of the total environmental phenomenon or global resource.

Natural environments Environments not disturbed by humans or domesticated animals.

Naturalness The absence of environmental disturbance by settled people.

Remoteness Distance from the presence and influences of settled people.

Semi-natural environments The basic vegetation has been altered but without intentional change in the composition of species, which is spontaneous (i.e. overgrazed areas).

Subnatural environments Some changes, but the structure of vegetation is basically the same.

Tourism urbanization Urbanization that has been generated by tourism. It consists of both resort infrastructure geared directly for tourists as well as infrastructure and housing that is required for the tourism industry labour force.

6 Planning and managing tourism impacts

Learning objectives

By the end of this chapter, students will:

- Be able to identify the roles of international organizations and their ability to influence tourism development and impacts at the global and supra-national levels.
- Understand the administrative differences between national, regional and local or community governments that can affect tourism development and impacts.
- Be able to identify techniques that can be used by communities and enterprises to plan and manage tourism development and tourism impacts.
- Recognize the different roles of government, business, the public and non-profit/non-governmental interests and organizations in tourism and community development.
- Realize the political nature of tourism policy and planning efforts, and the importance of networking and organizing to bring about political actions.

Introduction

The social, cultural, economic and environmental impacts of tourism, as we have discussed, are both good and bad. In some cases the positive impacts (such as jobs for the unemployed or the conservation of endangered cultural and natural environment sites) may be clearly evident. Negative impacts can also be easily seen in some destinations (such as soil erosion from overuse and rampant poverty adjacent to international luxury). However, as discussed in the first two chapters, the evaluation of most tourism impacts requires value judgements which are often contested by different interest groups. What is regarded as a positive effect of tourism by one group or individual may be seen as a negative impact by another. Despite this challenge, evaluation is necessary to mitigate problems and to push tourism development to higher quality levels. Quality is also a contested concept, but we can hope this to mean forms of tourism that better meets the needs of hosts, guests and the environment.

This chapter focuses on the planning and management of tourism. This includes both tourism development, where the goals are to change a small or declining tourism sector or to develop new locations, and limits on tourism, where the goal is to manage impacts from a tourism sector that are too large for its current context. The approaches and methods discussed here are, with some exceptions, not broken out into social, cultural, economic and environmental perspectives because the real world is multidimensional, with each impact area overlapping with the others. As such, planning and management should be holistic, multidisciplinary and comprehensive whenever possible. The complexity of tourism is one reason why tourism management plans produced by government agencies are not very common, even though tourism is often considered an important (and growing) sector within the economic, land use and natural resource management plans that are produced by national, regional and local government agencies. At the individual enterprise level, tourism planning has traditionally been narrowly confined to site development and marketing efforts. However, even here, growth and changes in the tourism industry and consumer concerns about the impacts of travel and tourism are forcing private sector tourism businesses to consider the broader social and natural environments that they operate within.

Ultimately the management and planning of the impacts of tourism falls under the scope of what is described as public policy. *Public policy* 'is whatever governments choose to do or not to do' (Dye 1992: 2). Following on from this approach Hall and Jenkins (1995) defined tourism public policy as whatever governments choose to do or not to do with respect to tourism. This definition covers government action, inaction, decisions and non-decisions, as it implies a deliberate choice between alternatives. For a policy to be regarded as public policy, at the very least it must have been processed, even if only ratified, by public agencies. This last point is significant as it indicates that the formulation of policy can occur outside of government. It also highlights the role of politics in managing the impacts of tourism. Various groups perceive and influence tourism policies in significant and often different ways. These include special interest groups (e.g. tourism industry associations, business groups, chambers of commerce, conservation groups and community groups), political leaders and politically influential individuals (e.g. parliamentarians and business leaders), members of the bureaucracy (e.g. employees within public tourism or regional development agencies) and others (e.g. academics, researchers and private consultants) (Hall and Jenkins 1995).

Ideally, the integration of approaches to the impacts of tourism occur within specific public sector organizations, as these are the ones usually charged with managing and planning for such impacts. However, this can be an extremely difficult thing to do, especially when public agencies may have only a limited mandate with respect to tourism, as opposed to their primary missions, which may be focused on resource management or economic development. Because tourism affects so many different parts of government and cuts across the interests and activities of so many different agencies, there is an increasing desire to establish a *whole-of-government* approach towards tourism. In part, this implies greater coordination between different elements of government and relevant non-government stakeholders (those affected by the issue). There are various ways to describe this priority, such as joined-up government, connected government, coordinated government, policy coherence, networked government, horizontal management and whole of government. The distinguishing characteristic of the whole-of-government approach is that there is an emphasis on goals and objectives shared across organizational boundaries, as opposed to working solely within an organization. It therefore encompasses the design and delivery of a wide variety of policies, programmes and services that cross organizational boundaries and can be the result of either 'top-down' decisions or local and community initiatives ('bottom-up'). In Australia, a whole of government approach has been defined as 'public service agencies working across portfolio boundaries to achieve a shared goal and an integrated government response to particular issues. Approaches can be formal and informal. They can focus on policy development, program management and service delivery' (Management Advisory Committee 2004: 1).

The whole of government approach is regarded as having great potential for what are described as *wicked problems* (discussed further below). Wicked problems are ill-defined planning and design problems (Horst and Webber 1973) – sometimes also described as 'messy' problems (Hall 2008c). The environment, in general, is a good example of a wicked problem, as is climate change, in particular. Figure 6.1 roughly outlines the spatial scale (local to global) and the visionary timelines (days to decades) of tourism development issues and related policy formations with respect to dimensions of global environmental change (see also Figure 2.8 for a version of this with respect to climate change). For example, the impacts of tourism are a daily phenomenon, whereas global environmental change, including climate change, occurs over decades and centuries. Destination planners, individual tour operators and tourists need to be aware of, and take action on, the specific impacts associated with a trip, as well as the cumulative impacts of their trip (such as reducing the amount of greenhouse gas emissions they produce), while the global environmental changes that the trip contributes to can only be seriously addressed through research, policies, innovations and interventions adopted at national and global scales. The figure also identifies the limits of influence of different levels of administrative and organizational institutions as they are discussed in this chapter.

The environmental dimensions of tourism-related change require different organizations that operate on different timescales and have varying jurisdictions, responsibilities and sets of values to work together in order to manage an impact (Hall 2008c). Because many tourism policy issues are international in scope and a number of policy and planning areas is being maintained not just by territorial state-bounded authorities, as in much of the past, 'but rather by a network of flows of information, power and resources from the local to the regional and multilateral levels and the other way around' (Morales-Moreno 2004: 108).

The combination of wicked problems and the desire for a whole-of-government approach to try to deal with such issues has led to the development of the concept of *governance*. Kooiman (1993: 258) defines governance as: 'The pattern or

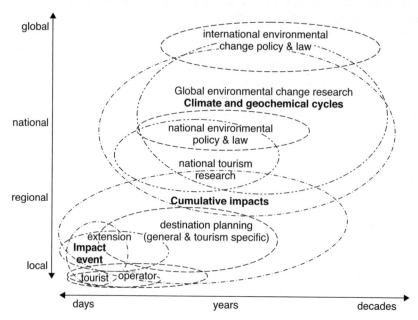

Figure 6.1 Dimensions of tourism impact scale and time frame for management and planning.

structure that emerges in a socio-political system as a "common" result or outcome of the interacting intervention efforts of all involved actors. This pattern cannot be reduced to one actor or group of actors in particular' and can be best understood through terms such as *steering* and *guidance*. This theme is also picked up by Morales-Moreno (2004: 108–9) who argues that, 'we could define governance as the capacity for steering, shaping, and managing, yet leading the impact of transnational flows and relations in a given issue area'. As the chapters in this book have demonstrated, tourism is clearly one such policy area that is marked by substantial transnational flows of people, capital and their associated impacts, as well as new sets of economic, political and socio-cultural relations. To understand some of the issues that emerge in managing the impacts of tourism it is important that we investigate how they may be managed at various scales.

This chapter first examines travel and tourism at the global (worldwide) and supranational (above individual countries) scales. Tourism-related governmental, non-governmental/non-profit and trade/business organizations exist at this level, though their administrative authority to influence policy at the national and local level is relatively limited as sovereign authority to legislate laws and enforce policies exist at the national and subnational (local or community) levels. Because these are the laws under which organizations operate, business strategies are also oriented toward national and local policy orientations, even for multinational corporations. However, multilateral international agreements and laws, with respect to tourism-related issues such as the environment and trade, are increasing in number and do have an influence on the actions of individual countries.

The second part of this chapter, and much of the chapter's focus, is on the management and planning of travel and tourism at the local community and site levels because this is where tourist attractions and tourism business activities directly interact with both tourist guests and local hosts; this is where the impacts of tourism are directly seen, experienced and mitigated. To address this, we first review different models or approaches to local planning, starting with a focus on natural protected areas and moving to working with communities and under-represented populations. This is followed by a summary of implementation tools, which are grouped into those that focus on tourism supply (attractions and destinations) and those that focus on tourism demand (the tourist).

Global and supranational management

Although a significant part of travel and tourism is international and even global in its operations, marketing and impacts, there are few international agreements or regulations that directly manage tourism activity. International law can be divided into 'soft' policy rules and declarations and 'hard' treaty requirements (Hall 2008a, c).

An example of an extremely influential soft policy declaration is the Brundtland Report, which was issued by the UN's World Commission on Environment and Development (WCED) in 1987 (see also Chapter 2). It defined the concept of *sustainable development* and declared it as a goal for the global community. Even though tourism was barely mentioned in this document, the tourism industry, government and academic leaders have taken up *sustainable tourism* as a major guiding principle in managing the impacts of tourism development. (We will return to sustainable development in the discussion of national and local tourism management practices later in this chapter.)

The principal organization representing government interests in travel and tourism is the World Tourism Organization (UNWTO), which became a specialized agency of the United Nations in 2003. UNWTO has established many soft regulatory mechanisms that shape tourism development at the global level. It is primarily responsible for creating international standards for the collection of tourist departures, arrivals and expenditures (as covered in Chapter 1), although it has increasingly become active in the development of tourism plans, encouraging free trade in tourism services, and promoting tourism as a sustainable form of development. Most countries are members of UNWTO, along with representatives from the private sector, some educational institutions, local tourism associations and authorities which pay to join as associate members. The organization also identifies major trends and issues faced by the industry, and provides consulting services for member countries. None of UNWTO's rules have 'hard' legal authority and are instead recommendations that member countries should adopt in their own policy-making.

Several regional organizations serve functions similar to those of UNWTO. Most prominent among these has been the Organization for Economic Co-operation and Development (OECD) and the Pacific Asia Tourism Association (PATA). Although having a far wider mandate on economy issues, the OECD does collect and assess tourism data on Europe, North America, Japan, Australia and New Zealand, and also makes recommendations with respect to using tourism as a development tool. PATA is focused on most of the countries bordering the Pacific Ocean, as well as South and Southeast Asia, and has both governmental and private business members. Both organizations collect and analyse tourism-related data for their member countries and organizations, and address regional issues in tourism development.

In addition, most of the multilateral regional bodies that exist around the globe have tourism sections or working groups that address issues of movement, trade, air and other transportation, and regional tourism marketing. These organizations include the European Union, the Arab League, the Union of South African Nations, the Association of Southeast Asian Nations, the Latin American Parliament, and the Organization of American States. The European Union's Schengen Agreement, which removed border checks between EU member states, is the most integrated travel agreement among these regional organizations, although less liberal arrangements have been adopted by other groupings. Table 6.1 provides examples of multilateral frameworks that enable tourists to move between countries.

A variety of private sector trade organizations have been established to represent different business interests related to travel and tourism. The two most influential of these is the World Travel and Tourism Council (WTTC) and the International Air Transport Association (IATA). WTTC states that its membership is comprised of the CEOs of the leading 100 (or so) global tourism and travel companies. The companies that comprise the WTTC membership include travel agencies, global hotel chains and many international airlines. As a trade organization, WTTC identifies the major concerns of its members and lobbies governments to adopt rules and regulations that recognize the economic importance of travel and tour-ism. They generally encourage deregulation and permissive foreign investment and ownership regimes. WTTC also tries to develop uniform industry positions on issues such as cross-border processes and environmental and sustainable development policies and assessments.

IATA is responsible for the designation of the universally adopted three-letter airport codes and two-letter airline codes, as well as the accreditation of travel agents throughout the world, except in the United States where accreditation is done by the Airlines Reporting Corporation. In the past, IATA was instrumental in standardizing international airfares, though that role has largely ended with airline deregulation in the early 1980s and the rise of low-cost airlines, many of which are not members of IATA. In addition to air passenger carriers, IATA also represents the interests of air cargo carriers. Like WTTC, IATA develops policies and lobbies government in the interests of its airline members.

UNWTO, WTTC and IATA exist primarily to support the interests of their members. This does not mean, however, that they do not respond to global issues

Table 6.1 Examples of multilateral mobility frameworks

Multilateral grouping	Members	Mobility	Common visa
Central America Four (CA-4)	El Salvador, Guatemala, Honduras, Nicaragua	Citizens of the four countries do not require a passport to travel between any of the four countries; a national ID card is sufficient	Central American Single Visa (Visa Única Centroamericana)
East African Community	Kenya, Tanzania, Uganda	An East African passport can be issued to members of each country for travel within the community. Visas are not required for nationals of the community who use their national passports	East African Single Tourist Visa
ECOWAS (Economic Community of West African States)	Benin, Burkina Faso, Cape Verde, The Gambia, Ghana, Guinea, Guinea Bissau, Ivory Coast, Mali, Niger, Nigeria, Senegal, Sierra Leone, Togo	Citizens of community members do not require a passport to travel between community countries; a national ID card is sufficient	No
European Union – European Economic Area, Schengen Agreement	Most European Union states and Iceland, Norway and Switzerland Ireland and the UK only participate in the cross-border policy cooperation measures	The 1985 Schengen Agreement among European states allows for the abolition of systematic border controls between participating countries; includes provisions on a common policy on the temporary entry of persons (including visas), the harmonization of external border controls and cross-border police cooperation	Schengen Visa – citizens of non-EU or non-EEA countries who wish to visit Europe as tourists can obtain a common Schengen Visa

Table 6.1 Continued

Multilateral grouping	Members	Mobility	Common visa
Nordic Passport Union	Denmark, Faroe Islands, Finland, Greenland, Iceland, Norway, Sweden	Any citizen can travel between countries without having passports checked; an ID card is sufficient. Other citizens can also travel between the Nordic countries' borders without having their passport checked, but still have to carry a passport or another kind of approved travel identification papers	Yes, within the context of the Schengen Visa (except the Faroe Islands)
SADC (South African Development Community) (expected to be introduced in 2008)	Angola, Botswana, Democratic Republic of the Congo, Lesotho, Madagascar, Malawi, Mauritius, Mozambique, Namibia, South Africa, Swaziland, Tanzania, Zambia, Zimbabwe	Protocol on the Development of Tourism has the objective of enabling the international and regional entry of visitors	SADC Univisa – originally intended just for main source markets, it is likely to apply to all long-haul markets by the 2010 FIFA World Cup in South Africa

Source: Hall 2008a: 43.

Photo 6.1 The Golden Triangle border area of Southeast Asia shows how countries can work together in the name of tourism. This photo is taken from Thailand, with Myanmar on the left of the Mekong River and Laos on the right. Special visa agreements allow tourists to visit all three countries in the border area. The red building in Myanmar is a casino, where visitors are required to confirm their ability to gamble a certain amount of money. Laos allows visitors from Thailand on to one of its islands, filled with tourist shops, without any visa formalities. Large tourism-related projects exist on the Thai side and are under construction on the Laotian side. (Photo by Alan A. Lew)

and efforts to address the impacts of travel and tourism. One of the major issues for all three organizations has been to encourage governments to reduce barriers to international travel. This includes outright bans on travel to some countries (such as US restrictions on travel to Cuba), visa restrictions that limit the amount of time a tourist can stay in a country, burdensome passport and visa requirements, and airspace restrictions on planes flying over some countries.

Considerable progress was made in many of these areas in the period of global deregulation in the 1980s and 1990s, fostering a tremendous growth in international travel and tourism that continues today, despite some setbacks. UNWTO and WTTC were both instrumental in the development of the Tourism Satellite Account (TSA) system (see Chapter 3), which is designed to show the true size of the travel and tourism economy. WTTC today is the primary producer of global TSA data, which they claim show travel and tourism to comprise 9.9 per cent of global GDP, and 8.4 per cent of worldwide employment (2008 forecasts in WTTC 2008c).

Deregulation and the expansion of travel and tourism, however, do not really address the impacts of tourism, except to encourage more of it. These organizations responded to the rising interest in sustainable development in the 1990s, with WTTC starting the Green Globe environmental certification programme (which is now a separate entity) in 1994, and UNWTO adopting a Global Code of Ethics for Tourism in 1999. The ten articles of the code of ethics include (UNWTO 1999):

1 Tourism's contribution to mutual understanding and respect between peoples and societies
2 Tourism as a vehicle for individual and collective fulfillment
3 Tourism as a factor of sustainable development
4 Tourism as a user of the cultural heritage of mankind and contributor to its enhancement
5 Tourism as a beneficial activity for host countries and communities
6 Obligations of stakeholders in tourism development
7 The right to tourism (as a form of leisure and recreation)
8 The right to tourist movements (travel)
9 The rights of the workers and entrepreneurs in the tourism industry
10 Implementation of the principles of the Global Code of Ethics for Tourism.

Although UNWTO has defined an implementation and enforcement strategy, in reality they are not in a position to force any member to adopt their code. Adoption and enforcement can only occur at the national, local and company levels (see Case 6.1), and need to be customized to those contexts.

More recently, all three of the international organizations have tried to address the challenge of global climate change. IATA, for example, adopted a strategy in 2008 to address airline carbon emissions (see also chapters 5 and 6). The measures they proposed include (IATA 2008b):

• Technology – accelerating the development of cleaner and alternative fuels, advanced traffic control technology, and new airframe and engine designs
• Operations – more efficient and conservation-oriented aircraft operations through inspections and training programmes
• Infrastructure – increasing airport and airspace efficiencies, flexible airspace access, and uniform air traffic control systems
• Economics – tax credits to encourage research and development into new technologies, an effective and voluntary global carbon trading system, carbon offset programmes for passengers.

UNWTO, WTTC and IATA do not have the powers to enforce any of these principles with respect to the actions of national governments or businesses. Instead, they rely on persuasion and diplomacy, as even the threat of expulsion from a body may not obtain the desired results. This is partly why such organizational goals tend to be general and idealistic. What these associations can do is to support relevant research and educate their members, governments and the travelling public. This will often include convincing legislators and parliamentarians that their industries have effective self-regulatory mechanisms, and do not require formal regulations. Thus IATA strongly encourages its members to voluntarily work toward its 'Greener Future' goals. If the members of these organizations are not seen to be effective in regulating themselves, then government regulations may be necessary. However, the effectiveness of such self-regulation in the longer term is debatable (Williams and Montanari 1999; Ceron and Dubois 2003; Wiley et al. 2008).

Case 6.1: Managing ecotourism impacts

Tourism is an economic activity that attracts entrepreneurs who often enjoy the benefits of tourism itself, such as meeting new people and seeing and experiencing new places. It is not surprising, therefore, that many tourism business interests share the concerns of tourists, hosts and the broader society over the potential negative impacts of tourism development. Some tourism entrepreneurs are proactive in trying to address these concerns through their business practices. Ecotourism companies are one example of this.

Ecotourism is sometimes considered a branch of sustainable tourism that focuses on natural area, though cultures within a natural area are a major component. Lew (1998a) surveyed 40 North American tour operators who conducted ecotours to the Asia Pacific region. The survey focused on ecotourism management policies that may differ from mass forms of tourism. While no single company had adopted all of the potential ecotourism policies that have been discussed by researchers on the topic (Fennell 2001), many of the policies had been adopted; the most popular were:

Use guides native to visited area	77.5%
Have an education programme for local guides	65.0%
Provide a pre-arrival information packet	60.0%
Providing a % of tour profits to local organizations	47.5%
Participate in local cleanup programmes	42.5%
Pack-it-out requirements	37.5%
Other activities to support sustainable development	40.0%

Two-thirds of the companies used local tour guides exclusively for their ecotours. Some of the 'Other activities' that were listed as supporting sustainable development included: donations to local charities and clinics and to fund conservation; teaching adult education classes and supporting libraries; providing medical services; planting trees; and using all recyclable materials. Over 90 per cent of tour participants were willing to donate to local causes, and the companies estimated that their ecotours cost an average of 11.1 per cent more than non-ecotours.

Photo 6.2 Canopy walk at the Poring Hot Springs in Mt Kinabalu Park in Sabah, Malaysia. Mt Kinablu (4,095 m; 13,435 ft) is the highest peak in insular Southeast Asia. Canopy walks are popular attractions in tropical rainforests, though they are also found in some temperate forests. They are considered a form of ecotourism because they provide an experiential education for tourists about the importance of the upper tree canopy, while protecting the ground-level flora and fauna from human intrusion. For many tourists, however, they are as much a thrill experience as an educational one. (Photo by Alan A. Lew)

Most of the companies addressed tourist behaviour in some manner. Though the policies generally varied depending on the sensitivity of the destination, responses included:

We strictly enforce sensitive behaviour on our tours	42.9%
We explain proper behaviour, but leave it up to the individual	33.3%
We only explain proper behaviour in the most sensitive place	11.9%
We seldom ever direct tourists in how to behave	11.9%

Although tour group sizes ranged considerably, from just one person to over 100 for one destination, the overall average group size was 11.4, and the general policy is to limit ecotour groups to a maximum of 15 participants. This is possibly the single most significant difference from mass tourism. Reasons cited for why ecotour groups were limited to this size included: to reduce or lessen environmental damage and cultural impacts in sensitive areas; to maintain the privacy of the destination populations; to enable better personal service and group control by guides; and to facilitate travel logistics in remote areas.

Source: Lew 1998a.

_____

International treaties

Some issues rise to a level of such importance as to force efforts to establish binding international agreements that require countries to adopt certain policies and laws. Examples of the topics covered by these 'hard' international laws, which are enforced through treaty agreements, are: nuclear and chemical weapons proliferation; intellectual property, copyright and patents; pollution, greenhouse gases and global climate change (e.g. the Kyoto Protocol); international crime and cyber crime (e.g. the International Criminal Court); and maritime law, including rights to offshore resources and territorial boundaries (e.g. the Law of the Sea). The Geneva Conventions, consisting of four treaties on the conduct of war, are probably the best known of the international laws.

Most of these issues are addressed through various structures related to the United Nations, and guidelines of international behaviour established by the International Labour Organization (ILO), the World Health Organization (WHO), the World Intellectual Property Organization, the International Telecommunication Union, UNESCO, the World Trade Organization (WTO) and the International Monetary Fund (IMF). The treaties and agreements are enforced through the threat of international economic sanctions that can be imposed by the United Nations.

For example, the ILO has been one of the leading UN agencies supporting indigenous and tribal human rights, adopting the *Indigenous and Tribal Populations Convention* (C107) in 1957, which was the first international legal instrument to recognize the rights of indigenous peoples (Anderson 1996). These rights were a major emphasis in the UN's 1992 *Rio Declaration on Environment and Development* (also known as the Earth Summit), which was a major document outlining the principles of sustainable tourism. Despite the human rights issues that tourism development in traditional and indigenous communities often raises, economic imperatives frequently predominate.

It is not surprising, therefore, that the treaty that most affects travel and tourism is probably the economically focused General Agreement on Tariffs and Trade (GATT), which is overseen by the World Trade Organization (WTO). The GATT treaty process started after the Second World War, with the goal of regulating and liberalizing international trade. It has developed rules to reduce tariffs (taxes governments place on imported goods), as well as to establish policies that govern international trade involving intellectual property, services, capital and agriculture. Disputes are taken to the WTO, which has the power to apply economic sanctions on countries that are found to violate the treaty rules.

The trade liberalization policies of the GATT, and particularly the General Agreement on Trade in Services (GATS) (see Chapter 1), are designed to encourage international trade by requiring member countries to allow direct foreign ownership of businesses on their soil, such as hotels and travel agencies, and to remove barriers that can prevent this, such as certain visa restrictions (see Case 6.2). The GATT treaty, therefore, impacts tourism by promoting it as an economic activity in member countries, but does not have any authority to control or limit the social and environmental impacts of tourism, although these issues are increasingly becoming part of the GATS debate in the area of tourism (Arkell 2003; Hall 2008a). Not all countries of the world are members of the WTO and the GATT treaty. As of 2008, Russia and many of the predominantly Muslim countries from Central Asia through North Africa were not members of the WTO. In addition, there are substantial differences between what developed and developing countries are seeking through GATS, with developing countries wanting more gradual liberalization strategies out of concern over the extent to which international destinations in their countries may be subject to the dominant market position of international airlines and tour operators (PATA 2004).

A good example of what developed countries want to have liberalized for tourism under GATS comes from a December 2000 communication to Members of the Council for Trade in Services, in which the United States presented its liberalization agenda for three sub-sectors: hotels and lodging; duty-free services; and the meetings, incentives, conventions and exhibitions (MICE) industry. The United States argued that WTO members should place 'no limitations' on market access and national treatment, even though 'some members, including the United States, already have done', and proposed that all members 'consider undertaking additional commitments relating to travellers and international conferences' (WTO 2000: paragraph 5). In contrast to the position of the United States, the developing countries had a different agenda with more of a focus on such matters as the mobility of workers and concerns over market dominance by foreign operators which can be illustrated by reference to the Columbian proposals to the WTO (2001b: paragraphs 4–10), which included:

- Full commitments should be adopted under the consumption abroad and commercial presence modes of supply, which would promote two-way tourism.
- Restrictions involving an economic needs test to gain market access to tourism-related activities should be eliminated.
- Given that the provision of such services depends on the cross-border movement of persons, Members should endeavour to ensure that their migration authorities facilitate such movement.
- Market access conditions under the presence of natural persons mode of supply should be improved in order to facilitate the temporary entry of natural persons supplying services in this sector.

- The competent authorities should take account of professional qualifications related to tourism services acquired in the territory of another Member, on the basis of equivalency of education and using qualification recognition methods.
- Anti-competitive behaviour by dominant operators can result in imbalances in the framework of liberalized trade in services.
- The existing classification of trade in tourism and travel-related services should be revised to include all services characteristic of this sector.

With respect to sustainable tourism and GATS, Dunlop (2003: 10) noted that the European Union, along with the United States, suggested that sustainable development needed to be considered within the proposed United Nations World Tourism Organization (UNWTO) Annex to GATS with respect to tourism, with the EU stressing 'the importance of access to high-quality environmental services – [as] a key offensive negotiating interest for the EC (and US) in the GATS negotiations', which would have significant impacts in a wide range of countries with respect to tourism trade with the EU (Hall 2008a). In effect, this would require efforts toward environmentally sustainable tourism by any country that sought more open tourism-related trade relations with the European Community, with the EC members being in a good position to sell those services to the countries that wished to trade with it. In addition, the EU sought to use any UNWTO Annex to eliminate restrictions on foreign direct investment in tourism, which some developing countries fear could be extremely damaging to their domestic companies. The differences in opinion between countries mean that some developing nations are arguing that no agreement is better than a bad one. As a result, 'many developing countries are turning to regional and bilateral free trade agreements, which they believe allow them to liberalise trade and services in a more progressive manner without unnecessary external pressure and in a way more suited to local conditions' (PATA 2004: 4).

Case 6.2: The international hotel industry's response to China's WTO accession

The significant impact of the General Agreement on Tariffs and Trade (GATT) and the World Trade Organization (WTO) on tourism development worldwide can be seen in the People's Republic of China, which was admitted to WTO in 2001. Over a six-year period, China was required to open its economy to foreign investors and to adopt international accounting and regulatory practices to ensure greater business transparency. Within two years, by the end of 2003, foreign service suppliers were permitted to engage in the retailing of all products; a year later foreign firms had the right to import and export most goods; and at the end of 2006 foreign firms were allowed to distribute almost all goods domestically. As a result of this timetable, 2007 was the year that major international hotel and travel agencies were first able to enter the Chinese market as sole proprietors, instead of as joint venture partners with domestic Chinese firms.

During this same time period, China was also transformed from a country where the government owned and ran most companies (in 2001), including hotels and travel agencies, to one where over 70 per cent of the country's GDP came from the private sector (in 2007). Some 46 million people lost their government jobs in the process, with most of them moving to new private companies. Having transformed its economy from Communist government ownership to mostly private ownership, China's next phase of economic development has been toward higher-quality value-added production, to ensure that wages and standards of living continue to rise.

The most significant impact of international tourism on the Chinese landscape has been in the constructions of hotels. Joint venture hotels were the first post-Second World War architectural buildings to dot the Chinese urban landscape after the country opened up following the death of Chairman Mao Zedong in 1976. These hotels were majority owned by Chinese government entities, but mostly built and managed under lease by international hotel companies. The purpose was to transfer modern construction, management and labour skills to the under-developed Chinese economy.

By the time that China was admitted to WTO in 2001, its accommodation sector was characterized by a small number of large, upscale, international chain hotels in major cities and a large number of smaller, older and much cheaper state-run hotels, most of which were in serious need of remodelling and none of which were part of a chain affiliation or ownership. Chain ownerships are considered more efficient in operations, bookings, management and brand development than single-owner properties, and China's domestic hotel industry was considered eco-nomically immature in comparison to international standards.

Jinjiang Inns was the only modern mid-priced domestic hotel chain established in China prior to 2001, though Home Inns was founded in 2002. By 2007 these two chains had 250 and 171 hotel properties, respectively, and both had plans to more than double by 2010 (Lew 2007a). This is a relatively small number of hotels to fill the vast and growing demand for mid-priced accommodation marketed between the low-end state-run facilities and the high-end international chains. As a result, China's accession to WTO has prompted a boom in mid-priced hotels built by international hotel chains. Both Starwood and Marriott announced plans to quadruple their number of hotels in China, from 27 and 25, respectively, in 2007, to over 100 each by 2012. Though mostly high-end, they also include Marriott's mid-range Courtyard by Marriott. Also in 2007, Accor, Europe's largest hotel chain, announced that it would expand from its 50 hotels in China to 180 by 2010. Accor's hotels include the higher-end Sofitel and the budget Ibis and Motel 6 chains. Bass (Inter-continental and Holiday Inn), Cedant (Ramada, Super 8 and Days Inn) and Shangri-La are other major international hotel operators in China that are expanding their chains throughout the country, especially along the rapidly expanding limited access highway network that is tying the country together.

Source: Lew 2007a.

International conservation organizations

Not all international organizations are focused primarily on the liberalization of economies to support tourism development, as are those discussed above. Several very important international organizations are focused on the conservation of cultural and environmental resources. Examples of governmental organizations include the United Nations Education and Scientific Organization (UNESCO), the United Nations Environment Program (UNEP), the World Conservation Union/International Union for the Conservation of Nature and Natural Resources (IUCN), the International Council on Monuments and Sites (ICOMOS), and the World Heritage Centre. The objectives of these quasi-governmental organizations are generally not specifically for tourism, though most of their activities directly impact tourism (see Chapter 4). Non-governmental and non-profit organizations (NGOs) are more likely to work specifically on tourism issues. Some of the more prominent of these are: Tourism Concern, the World Wide Fund/World Wildlife Fund (WWF), and the Ecumenical Coalition on Tourism (ECOT).

These conservation and tourism organizations work at a variety of levels, from establishing global agendas and guidelines to working on local conservation programmes. The following discussion focuses on the World Heritage List, which includes both natural sites and cultural sites. The designation of a place on the World Heritage List is considered a major achievement as it reflects both an international level of appreciation for a resource, as well as a high level of recognition of quality in its management. (The impacts of such a designation were discussed in Chapter 4.)

Natural protected areas

IUCN defines a natural protected area as, 'An area of land and/or sea especially dedicated to the protection and maintenance of biological diversity, and of natural and associated cultural resources, and managed through legal or other effective means' (IUCN 1994). The designation of a place as a protected area does not ensure its full protection. The level of protection varies from country to country, and from one site to the next, due to differences in legal authority and status. Management objectives also vary considerably, and often include conflicting mandates for development and conservation. The effectiveness of planning and management efforts also varies due to the complexity of environmental and political challenges that each protected area faces.

For tourism development in natural protected areas, the management strategies must support protection of natural and cultural resources, while allowing for adequate and appropriate access for tourism and recreation. This is a common mandate for national parks, for example (see Chapter 2), and presents planners and managers with the challenge of determining trade-offs among tourism development, the protection of resources and the interests of the local community. IUCN designates six levels of protected areas, based on their primary management objective (see Table 5.6).

Photo 6.3 A cruise ship full of tourists at College Fjord, near Prince William Sound, Alaska. The glaciers were named for different Ivy League colleges and are under the management of the Chugach National Forest, which has designated this area as the College Fjord Wilderness Natural Area. Wildlife is abundant in the largely uninhabited area, with bears on the shorelines and harbour seals basking on floating glacial ice, which also protects them from prowling orca whales. The Alaska Department of Environmental Conservation works with the Alaska Cruise Association to limit cruise visits to minimize environmental impacts and maximize visitor experiences. (Photo by Alan A. Lew)

IUCN protected area levels are relatively flexible to account for the wide range of protected area management situations around the world. In addition, some form of recreation and tourism is possible in all of the categories, except for those under strict scientific research controls. Cultural resources, however, are only touched upon peripherally by the IUCN guidelines, even though traditional cultural issues are often an integral part of the management of natural areas throughout the world, and some national park systems are equally responsible for both nature and historical sites. Marine Protected Areas (MPAs) were also not considered in the original IUCN categories. A challenge of MPAs is that, more so than most land-based preserves, their resources migrate and extend far beyond their administrative and political boundaries.

The range in management objectives for protected areas means that there is no single way of organizing the relationship between tourism and protected resources. Different resources will be attractive to different tourist and recreation market segments (see Case 6.3), and it is important for the protection of the resource that it is marketed to tourists that can appreciate and respect its management objectives.

Case 6.3: Market segmentation for Denmark

Not all tourists are the same. They have different interests, abilities, motivations and resources. Similarly, not all tourist attractions are the same. A mismatch between a tourist's interests and a destination's attractions can result in bad experiences for both the guest and the host. For example, the tourist may be interested in nightlife activities, such as dancing and live music, but the destination may essentially close down after dinner. The result could be an upset guest at the destination and negative word-of-mouth or online reviews after the holiday.

It is, therefore, important to market a destination to the tourist market segment that would be most interested in its attractions. Proper *market segmentation* and appropriate marketing also enable a more efficient allocation of marketing resources. Market segmentation is often determined through surveys of visitors. The goal of these surveys is to determine:

- Which tourists most appreciated and *enjoyed* the destination?
- Which tourists spent the most *money* at the destination?
- Which tourists spent the most *time* at the destination?
- What *activities* did different tourists participate in, and which did they enjoy the most?
- Which tourists are most likely to *return* to the destination?

In addition to asking these types of questions, the survey needs to obtain background data on the tourists to correlate with their interests. The types of background data include: demographics (such as age, gender, occupation, education level, income level, ethnicity, home location, past visits), interests (such as comfort/safety or adventure/risk, contemplative or social, or educational), activities (such as nature-based or cultural, and group or individual) and trip characteristics (such as accommodation type, travelling partners, transportation and itinerary taken). Demographic data are collected in every survey. Collecting other forms of data depends on available resources (time and money) and survey objectives.

A survey of 7,600 domestic and international tourists to Denmark in 2004 used factor analysis to identify seven statistical groupings among 22 motivations that the respondents gave for visiting Denmark. This means that there are seven basic categories that tourists in Denmark can be divided into based on their interests in travelling there. These groupings were:

1 *Nature*: nature (beach, forest), clean country, safe place to stay, few tourists.
2 *Local ways*: people in general, possibility for enjoying Danish food, shopping possibilities, other motive.
3 *Attractions*: children friendly, attractions and amusements, possibility of activities.
4 *Value for money*: good or cheap transport to destination, price level or inexpensive, possibility for spa, health, fitness.

5   *Physical activities*: possibility for bicycling, possibility for angling, possibility for yachting.
6   *Museums and culture*: museums/cultural, historical heritage, special events, theatre and musical festival.
7   *Visit friends and golfing*: visiting friends and family, possibility for playing golf.

Each of these could provide a separate marketing theme with related attractions. The country of Denmark could decide to focus on some themes more than others in any given year. In addition, the marketing themes could be targeted to media outlets that focus on individuals with those interests. Such targeted marketing can be further enhanced by cross-tabulating these seven factors with demographic data, including place of residence, media preferences and income levels.

In this particular study, one of the cross-tabulations undertaken was to compare the factors with the tourist's primary destination in Denmark. This showed that the nature factor most correlated with the city of Bornholm, the local ways and the museums and culture factors were most related to Copenhagen, and the value for money factor was most associated with the rest of Denmark. This provides further insight into how different geographic regions in Denmark might best market themselves so that both the hosts and guests have the best tourism experience.

Source: Zhang and Marcussen 2007.

National and regional management

Unlike most international organizations, whether government-based, industry-based or NGOs, countries have the most substantive authority to monitor, manage and regulate all aspects of land and life within their jurisdictional borders. They are able to legislate laws, establish regulatory authorities to administer laws, and grant variations of similar authority to lower, local level governments (such as provinces/states, cities and counties). Governments shape the economic climate for the tourism industry, help provide infrastructure and educational needs, establish the regulatory environment in which businesses operate, and usually take an active role in tourism promotion and marketing. However, different levels of government tend to have different sets of objectives to achieve through tourism development. This is a situation that makes the study of tourism policy and decision-making all the more complex, because the aims of regional and local government may diverge from those of central government. The International Union of Tourism Organizations (IUOTO 1974), in its discussion of the role of the state in tourism, identified five areas of public sector involvement in tourism: coordination, planning, legislation and regulation, entrepreneurship, and stimulation, including government-sponsored destination promotion. To these, two other functions can be added – a social tourism role, which remains very significant in European tourism although it is not as important in North America; and a broader role of protection of the public interest (Hall 1994). It needs to be remembered that different countries and levels of government will have different foci with respect to the various ways that the state is involved in tourism, which will affect not only their approach towards tourism and its impacts, but also the design of national and regional public agencies responsible for tourism.

Many countries have a national-level tourism body, although these vary considerably in the scope of their mandates, their degree of authority and their source of funding. Table 6.2 provides a summary of these approaches.

Some have a national tourism organization (generically referred to as a national tourism organization, or NTO), that collects tourist arrival and expenditure data and sets policies that impact the marketing and development of tourism facilities and sites. The China National Tourism Administration (CNTA) is an example of this type of government body. In addition to publishing statistics, it oversees the hotel star ratings for the country, the certification of tour guides, national-level marketing efforts, and works with other government agencies to identify natural and cultural resources and transportation infrastructure for development and promotion. These national-level NTOs can have considerable budgets and the authority to influence a country's tourism development by establishing priorities and guidelines for the development of tourist sites.

Many countries combine their tourism administration with other administrative portfolios, such as natural resources, cultural resources or economic development. Some countries assign their national tourism interests to quasi-governmental or public–private sector organization. In the United States there are two organizations that oversee tourism at the national level. The Office of Travel and Tourism Industries (OTTI, formerly known as ITA) of the US Department of Commerce is

Table 6.2 Approaches to national tourism organization (NTO) design

Type	Nature	Example
1 Unified	Tourism policy, marketing and development combined into a single agency	China National Tourism Administration
2 Policy/ marketing split	Different government agencies are responsible for developing tourism policy (including undertaking research) and national tourism marketing and promotion	Tourism Australia, Tourism Canada, Tourism New Zealand
3 Policy only	A government body with a degree of autonomy is responsible for policy and research but there is no single government agency responsible for national tourism promotion	Office of Travel and Tourism Industries, USA
4 Integrated	Tourism functions are integrated within a larger government agency that has other areas of responsibility. Although there may be a tourism-specific section it is not an autonomous body	Innovation Norway

the primary federal government agency, while the Tourism Industries of America (TIA) organization, a private sector body, is the principal trade organization. The functions of the OTTI are (OTTI 2008):

- management of the travel and tourism statistical system for assessing the economic contribution of the industry and providing the sole source for characteristic statistics on international travel to and from the United States;
- design and administration of an international promotion program and export expansion activities;
- development and management of tourism policy, strategy and advocacy; and
- technical assistance for expanding this key export (international in-bound tourism) and assisting in domestic economic development.

OTTI collects and publishes basic US inbound arrival and outbound departure data, conducts some overseas marketing campaigns (though this is very limited) and represents the United States overall at overseas travel trade shows. TIA takes the data from OTTI and produces value-added reports for its members (some are also publicly available), which are also used to support their educational and lobbying efforts. TIA has been the most influential group in lobbying for user-friendly changes to the US Transportation Safety Administration (TSA) airport security methods. TIA does some industry-wide marketing, but not a lot, and it oversees the annual International Powwow, which is the largest travel trade sales event in the United States, bringing together tour and destination providers with travel agents and tour sellers. In the United States, as with many other federal systems, however, most direct tourism marketing is conducted at the state and local level, rather than the national level.

Other national-level authorities can have significant impact on the tourism milieu and overall image of a country. These include a country's aviation authority, its highway and railroad construction authorities, and its national park and public lands management agencies. Many of these governmental agencies recognize their relationship to tourism, but typically have other mandates that take higher priority (Hall 2008c). Depending on a country's national goals, which are defined by political leaders, and the individual resource site in question, these agencies may or may not work with private sector tourism interests to promote or protect natural and cultural resources. However, through their budgeting process and administrative rules and regulations, they often determine the access and quality of sites that are visited by tourists.

Because national-level government agencies tend to have the greatest political power to influence tourism development and resource management, this is also a level at which many non-governmental organizations operate. Those that impact and influence tourism include environmental groups, consumer organizations, trade unions, trade associations, professional associations and many other special interest groups. Most of these groups operate both at the national level, by lobbying the national government, and below at regional and local levels.

For most countries, 'regional authorities' generally refers to a level of government just below the national level. This will vary from one country to the next, and will be more formally structured in some countries than others. In the United States, for example, regional authorities exist at either multi-state or multi-city levels. Multi-state authorities, such as the Tennessee Valley Authority, are not very common, though multi-city regional planning authorities are found in large metropolitan areas, mostly focused on transportation. In contrast, the EU has deliberately encouraged and supported the development of a number of cross-border organizational structures that bring various levels of government from different countries together in order to encourage European integration – of which tourism is an extremely significant component. Local governance usually refers to the municipal or county level.

The primary source of legal authority for tourism-related activities is either at the national level, for most countries, or at the province or state level, in a federal system of government. However, even in federal systems the powers of states and provinces are determined by a national constitution. Depending on the degree and form of legislative and administrative authorization that a country or a state government grants to the regional and local-level authorities, these authorities may be able to influence tourism policy, development and management through a combination of stimulus and regulatory measures. Examples of the types of influences that local governments may exert include the following (adapted from Hall 2008c), although note that regional and local governments vary considerably in their authority to undertake these actions (see Case 6.4).

- Coordinate governmental and non-governmental stakeholder interests and efforts.
- Plan for tourism development and management.
- Legislate land use policies and regulations that allow or limit development.
- Create entrepreneur education and financial support programmes, which can be targeted to meet specific needs.
- Provide tax, infrastructure and other support incentives to encourage particular forms of development or conservation efforts.
- Create funding sources through their taxing authority, such as designating special hotel and restaurant taxes that are paid more by tourists than by locals.
- Fund tourism marketing and promotion campaigns through local tax revenues.
- Provide for social tourism programmes that provide leisure experiences for targeted underprivileged populations.
- Protect sensitive natural or cultural resources, including landscapes, from inappropriate development or use.
- Adopt budgets and operational practices that support some forms or locations of development or conservation over others.

Consumer organizations that represent travellers and tourists typically operate at the national level, though exceptions exist, especially at the local level. These organizations typically arise to address especially egregious treatments of travellers. An example is the International Cruise Victims Association (ICV), which mostly operates in the United States on behalf of cruise passengers who have been subject to crimes or have died from accidents on cruise ships in international waters where no country has jurisdiction. Also in the United States, several formal and informal organizations exist to lobby for the rights of airline passengers, including the Aviation Consumer Action Project and the Consumer Federation of America. Trade groups, such as IATA, also provide some consumer information, though usually less confrontational and avoiding controversial issues. Governments are another important source of consumer information, which can sometimes support certain policy orientations. The US Department of Transportation has an Aviation Consumer Protection Division with an extensive website to assist passengers with industry complaints (USDOT 2008), and many countries issue official travel warnings and country advisories for their nationals. Among the more influential of these travel warnings are those issued by the US Department of State, Foreign Affairs and International Trade Canada, and the UK's Foreign and Commonwealth Office.

In the EU the European Consumers' Organization (BEUC) is partly funded by the European Union and has been very active in promoting consumers' interests to various European bodies and to the tourism industry since they were founded in 1962. Most of their actions are warranted, as they represent a voice that is often missing. For example, in May 2008 the European Commission released a report on the purchase of airline tickets online, which found that there were serious issues with respect to consumers being misled or confused. The Commission found breaches in 137 out of the 386 websites checked in September 2007 in 13 EU states (Kuneva 2008). The main issues identified were misleading prices – found on 58 per cent of the checked websites; unfair contract terms, such as missing or wrong language versions; content of the small print or pre-checked boxes for optional services – found on 49 per cent of the sites; and problems with non-availability of advertised offers on 15 per cent of sites (Kuneva 2008). The websites represent 80 airline companies but they could not be named due to legal restraints in most EU member states that do not permit the publication of the names of companies under investigation. In response to the Commission's findings the BEUC said in a press release that the names of all the investigated companies in all member states should be made known. 'What is really missing now for European consumers is the names of the companies which are not playing by the rules. It is only armed with this information that consumers will be able to fully play their role in the market by turning their backs on these websites!' (BEUC 2008).

Although there are a number of environmental and social issues groups that arise in response to particular tourism developments, there are very few public interest groups with a broad interest in tourism. One of the most notable of such groups is Tourism Concern, a UK charity that is independent and non-industry based. The organization has a membership of over 900 people and works in over 20 destination countries. Tourism Concern's vision is for 'A world free from exploitation in which all parties involved in tourism benefit equally and in which relationships between industry, tourists and host communities are based on trust and respect' (Tourism Concern 2008: 5). Tourism Concern has been involved in numerous social welfare campaigns, including support for tourist boycotts of Myanmar and promoting improved working conditions for employees in the tourism industry in developing countries. According to Tourism Concern (2008: 8), Tourism Concern's work is grounded in a rights-based approach to development.

Tourism generates positive benefits as well as negative effects. Communities who have greater control within tourism development are able to direct development according to their priorities. In a rights-based approach:

- Local communities must have the right to participate in the decision making about tourism development where they live.
- Tourism industry operators and governments must be accountable to the people whose land and cultures are being utilised for the benefit of the tourists and tourism industry operators.
- Strategies must be prioritised to empower people to be better able to have a say in the development of their communities and country and the capacity to shape tourism development for equitable benefit.
- Attention must be given to marginalised and vulnerable groups such as women, children, minorities, illegal workers and indigenous people working or affected by the tourism industry.

Case 6.4: Sustainability trumped in Aberdeenshire

Sustainable development and sustainable land management are important values driving policies for public agencies around the world. What happens, though, when the promise of jobs and economic development conflict with those of sustainability? While there are many case studies from less developed countries of moneyed interests overcoming local opposition through their influence at the national level, this situation is also common in the developed world.

In November 2008, American developer Donald Trump was given planning approval to build a £1 billion golf project in Balmedie, Aberdeenshire, Scotland, by the Scottish government. The tourism project was regarded as a major economic boost to the region, which is gradually moving away from oil as its economic base. The region's economic development agency Aberdeen City and Shire Economic Future (ACSEF) stated locally that there was over £1.4bn in potential developments, with £600m of investment already planned. The development manager, Rita Stephen, said it secured future financial growth for the area and signalled 'a new era for the region that will secure our long term vision for economic growth, diversifying the economy and enhancing our quality of life. Aberdeen city and shire must now be one of the most interesting and dynamic city regions in the UK with exciting prospects for residents, visitors and those wishing to relocate here' (BBC News 2008d). However, the project met substantial opposition, highlighting some of the difficulties of effectively integrating tourism economic development objectives with other planning goals.

The Trump plan to build two championship golf courses, a five-star hotel, 950 holiday homes, 36 golf villas and 500 private homes in the coastal area, which first came to public attention in January 2006 (BBC News 2006) was originally rejected by the Aberdeenshire Council infrastructure committee in November 2007 at which point the Trump Organization threatened to take the proposed development outside of Scotland. A spokeswoman for Mr Trump told BBC Scotland: 'We are surprised by the decision, it would have been a great development.. . . We are considering an appeal, and also considering doing something very spectacular in another location. Sadly, it will not be in Scotland.' The Trump Organization's head of international development, George Sorial, said: 'Obviously we are very disappointed. It is our position that the council has failed to adequately represent the voice and opinion of the people of Aberdeen and the Shire who are ultimately the losers here . . . I think we have been very frank all along – we do have options elsewhere in the UK and we will sit down now and look at that' (BBC News 2007c).

Opinions for and against the project in the region clearly ran high, with one councillor who voted to reject the development claiming she was assaulted while other councillors sought to find means to overturn the decision. In early December, Martin Ford, the infrastructure committee's chairman who had used his casting vote to reject the application, stated that the resort plan was 'dead':

> There's no possibility that I can see that we can go back to re-discuss an application which has been dealt with. As far as I understand it, it's dead.. . . It was unacceptable for quite a lot of significant reasons. It broke a whole raft of planning policies in relation to environmental protection, housing in the countryside and it had wider implications in terms of the council's whole approach to biodiversity conservation and the environment.
>
> (BBC News 2007d)

However, within two weeks Martin Ford had lost his position as chairman of the infrastructure committee and the Scottish government decided to intervene to decide whether or not the development should go ahead, in light of the Trump Organization's meeting with the First Minister of Northern Ireland to discuss a golf course development there (BBC News 2007e). A Scottish government statement, issued on 4 December, stated: 'Ministers recognise that the application raises issues of importance that require consideration at a national level. Calling the application in allows ministers the opportunity to give full scrutiny to all aspects of this proposal before reaching a final decision'. A decision agreed with by the head of the Aberdeenshire Council, Anne Robertson, who stated, 'What is important in all this is securing the economic future of the north-east of Scotland. The Scottish Government quite rightly feels this application raises issues of such importance that they require scrutiny at a national level. If the decision of ministers to call this application in keeps it alive, then we welcome this intervention' (BBC News 2007e).

The Scottish government inquiry commenced in June 2008 with the golf course proposal receiving significant support from business and tourism interests. In contrast conservationists and local campaigners opposed it in great part on the basis of its environmental impacts, particularly given that part of the course would be built on sand dunes which are part of a Site of Special Scientific Interest (SSSI), and housing concerns (BBC News 2008e). In November it was announced that the plan had been approved, with Scottish First Minister Alex Salmond, who is also the Member of Scottish Parliament (MSP) for Gordon in Aberdeenshire, citing 6,000 possible jobs as a result of the development. Other favourable comments came from Aberdeenshire Council leader, Anne Robertson: 'I truly believe a development of this type will bring significant benefits to the area, particularly in terms of jobs and tourism' and Ken Massie, VisitScotland's regional director for Aberdeen and Grampian, saying 'We welcome the decision as the development will make a significant impact on the economy'. In contrast, opponents were dismayed; Green MSP Patrick Harvie stated:

> The fact that this project has been approved in its entirety demonstrates a blatant disregard for the legal protections these dunes are under. We can only hope that the current economic crisis tears a hole in his business plan, and that he fails to get the money he needs to fund the project. Even the credit crunch must surely have a silver lining.
>
> (BBC 2008d)

Aedan Smith, head of planning and development at the Royal Society for the Protection of Birds (RSPB) Scotland, one of Scotland's largest environmental groups said: 'The development will cause the destruction of a dune system, with its precious wildlife, on a site which is protected by law and should continue to be available for future generations to enjoy. We, and the thousands of other objectors, consider that this is too high a price to pay for the claimed economic benefits from this development.'

Such statements clearly illustrate the difficulties of integrating different values and perspectives on tourism-related land use and finding a 'balanced approach' that provides for sustainable development. However, the elements of this case study in terms of the difficulties in finding 'win–win' situations are repeated all over the world.

Source: Trump International Golf Links Aberdeen Scotland:
http://www.trumpgolfscotland.com/intro.asp

Community and enterprise: planning and management models

The Recreation Opportunity Spectrum (ROS)

In large protected areas with many different natural and cultural resources, it is often necessary to plan for different management objectives in different parts of the administrative area to meet a range of conservation and recreation mandates. The *Recreation Opportunity Spectrum* (ROS) is the most widely used method for resource planning and visitor management in large protected areas. The ROS approach seeks to match the recreation interests and behaviour of the public to the most appropriate places in the protected area, and to keep inappropriate behaviour away from more environmentally sensitive areas. The major areas of assessment in undertaking a traditional ROS management plan include (More et al. 2003):

- *The Physical Setting*: different areas and sites are scaled from primitive undeveloped areas to highly developed urban areas. The Physical Setting also includes assessments of:
  - *Remoteness and accessibility*: different areas are scaled from far from roads to easily accessible.
  - *Evidence of human settlement*: different areas are scaled from no human structures to many buildings.
- *The Social Setting*: different areas are scaled from very isolated with almost no other people present to large numbers of people present. The Social Setting also includes assessments of:
  - *Activities available*: different areas are scaled from remote hiking to well-organized and intensive recreation.
  - *Experiences available*: different sites are scaled from physically challenging to easy activities.
- *The Managerial Setting*: different areas are scaled from no administrative signage to clearly posted rules.

Each of these areas of assessment has a range of categories and criteria for evaluating conditions in the areas under study. For protected areas the entire administration territory should be assessed and rated. *Geographic information systems* (GIS), which include databases associated with digital maps, have made this process more flexible and precise, and allows for the generation of alternative planning scenarios.

In the traditional ROS approach, the result of this analysis would be mapped showing the protected area and the management objectives for different parts of that area. The primary recreation opportunity classes (also known as management categories or zones) are:

- *Primitive management areas* These areas are unmodified natural environments of fairly large size where interactions between recreation users is very limited and the managerial setting is free from evidence of management restrictions and controls.
- *Semi-primitive motorized and non-motorized* These areas are mostly natural or natural-appearing environments where interaction between users is low and management controls and restrictions are present, but are minimized.
- *Semi-developed natural (or roaded)* These areas are predominantly natural-appearing environments where human evidence is in harmony with the

environment and interaction between users is low to n
modifications are in harmony with the environment and m
is permitted for maintenance purposes only.
*   *Developed natural (or rural)* The natural environment
    substantially modified by humans, with sights and sound:
    readily evident, and where interaction between people is
    high.
*   *Highly developed (or urban)* These are areas that are urbaniz
    though they may be in natural settings, where much modification has been
    undertaken to enhance recreation activities. Non-native vegetation, often
    maintained by humans, is common and the sights and sounds of humans are
    widespread with much interaction.

After an assessment is completed, each zone is assessed for inconsistencies and
conflicts that may require active intervention and mitigation measures, instead of
just management maintenance.

The original ROS approach was developed in the 1970s for the US Forest
Service to assist in the management of its vast land areas in the western United
States. The traditional areas of assessment and resulting categories clearly reflect
this history. However, the basic approach has been modified for use in different
environments and under different types of land ownership. The version above was
actually modified for use in the New England states, for example. Other
modifications have been suggested specifically for tourism and for ecotourism.

The *Tourism Opportunity Spectrum* (TOS) for adventure travel destinations,
suggested by Butler and Waldbrook (2003), assessed factors of: (1) access, (2)
non-adventure activities, (3) the tourism plant (or infrastructure), (4) social
interaction, and the tourist's degree of acceptability of (5) visitor impacts and (6)
regimentation or management control. The resulting categories include:

*   hard adventure
*   medium adventure, and
*   soft adventure.

The *Ecotourism Opportunity Spectrum* (EOS) (Boyd and Butler 1996) assesses
a similar range of variables and identifies areas for Eco-Specialists, Intermediate
Ecotourists, and Eco-Generalists. Neither of these proposed methodologies,
however, has been widely adopted in tourism, in part because so much of tourism
is based in more developed areas where ROS-type approaches are harder to apply,
although some of their ideas can be readily incorporated into ROS plans for nature-
based tourism areas.

The ROS approach, along with its modifications to adapt it to tourism environments, assumes that there is a level of visitor or recreation use beyond which the
impacts on the environment or society is not acceptable. This implies that it is
possible to determine an environmental or social *carrying capacity*. However,
many years of efforts to develop and implement the carrying capacity concept has

Photo 6.4 The Stone Forest (*Shilin*) National Park, near Kunming, China. This is the largest area of this distinctive limestone rock formation, which is also found in smaller sizes elsewhere in this part of Yunnan Province. The park is also designated as a 'geopark', which is a UNESCO designation that is primarily based on geologic features, but also includes sustainable social and economic development principles. This part of the park, which receives the most visitors, is hardened with designated pathways and landscaping. More remote areas have less of these forms of site hardening. (Photo by Alan A. Lew)

found there is little evidence that changes to the level of use has any direct relationship to impact measurements (McCool 1994; Sayre 2008). Part of the reason for this is that the impacts that occur at any given use level are highly dependent on the degree of site hardening, visitor management (over time and space) and visitor education, as well as in changing expectations of what an acceptable impact might be. More recently, however, the concept of environmental carrying capacity has formed the basis for ideas related to the *ecological footprint* (EF) of resource use, which is also a measure of the intrinsic sustainability of a given area and level of consumption. EF analysis applies the ideas of carrying capacity, but focuses on how large an area is required to support a community, firm or region, given certain assumptions about biological productivity and consumption patterns, rather than asking how many individuals a given area can support. Therefore, the ecological footprint of a specified population is the area of land and water ecosystems required on a continuous basis to produce the resources that the population consumes, and to assimilate the wastes that the population produces. The EF is measured using space equivalents, such as hectares of land or water surface, which translates into the size of one's ecological footprint. In simple terms, EF produces a net ecological budget, expressed in areal terms, by dividing human consumption (demand) by ecosystem productivity (supply) (Hall 2008c). Environmental degradation occurs when consumption/demand outstrips ecosystem productivity/supply.

In a study of tourism to the Seychelles, Gössling et al. (2002) found that the average tourist's holiday required more than 1.8 ha of world average space to maintain the necessary resource flows and off-set greenhouse gas emissions. The combined ecological footprint of all leisure travel to the islands was over 2,120 km². This can be compared to the total land area of the Seychelles of 455 km². It was also found that an average ten-day holiday in the Seychelles corresponded to 17 to 37 per cent of the annual EF of a citizen of a country in the developed world. 'A single journey to the Seychelles thus required almost the same area as available per human being on a global scale' (Gössling et al. 2002: 206). One approach toward obtaining a broader view of alternative management scenarios for natural protected areas is the use of the Environmental Impact Assessment (EIA) (see also Chapter 2). Like the ROS, the EIA approach has been widely adopted as a planning tool, and even required by most countries around the world. In addition, it has been used to mitigate development impacts in both remote and urban environments. The purpose of the EIA is to identify potential positive and negative impacts of a large development on the environment. The EU requires that the following seven areas be included in an EIA (Watson 2003):

1   Description of the project, including its site characteristics and its environmental impacts during construction, during operation, and through and after decommissioning.
2   Alternatives that have been considered to the proposed project.
3   Description of the environment, including a list of those aspects that may be affected by the development, such as fauna, flora, air, soil, water, humans, landscape and cultural heritage.
4   Description of the significant effects on the environmental aspects listed in step 3.
5   Mitigation of significant impacts identified in step 4.
6   Non-technical summary of the EIA written for the lay-person that excludes jargon and complex diagrams.

7   A discussion of technical challenges and weaknesses in the knowledge that the EIA was based on, with suggestions on areas for further research and development.

The approaches for addressing development impacts and managing resources discussed so far (ROS, TOS, carrying capacity, EF and EIA) primarily rely on scientific methodologies and expert assessments. This is because they were originally developed for managing natural (ie. non-human) ecosystems. When the human component is added to the mix, the issues and their solutions become much more complicated and political. Other approaches to visitor impact management have been developed to address the social and economic issues that arise in the more complicated planning environments. These include Limits of Acceptable Change (LAC), Visitor Impact Management (VIM), Visitor Experience and Resource Protection (VERP), Visitor Activity Management Process (VAMP), and the Tourism Optimization Management Model (TOMM).

All these models are variations of the traditional *rational planning* approach, as developed by urban and regional planners (Hudson 1979). Other terms for this approach are *synoptic planning* and *comprehensive planning,* though the latter also has some additional connotations. The actual steps in the rational planning process are described in different ways by different authors, though they can be generalized as:

1   *Identify a problem or an issue* Problems may be brought forward by the public, including special interest groups, by politicians, or identified by an administrative agency through their managers, staff or committees. For tourism, this may be related to the development of a specific project, such as a major convention centre, or the revitalization of a heritage retail district, or the need to manage tourism that has evolved in a previously ad hoc (unplanned) manner.

2   *Conduct background studies and collect data* Background studies are conducted to better understand the problem or issue. They are usually undertaken by planning staff or consultants, and may involve discussions with different stakeholders. For tourism, this may involve a citizen participation process (discussed below), economic studies to understand the role of tourism in a community, and architectural or environmental audits to assess resource conditions.

3   *Identify a goal or goals to solve the problem(s) or address the issue(s)* This is usually done by planning staff in consultation with stakeholders and political leaders. Typical tourism-related policy goals include:
    • *Social goals*: education; cross-cultural understanding; increase opportunities for recreation; and increase community identity and pride.
    • *Economic goals*: maximize local income, employment and government revenue; and support, through tourism, for other economic sectors, such as agriculture and fisheries.

- *Environmental goals:* protect natural resources; and minimize resource use and waste.
- *Cultural goals*: protect historical resources; and maintain cultural traditions.
- A *visioning process* may be included in the goal formulation. Visioning uses an open-ended approach that encompasses all possible future scenarios (see Chapter 7) to identify the broad desires for a community, business or other group. Goals are then formed to narrow down the broader vision to more specific and achievable objectives.

4 *Identify objective criteria for assessing alternative plan scenarios* To provide an unbiased assessment, it is important to set the evaluation criteria before creating the scenarios in step 5. If this is done after the scenarios are developed, then they will be influenced by personal preferences for one scenario over another. For tourism, specific criteria would reflect the social, economic, environmental or cultural goals identified in the previous step.

5 *Develop alternative plan scenarios, including policies and guidelines, to achieve the goals* Usually a minimum of three alternatives is developed, including more intensive and less intensive levels of change. There might also be a no-change scenario, which is often required for an EIA. Each alternative comes with its own set of policies to achieve the goals defined in step 3. The policies are what make the scenarios different from one another.

6 *Assess alternative plan scenarios using the predefined assessment criteria* This typically involves presenting the alternatives to political decision-makers, stakeholder groups and the general public, as well as to specialists. Comments and possible modifications are obtained through this process. Depending on the scale of the management plan, this can take from a couple of months to a year.

7 *Select the preferred plan from the alternative scenarios, typically following a public comment period* Because this is a political process, the final decision is typically done by either elected officials or political appointees. Sometimes the final alternative may be put to a public vote process, though that is rare.

8 *Implement the plan* The selected plan becomes the official policy statement for the government or administrative agency. It guides the creation of administrative rules, budget allocations, decision-making and physical (construction) development.

9 *Monitor, evaluate and revise the plan and its implementation* This is normally done about every three to five years.

10 *Identify new problems and begin the process again* Return to step 1 for problem identification.

The rational planning process, as it is described here, is for large, often community-wide, planning issues and development projects. A *comprehensive planning* process would look at several different issues (such as open space, housing and economic development) and apply the approach to each of these areas. Alternatively, it can be scaled down to smaller sites and issues, though the goal of objective analysis should be maintained throughout.

The result of this process is what is called a *policy plan*, as it is based on the policies that are adopted in step 7. The goals are established early on, and specific rules and regulations to implement the policies may change over time. The policies, however, are the most important element that comes out of the planning process and it guides later decision-making. Examples of tourism policies include:

- *Market policies*: focus on a mass tourism market; or focus on speciality and niche markets (such as upscale tourists or ecotourists).
- *Resource policies*: develop additional outdoor recreation resources (such as water sports or hunting); or develop new facility-based recreation (such as team sports or casino gaming); or adopt regulations to better protect cultural resources.
- *Tourism development policies*: define and enforce a preferred scale, location or concentration of development; and implement a preferred growth rate.
- *Investment, education and employment policies*: identify roles for government, private sector and educational institutions.

The *Limits of Acceptable Change* (LAC) process was essentially developed by applying the rational planning process to US National Forest lands. The steps are virtually identical to those of rational planning, but differ significantly from the ROS approach in that the focus is on the *preferred conditions*, which are achieved through the establishment of goals. The acceptable change components are reflected in the monitoring process, where, once the preferred goals are achieved, management practices focus on maintaining the restored conditions. The *Visitor Activity Management Process* (VAMP) model was developed by Parks Canada and is similar to LAC except that it includes a larger public participation, service and experience component (Eagles et al. 2002). The *Visitor Experience and Resource Protection* (VERP) approach, developed for the US National Park Service (US NPS), focuses more on the relationship between the protected area and the larger community, including private sector businesses, in which the protected area exists.

The *Visitor Impact Management* (VIM) process was also developed for the US NPS. Again, this process is very similar to the rational planning model and the LAC processes up to the development of alternative scenarios. Instead, the focus of the VIM is on identifying scientific indicators of change caused by visitors (the US NPS prefers the term 'visitors' instead of 'tourists'). These indicators are then compared to existing actual conditions, and management strategies are identified to achieve management goals. Monitoring these indicators also identifies areas for intervention as they arise after the plan has been implemented. And finally, the *Tourism Optimization Management Model* (TOMM) approach is similar to the VIM model, except that it is intended more for human communities. It establishes ranges of acceptability (instead of absolute limits) and applies different levels of intervention depending on the degree to which these ranges are contravened (Hall and McArthur 1998; Mendelovici 2008). Tourism indicators are divided into socio-cultural, environmental, economic and experiential. Table 6.3 shows the

Table 6.3 Tourism Optimization Management Model indicators for ?
Australia

Socio-cultural indicators:

1 *Influence tourism*: percentage increase in residents stating that t
tourism-related decisions
2 *Access to areas*: proportion of residents that feel they are able
recreational opportunities that are not frequented by tourists
3 *Local employment growth*: percentage increase in the proportion of r...
derive all or some of their income from tourism
• Secondary data (collected by other agencies)
  • Number of petty crime reports committed by non-residents per annum
  • Proportion of the community who perceive positive benefits from their
    interactions with tourists
  • Number of traffic accidents involving non-residents per annum

Environmental indicators:

1 *Hooded plovers*: number of hooded plover pairs (a wildflower that tourists come to
see)
2 *Waste management*: percentage of total waste diverted from landfill on Kangaroo
Island
• Secondary data (collected by other agencies)
  • The number of seals at a designated tourist site
  • The number of osprey at a designated tourist site
• Potential data (not currently collected)
  • The proportion of visitors to the Island who visit natural areas zoned specifically
    for managing visitors
  • Net overall cover of native vegetation at specific sites
  • Energy consumption / visitor night / visitor
  • Water consumption / visitor night / visitor

Economic indicators:

1 *Average nights*: annual average number of nights stayed on Kangaroo Island
2 *Residents income*: percentage of residents deriving most of their income from
tourism
• Potential data (not currently collected)
  • Growth in tourism yield
  • Seasonal fluctuation in the number of visitors

Experiential indicators:

1 *Experience with wildlife*: proportion of visitors who believe they had an intimate
experience with wildlife in a natural area
2 *Match to marketing*: proportion of visitors who believe their experience was similar
to that suggested in advertisements and brochures
3 *Overall satisfaction*: proportion of visitors who were very satisfied with their overall
visit

Sources: based on Hall and McArthur 1998; Mendelovici 2008.

ators developed using the TOMM approach on Kangaroo Island in South
stralia.

Environmental Management System

The *Environmental Management System* (EMS) approach differs from those
discussed above in that the emphasis is on reducing and minimizing the environmental impacts of an organization or enterprise. There are four basic steps in
the EMS approach: planning, implementation, monitoring and reviewing (USEPA
2008).

• The planning step involves identifying the organization's environmental
   relationships, which often includes estimating a carbon or environmental
   footprint, and setting organizational goals and a timeline for the EMS.
   Designating an appropriate planning team and leadership is also part of this
   step, as is establishing a process of communication across the organization.
• The implementation step includes identifying aspects of the organization that
   have significant environmental impacts, such as emissions, waste, use of
   raw materials and fossil fuel use. The legal requirements for these must be
   understood, as well as the views of different stakeholders toward them, both
   within and outside of the organization. Goals for different areas within the
   organization, and measures and procedures to implement and achieve those
   goals, are the final part of the implementation step.
• The monitoring phase consists of an internal audit process to assess the
   effectiveness of the implementation procedures. Auditors must be trained and
   the whole auditing process will evolve over time. No plan is perfect and the
   audits should be designed to identify shortcomings so these can be addressed
   in a larger, less periodic, review process.
• The review process should be undertaken when enough knowledge is gained
   from the internal audits to identify the major shortcomings of the original
   implementation plan so that these can be addressed. Ideally the whole process
   should be cyclical, with the review leading to a new plan, implementation and
   audit procedures.
• The EMS approach is probably not suitable for a community or regional
   planning process, but has many other potential applications from individual
   businesses of all sizes, to government agencies (including NTOs), land
   developers, non-profit attractions (museums, historic sites), educational
   institutions and cruise ships. In addition, an EMS approach can be expanded
   beyond environmental impacts to incorporate social and cultural issues. Such
   a planning process could be an important building block for larger community
   and regional policies to address tourism and sustainable development.

Less than rational planning

These various planning models show the importance of human factors in assessing and managing natural, recreational and tourism resources. However, the basic rational planning approach does not always effectively incorporate this human component. Even more scientifically objective planning models that focus on non-human resources require that some set of goals be established, which are expressions of human values, and include political and economic trade-offs. Criticisms of the rational planning approach, which apply to all the visitor impact management models described above, include:

1 It assumes predictable and rational human behaviour and environmental conditions, when human behaviour is often political and irrational, and environmental changes often occur in unpredictable ways.
2 It assumes unlimited knowledge (synoptic) and problem-solving capabilities on the part of the planner, including the ability to identify all alternative scenarios (for approaches that build scenarios), whereas data collection is often limited by time and cost, resulting in recommendations being based on less-than-perfect knowledge.
3 It assumes scientifically valid objective variables and facts in a closed system, while the very act of selecting indicators to measure reflects value judgements and crisis situations and unpredictable opportunities are likely to arise from unforeseeable sources, such as climate changes, political changes and new technologies.

In reality, planners and managers, whether they work for governments, NGOs or business interests, almost always face situations of limited knowledge, limited time horizons, limited budgets and unpredictable influences. As an alternative to rational planning, an *incremental planning* approach intentionally assumes these limitations. Goals are still important, but the political processes and community values are made more explicit, and methods are kept open and flexible to meet changing circumstances (Lindblom 2003). The overall process of incremental planning is similar to rational planning and it still assumes an ability to adequately explain, predict and control, though within limitations. In its implementation, however, policies are allowed to be more flexible with less stringent time lines (see also Bendor 1995).

In general, rational planning models are more appropriate for *tame* problems (Rittel and Webber 1973), including the construction of non-controversial buildings and infrastructure that can be addressed by engineering solutions. Incremental approaches are more appropriate for *wicked* or *messy* problems (as discussed previously), which are frequently social and characterized by significant changes, where the problem is hard to formulate or define. Wicked problems may also be difficult to quantify, and have no 'right or wrong' policy choices, but rather 'better or worse' ones. Other characteristics of wicked problems are those with no opportunity for trial and error, those with few established methods or rules, and those where there is uncertainty as to when a problem is solved. The degree of necessary incrementalism will vary considerably from one project or community to the next.

A *transactive planning* approach may be more appropriate than assuming rationality but acting incrementally. Transactive planning (also known as *post-rational planning*) is based on *social learning theory* and transactive (inter-personal) interaction (Healy 2003). With this approach, the building of trusted relationships is more important than specific planning steps, which makes it more like an incremental approach. The ultimate goal is to create self-learning (or intelligent) institutions that will be self-sustaining after the planning process has ended. This typically results in the creation of community-based development and oversight organizations. Tourism examples of this include Australia's ecotourism certifications programmes, and community-based tourism groups in Costa Rica (Lew 2007b) (see also Case 6.5). Transactive approaches, however, are highly time-consuming and require long-term commitments by all stakeholders.

Advocacy and sustainability

As important and functional as all these planning and management models are, they do not directly address what may be the most important aspect of the planning process: goal formulation. When a goal is recognized and named, it beomes an expression of the values of a community. It expresses how the community sees itself now, and what it wants to become in the future. The voting and election process (from local to national levels) is one way that community-wide values are identified. Elected politicians, and political appointees, set many of the goals for a community. The models above describe how those goals are turned into policies for implementation. However, there are situations where it is important for planners to be directly involved in the goal decision-making process.

The *advocacy planning* approach evolved from social movements of the 1960s and 1970s (Davidoff 2003). The role of the community advocate is to represent the interests of the weak (the poor, the disenfranchised and environmental causes) against the strong (often, but not always, the business community). Community advocates use their technical knowledge to empower under-represented communities to have their voices and values heard. *Environmental activism* is another form of community advocacy. *Sustainable development* is also a form of advocacy planning, though typically less confrontational. The advocacy approach has been criticized for inhibiting efficient planning, though it can bring social equity and environmental sustainability issues to the forefront of the planning process.

Although some would deny it, sustainable development approaches to tourism have broadly predefined goals that make it more of an advocacy approach than a purely rational one. According to the World Commission on Environment and Development (WCED 1987: 43; also known as the Brundtland Report) sustainable development is development which 'meets the needs of the present without compromising the ability of future generations to meet their own needs'. As was discussed in Chapter 5, it includes five basic principles or goals: (1) that planning should be holistic (people and environment) and strategic (long term); (2) that essential ecological processes should be preserved; (3) that cultural heritage and biodiversity should be protected; (4) that forms of production should be sustainable for future generations, and (5) that resource use should ensure fairness between nations.

The concept of *sustainable tourism* applies the sustainable development approach to tourism, with goals of balancing environment, community and economic values (which are sometimes referred to as the *Triple Bottom Line*). In practice, this means advocating for the environment and the community against exploitative business interests, which are often supported by government officials. For tourism, this has been especially so in relation to indigenous populations and sensitive natural environments in the less developed economies of the world. However, the concept of sustainable tourism has also been co-opted by the tourism business interests who rarely address, let alone advocate for, the core principles defined by the WCED's Brundtland Report.

Over three decades of trying to apply sustainable tourism development approaches have resulted in a number lessons. Among these are (Lew and Hall 1998):

- *Sustainable tourism is a value orientation in which the management of tourism's social and environmental impacts takes precedence over market economics.* Business profits are still important, and some environmental and social sacrifice is also necessary. But to ensure long-term sustainability, the balance must always tip in favour of under-represented social and environmental interests.

- *Implementing sustainable tourism development requires measures that are both scale and context specific.* Different implementation methods will be required depending on the scale of the issues (from global to site) and the environment in which they occur (social, political and physical). The range of strategies discussed in this chapter provides some guidelines to these scale and context issues.

- *Sustainable tourism issues are shaped by global economic restructuring and are fundamentally different between developing and developed economies.* In some less developed countries, poorly paid government officials are more highly susceptible to graft and corruption. This highly skews the power relationship in favour of business and global economic interests and against environmental and community interests. Rational planning approaches can be a waste of time if decision-making is based more on money influences than objective analysis. This is another reason why an advocacy approach is often more suited to ensuring adherence to sustainability.

- *At the community scale, sustainable tourism requires local control of resources.* Numerous research studies of the impacts of tourism on populations that are disenfranchised from the land that they occupy demonstrate the importance of this lesson to ensuring sustainable tourism development (Hall and Lew 1998).

- *Sustainable tourism development requires patience, diligence, and a long-term commitment*. It cannot be achieved by hiring an outside consultant to fly in, write and present a report, and fly out. The transactive planning model, as discussed above, with its creation of intelligent institutions, is more appropriate to ensuring a sustainable approach to tourism development.

Public participation and politics

As the transactive and advocacy planning approaches, as well as sustainable tourism, make evident, involving local citizens and residents in the planning process is essential for successful and sustainable approaches to community tourism development. An effective public participation process will improve the decision-making process by helping to ensure that all the crucial issues related to tourism development or management are addressed. In particular, public (or citizen) participation can also generate new ideas, identify public attitudes and opinions, and provide a sounding board for reviewing alternative proposals. Ideally, public participation techniques can help to build community consensus by disseminating information so people feel properly informed about changes that are proposed. The process can identify and ideally resolve conflicts, serve as a safety valve for pent-up emotions, and develop support, or minimize opposition, to a preferred alternative.

In developing a public participation process, it is important to identify and understand the diversity of stakeholders in the community. Stakeholders are those individuals and groups who have a special interest in tourism development in general, or in a specific project. These may include residents in the vicinity of a proposed project, landowners, business interests, non-governmental or non-profit organizations, government agencies at different levels, and possibly educational institutions. Sometimes there is a significant difference between long-term residents and newcomers, and there may be interest groups that are external to the community, including outside investors and frequent tourists, who have a stake in the planning outcomes.

There are three basic techniques used in public participation: committees, surveys and meetings or workshops. Citizen *advisory committees* or review boards are usually ad hoc (informal) and temporary, and are particularly well suited to situations where individuals who are considered community leaders on a particular issue can be clearly identified. They receive some agency staff support and only have advisory authority to policy-makers. Attitude and *opinion surveys*, conducted either in person or by mail, are especially suited to gain a broader opinion from across the community. Mailed surveys can also be an effective part of a public information programme, as educational material can be included with the survey instrument.

Meetings and workshops are very effective at educating the public, obtaining public input and addressing citizen concerns. They address a larger and broader population than an advisory committee might and can be more informative than a survey. They can take several forms including citizen training sessions, mini-conferences, collaborative design-ins, focus group interviews, design charrettes and neighbourhood meetings. The selection of a public participation approach or approaches depends on the goals for undertaking the process, the stakeholders that are to be involved in it, resources available to conduct a citizen process, and the level of existing citizen education and preparation on a topic.

An important factor in a public participation process is that planners, managers and decision-makers are open to changes and modifications of their personal preferred plans. The potential effectiveness of the process to educate both the public and the managers, of course, assumes that sufficient community support exists and that political or administrative leaders are not forcing a highly unpopular project or regulation on a community.

The political decision-making process is an expression of political power. When close relationships exist between government leaders and the business community, this can reduce the political power of the broader citizenry. Under-represented populations, such as the poor, can be especially disenfranchised in this process. Corruption and graft in some settings is especially challenging, and typically results in poor decision-making and the long-term destruction of public resources and trust, in exchange for short-term personal gain. Hall and Coles (2008b: 277) have argued, 'If (sustainable) tourism for poverty alleviation is such a key concern, critical issues such as corruption cannot be conveniently overlooked because of their ability to take value out of the international value chain and to contribute the sorts of unacceptably "challenging externalities" that have dissuaded major investors such as the World Bank from supporting tourism projects (Hawkins and Mann 2007)'. Furthermore, tourism consultants (often from overseas), though treated as experts, are biased toward the status quo system of corruption because they are almost always funded by government interests and, at a minimum, legitimize the planning and approval processes. Thus, tourism policy that lacks real public support will empower some stakeholders (usually higher-end business interests) and contain and disenfranchise others.

Case 6.5: Community development agreements

For most development projects it is the private developer who is responsible for conducting the public participation process related to their development. In part, this is because local governments do not have the personnel and financial resources to undertake such an effort. One possible outcome of a public participation process is a *Community Benefits Agreement* (CBA) (Gross 2005). A CBA is a private contract between a developer and the community in which a development project is proposed. The agreement stipulates the benefits that the developer guarantees to provide to the community, including hiring local applicants first, providing living wage pay-scales and salaries, and developing affordable housing.

The city of Los Angeles has experienced a couple of successful CBAs connected to tourism-related developments. In 1998 the Hollywood Highland Center, which is now where the Motion Picture Academy Oscar ceremonies are held, was proposed. The project would include a 4,000-seat theatre, several hotels, parking lots and 111,500 meters² (1.2 million ft²) of retail space. Traffic congestion, environmental pollution and increased crime were the major concerns of the area's residents. With assistance from the Los Angeles Alliance for a New Economy, the conflicting parties came to an agreement in which the developer guaranteed to fund traffic improvements, living wage pay-scales for all employees of the Center, a union-neutral hiring policy, and a local-first hiring plan. The level of community support that this contract agreement generated enabled the developer to obtain approval for the development, along with favourable tax subsidies, from the Hollywood City Council.

A similar process, but one that involved over 30 local community groups, developed a CBA for the Los Angeles Staple Center, home to the Los Angeles Lakers basketball team, as well as other professional sports teams. In addition to the sports arena, the 2001 project included two hotels, an apartment complex, a convention centre, a 7,000-seat theatre and retail shopping. After benefits to the community never materialized following the first phase of the project, despite initial verbal agreements, a long nine-month period of negotiations with economic justice groups, environmental groups, church organizations, health care organizations, immigrant rights groups and tenant rights groups ensued, resulting in a written CBA contract. Some of the major stipulations of that agreement included:

- $1 million for the improvement of area parks and recreation facilities,
- $25,000 a year for the first five years to fund a parking permit system to limit parking in residential areas to residents only,
- To ensure that 70 per cent of the jobs created at the project provide a living wage pay-scale,
- To give priority for hiring to people displaced by the project and to low income individuals living within three miles of the project,
- To set aside 20 per cent of the residential units as affordable housing and to provide $650,000 in interest-free loans to a non-profit group to develop additional affordable housing in the community,
- To cooperate with the coalition of community groups to create an Advisory Committee to assist the CBA's implementation and enforcement.

While these examples of the CBA approach were for large development projects, size can be a relative issue. Much smaller tourism projects can have just as significant an impact in more sensitive locations, including coastal and mountain areas and areas with very traditional cultural values. The CBA approach can help to guarantee that the promises of positive impacts from a new development actually materialize for those who need them the most: the poor and the environment.

Community and enterprise: implementation approaches

In general, the goals of a tourism management plan will be to increase tourism's positive impacts and decrease tourism's negative impacts, or maintain current conditions if they are deemed optimal. Location-specific policies would be included in the management plan that would guide the preferred approaches to achieving those goals. There are basically two broad approaches to implement those policies: *supply* approaches and *demand* approaches. Supply approaches manage the tourism resources, which are primarily at the destination site or area. Demand approaches manage tourists to increase or decrease their numbers or change their behaviour in relation to travel and tourism.

Supply approaches to managing tourism impacts include:

* increasing or decreasing the number of attractions and supporting facilities,
* increasing or reducing accessibility to sites and facilities,
* hardening, softening, protecting or otherwise modifying attractions sites, and
* spatially concentrating or dispersing sites and facilities.

Demand approaches to managing tourism include:

* marketing enhancements and educational campaigns,
* human resource and service development, and
* behavioural guidelines and regulations.

Supply: increase or decrease the number of sites and support facilities. Without mitigating measures, too many visitors to a popular site can overwhelm its physical capabilities. On the other hand, too few visitors can also lead to product decline from inadequate funding. Public agencies, non-governmental or non-profit organizations and private sector businesses can create new attraction and recreation sites if sufficient demand, funding or profit incentives are shown to exist.

One way that local governments can address these issues is though special modification to local land use regulations (Hall 2008c). Tax incentives, such as a tax reduction or a full tax holiday, on local property taxes and government-sponsored low-interest loans can provide financial incentives to either upgrade or build new attractions, hotels and convention centres, and other tourism-supported facilities and businesses. Larger development projects, such as a major convention centre, can require more significant action on the part of local governments, including *urban renewal* (see Chapter 2), which has the government buying out landowners to consolidate their properties into a single large parcel, which is then sold or leased to a developer. Private developers can also take this approach, though it is usually too costly, especially if an individual owner refuses to sell. Governments can also use their eminent domain powers to force a sale, which private developers cannot do. This approach is often employed when governments need to develop large infrastructure projects, such as those associated with airport development and road building.

Local governments also control urban growth through their *capital improvements programme* and budget process. This is the community's planning to allocate funds to build new roads, water pipes and sewerage works in undeveloped areas. These usually involve a multi-year planning process and can be used to influence where and what developments take place in the near term versus the longer term.

There may be instances where the preferred approach is to remove tourist facilities that are causing unacceptable impacts that cannot be mitigated in any other way. This is rare and would probably require some form of government intervention. Restroom facilities, for example, typically deteriorate over time leading to user complaints and environmental hazards. Such a facility could be closed to discourage human use of a site. It could also be moved to a less environmentally or socially sensitive location.

Entrepreneurial education and support programmes can be established to encourage more business activities that will grow the local tourism economy. These programmes can be spearheaded by either public or private sector interest groups, and may be conducted in cooperation with a community college or other adult education institution. Such programmes can be directed to specific types of businesses or niche markets that support the destination's image and needs.

Supply: increase or reduce tourist accessibility to sites and facilities. Tourists are more likely to visit sites that are easily accessible than those that are difficult to access. Roads are the major infrastructure that can be changed to help encourage or discourage access to specific sites. The provision or denial of utilities (especially electricity and water) and other services, such as public transportation, can also influence the number, size, quality and visitation rates of tourist attractions and facilities. Roads can also be downgraded to reduce the numbers of visitors by limiting access to only those with all-wheel drive vehicles, for example. Increasing the level of public safety at a site, such as increasing police, lifeguards or other security personnel, can increase tourism by making the place more comfortable for larger numbers of tourists.

Posting signs that prohibit access outright to some areas and explicitly encourage movement in other directions are design modifications that can be used to influence tourist behaviour, sometimes described as a demarketing strategy (Beeton and Benfield 2002) (see also Case 6.7). This can be done overtly, through properly placed signs, or covertly (Lew 1999). Covert structural designs encourage people to take certain paths, though they cannot absolutely prevent them from wandering from those paths. Clearly marked trails or paved pathways, for example, encourage people to follow them. Barriers of different types need not be overly obtrusive, while still discouraging movement in those directions. Brighter (well-lit), open spaces attract people, while dark and hidden directions discourage people. Consistent colours and symbols will also attract people, once they have become familiar with the coding system. Concentrations of people and retail activity will also attract people.

Supply: harden, soften, protect or otherwise modify existing sites. Attraction sites can be changed to address specific impact goals and policies. The design guidelines described in the paragraph above are an example of a soft approach to site modification. Soft approaches are more subtle and less intrusive. The more common need is to harden a site to withstand large numbers of tourists. This is typically accomplished through the construction of pathways that prevent visitors from directly tramping on the soil of a place, for example. This may be in the form of boardwalks, paved trails, tramways and suspended walkways, and visitor transportation systems. Other site-hardening tools include paved parking lots, animal-proof rubbish bins, boat docks and moorings, fencing and enclosures, and the addition of structural support for older buildings. Especially sensitive sites can be fenced off from direct visitor contact and should undergo more attentive monitoring. Some forms of site hardening are less obvious and intrusive than others, depending on their material and extent. An alternative to site hardening is to move visitor facilities, such as information centres and restrooms, to more durable and less sensitive locations.

Environmental and development regulations and guidelines can also be used to modify the nature of attractions and facilities to either promote tourism's growth or limit its environmental and social impacts (Hall 2006c). These can include landscape requirements, building design standards and incentive programmes. Landscape requirement may require or suggest a minimum number of trees or specific types of vegetation, such as the exclusion of non-native species. These stipulations can be required or can be encouraged through incentives that, for example, might allow for a larger hotel if certain landscape standards are met. Similar stipulations can be required or encouraged for sustainable construction that reduces energy consumption, or for business operations that utilize a certain percentage of local community products and human resources (employees).

Local government regulations or developer-imposed deed restrictions can require certain architectural design standards to conserve heritage districts (historic preservation) or desired thematic designs. Some retail districts, for example, require that all businesses maintain an adopted theme, such as a mining town theme or a waterfront wharf theme, with detailed guidelines on appropriate colours, signage and building ornaments (see Case 6.6). Assistance programmes can provide design and funding help to businesses to ensure theme consistency. Such programmes may be funded by public authorities, private entrepreneurial organizations or through some form of public–private cooperation.

Supply: spatially concentrating or dispersing sites and facilities. Above the scale of the individual site, it may be possible to cluster tourist attractions or facilities to limit their impact on residential areas, or alternatively, to disperse them in order to spread the economic impacts of tourism. Concentration often occurs in the hotel industry, with hotels, motels and restaurants clustering at transportation nodes and near popular tourist sites to take advantage of co-location. Some of these clusters occur naturally over time, such as the Waikiki district of Honolulu, which experienced rapid hotel growth after the Ala Wai Canal was built in 1920 to drain a sandy swampland (see Photo 3.5). Others are master-planned resort communities, such as the Nusa Dua enclave on the island of Bali, Indonesia, which was planned in the 1970s in a dry and underpopulated part of the island, and today has over a dozen four- and five-star resorts. Land use planning, zoning and regulations can encourage clustering, though it is clearly more suited to private sector developments than for geographically fixed cultural and natural attractions.

Photo 6.5 Melbourne's Chinatown is an example of a thematic ethnic retail district. Private businesses contribute to the theme through their signage and the products they sell. Government contributions, however, can also be seen in the themed gates and street lighting. Architectural controls may also help to conserve older buildings in the neighbourhood, and the wide, pedestrian-friendly sidewalks help to encourage visitors. (Photo by C. Michael Hall)

Alternatively, it may be desirable to disperse some or all tourist services. This can help to ensure that tourists are not overly concentrated in a single location, and may be more suited to some destinations where fixed attractions are naturally widely dispersed. In such a situation, tourists typically tour the sites over the course of several days, staying at different locations each night. However, the concentration of facilities in one place may not be suited to the needs of visitors to these kinds of touring regions (Stewart et al. 2001). Furthermore, Wall (1997) has suggested that there are two forms of dispersed tourism: lines, where attractions are distributed along a linear path, and areas, where they are distributed more broadly. An example of dispersed tourism development is the Munda Maya Trail (also known as La Ruta Maya), which is an ongoing effort to coordinate tourism development and promotion along a linear route that connects Mayan archaeological sites and crosses the countries of Mexico, Belize, Guatemala, Honduras and El Salvador (Ceballos-Luscarain 1990).

Case 6.6: Heritage themeing in Oregon

Heritage conservation and tourism development are closely related, but are not always friends. Heritage supporters tend to focus more on culture and authenticity issues, while tourism boosters tend to focus more on the economic bottom line. The experience of three communities in the state of Oregon illustrates how these divisions arise and some of their implications for different forms of tourism development and community identity (Lew 1988).

The ethnic festival

Junction City, in the Willamette Valley of western Oregon, was originally settled by ethnic Danish migrants from the Midwest region of the United States in the 1800s. They were later joined by other ethnic Scandinavian migrants. In the late 1950s, the new Interstate Freeway bypassed Junction City, which had previously been on the main highway between California and Portland, Oregon. This devastated the community's economy, causing huge job losses. In 1961, a local physician recommended that the city hold a Scandinavian Festival, to raise people's spirits and pride. The Chamber of Commerce provided some financial support for the first festival, but it was mostly taken up by local social organizations, such as the Danish Brotherhood, the Danish Sisterhood and the Sons of Norway.

Today, the Scandinavian festival draws over 200,000 visitors yearly over a four-day period, each day of which is devoted to Denmark, Norway, Sweden and Finland. The focus of the festival is on authenticity in the booths that line the streets of Junction City, and support for non-profit organizations, including schools. The Festival Committee interviews all groups before they are allowed to participate, and monitors their authenticity during the event. Despite the early involvement of the Chamber of Commerce, commercial interests are expressly discouraged in what is considered one of the leading ethnic festivals in the United

States. As a result, the Scandinavian theme is only slightly present in the town's commercial landscape outside of the festival event.

The cowboy town

Sisters is a small community (959 people in 2000) located on the relatively dry eastern slopes of the Cascade Mountain Range in central Oregon. It is at the confluence of three major highways that cross the mountains from the Willamette Valley, and was named after the Three Sisters volcanic peaks (Faith, Hope and Charity) that loom not far away. It was originally known for its several gas stations and restaurants, though the Sisters Rodeo has been an annual event since 1940. In the early 1970s, the Junior Chamber of Commerce converted two small buildings, one at each end of the small community, with 1880s Western cowboy façades.

The theme became popular and by 1977, the 1880s Western architectural motif was made mandatory for all retail businesses in the town by the town council, with design guidelines written into the land use ordinance and building code. The business owners and residents all felt that the cowboy theme was appropriate to the community, even though there was only one truly historic 1880s-style building, and made the town more interesting to serve the large second-home recreation and retirement populations in the surrounding region. In fact, the developer of a large second-home subdivision under development near Sisters in the 1970s paid for architectural sketches to show each business how they could be 'Westernized'.

The authenticity of heritage in Sisters is based more on how well buildings meet the architectural ideal of the cowboy façade, not how authentic the actual building is. The theme fits well with the other amenities of the town's high plateau region, including nearby winter skiing, hunting and fishing, and outdoor lifestyles. And nearly all the businesses in Sisters are members of the Chamber of Commerce, demonstrating the unity that the business community has for this approach to tourism development.

The historic downtown

Oakland is a small town with a historic central district, reflecting its earlier days as the 'Turkey Capital of the United States'. By the 1960s, however, it had become a bedroom community for people who commuted to work in nearby Roseburg, Oregon. At that time the community decided to apply to have the historic core of the town designated as an official historic district by the State of Oregon. Although most of the community was involved in this discussion, when the proposal was put forward only the retail district was included and all the residential buildings were removed. This caused a major split in the community between the business people and the non-business population. The proposal was withdrawn for ten years until a more encompassing and more acceptable version was developed.

However, the split between business and non-business interests evolved into a sense that the historic district should not be commercialized and Disneyfied (like

Sisters, Oregon). Residents felt that 'visitors' were welcome to share their cultural treasure, but they did not want 'tourists' ruining their community. These sentiments may be more prominent here because most of the residents were not dependent on Oakland for their employment. Retail businesses in the town have made some efforts to expand their markets, at least to adjacent communities, but some residents continue to be wary of the potential impacts of their efforts.

These three case studies show how history, geography, community stakeholders and larger social processes all impact the decision-making process and the way that heritage resources and tourism evolve in a community. These dynamics may be more straightforward in smaller communities, but similar processes occur in larger cities and urban neighbourhoods, as well.

Source: Lew 1988.

Demand: marketing enhancements and educational campaigns. The traditional way to increase tourism is to advertise. This approach, however, can also be used to educate potential tourists. Marketing campaigns serve two functions: informing potential tourists about the experiential opportunities at a destination, and creating or reinforcing a destination's image. It is important to match the marketing campaign with the actual resources and experiences available at the destination. Effective marketing also targets the most appropriate market segments, rather than broad, generic populations. Destination-wide marketing often involves some role for government bodies, including national, regional or local tourism organizations (NTO, RTO or LTO). However, these organizations typically include significant participation by local business owners, and their marketing budgets can be a combination of public tax funding and private funds. At the local level, chambers of commerce and convention and visitors bureaus may take the lead marketing role if no government LTO exists.

Marketing and destination image development includes both advertising and destination-based visitor centres that meet the informational needs of travellers. Both of these have a significant role in educating tourists and potential tourists about the attractions of a destination, as well as proper behaviour in order to avoid environmental and social transgressions where this is important (see Case 6.7).

One way that many local governments fund all or part of these efforts is through a special *bed, board and booze tax* (hotel, restaurant and bar sales tax) (Kerr et al. 2001). These typically range from 1 per cent to 5 per cent, and sometimes more, on top of the existing sales or value-added tax. In most communities this tax is specifically targeted to marketing and other tourism-related efforts, either exclusively or partially. Other communities, however, apply these taxes to their general fund, which does not guarantee an ongoing marketing effort.

Demand: human resource and service development. The most valuable form of tourism marketing is word of mouth, as people trust the opinions of their friends and relatives more than any form of overt advertising. As such, the tourist's actual experience and impressions of a destination are crucial in a place's marketing efforts. These are shaped by the quality of the attractions and facilities on offer, and the helpfulness and friendliness of the people they meet at the destination.

One way of encouraging quality in hotels, restaurants and tourist attractions is through certification programmes. Many NTOs have developed hotel rating programmes to provide tourists with a degree of consistency and predictability. The ratings criteria will vary from one country to the next, and may not match well with private star ratings available in guidebooks and on travel websites. Restaurant and attraction ratings are less common, though the China National Tourism Administration (CNTA) has star rating systems for these as well, while privately run evaluations, such as those of *Michelin Guide* in the case of restaurants, can also be extremely influential. In addition to encouraging overall levels of quality (based on the evaluation criteria), certification schemes can also be used to encourage more sustainable and environmentally sensitive practices. Green certification programmes exist at many levels, from the international Green Globe, mentioned above, to national and local programmes. These mostly apply to hotels, resorts and attractions, and can be used in their promotions and image development.

The human factor is also essential to the quality of the tourist experience at a place and human resource development is considered as important as marketing in the overall management of a destination image and experience. Human resource development occurs for both current employees and potential employees. It includes both those who work in the private sector and those in the public sector. And it includes the hospitality and travel industries (hotels, restaurants and visitor services), the travel industry (travel agents, and airlines and other transportation companies) and the tourism industry (destination planners, marketers, researchers and visitor services). All of these areas require different levels of skills and experience, including entrepreneurs, managers and employees (Baum 2006).

Sometimes these needs are addressed through a formal human resource development plan, which starts with an assessment of the current and future employment situation for tourism-related activities, and compares that to the current available labour force. Current and future training and education needs are then identified, along with options for addressing those needs, including both public education and private in-house training programmes. Other education issues include tourism sensitivity training for ancillary sectors, such as taxi drivers and retail clerks, and public education initiatives to remind the community about the importance of tourism and encourage an appreciation for the benefits that visitors bring.

Some communities will devote one day every year to enhancing employee and resident attitudes toward tourists. The education of employers, employees and the general public can also be an effective way to inform people about how they can behave in more sustainable and environmentally and socially sensitive ways.

Demand: behavioral guidelines and regulatory limits. Tourist behaviour can be changed both directly and indirectly through education efforts and mandatory regulations. Direct control over the number of visitors to a site or area requires the ability to control access. This can be done for heritage sites where a maximum capacity can be controlled through tickets and reservations. Tickets can be limited

either by their absolute number or by charging higher price{
required as one way of limiting the length of time that t{
thereby limiting their physical impacts. The Inca Trail to
uses a combination of these approaches to limit the numb
time on this popular trail.

It is also possible to segment visitors to only allow som
Visitation on some days or time slots can be limited to or
children, for example. Access to more sensitive environr_
to individuals with certified skills or equipment. Some forms of equipment
banned if it has the potential to cause damage, such as off-road vehicles.

Educating visitors is another way to influence their travel behaviour. They can
be encouraged to visit different sites as a way to disperse their impacts, and they
can be informed of particularly sensitive environmental and social issues related
to tourism. For example, in some developing countries young children will resort
to begging from tourists instead of attending school. Tourists to countries such as
Sri Lanka and Thailand have been informed either on their plane or at their port of
entry not to encourage this behaviour.

Codes of ethics for tourists and the tourism industry have been developed by
many different non-governmental organizations. The UNWTO Global Code of
Ethics for Tourism, which is aimed more at the tourism industry, is discussed
earlier in this chapter. As an example of a code for tourists, the Pacific Asia Travel
Association (PATA 2002) has adopted a Traveller's Code for visiting indigenous
cultures, the major points of which are (see also Case 6.1 and Chapter 4):

- *Be flexible* – and open to accepting other cultures in their own terms.
- *Choose responsibly* – for businesses that clearly support local cultural and
  environmental stewardship.
- *Do your homework* – so you are familiar enough with the destination so as to
  avoid offending local traditions and harming the environment.
- *Be aware* – of special events and religious and social customs that you are
  likely to encounter during your visit.
- *Support local enterprises* – by eating in local restaurants, staying in locally
  owned accommodation and buying locally made arts and crafts.
- *Be respectful and observant* – of local laws and norms of behaviour.

Many tourists already behave according to at least some of these principles,
simply because it makes for a more enjoyable experience. While there are
exceptions, most visitors do not want to be an 'ugly tourist' and are willing to learn
and be educated about the places they are visiting. This is an area that has barely
been touched on by travel and tourism providers. It has been mostly left to attraction sites to educate their visitors, but responsible transportation, accommodation
and marketing offices could also support such efforts. Airlines in particular,
because they bring people to the furthest and most different destinations from their
homes, could be a major force for managing tourism behaviour and mitigating
some of tourism's more negative impacts.

6.7: Managing tourism impacts on the Hopi
Indian Reservation

Unique cultures are a popular tourist attraction worldwide. As discussed in Chapter 4, however, tourism can change and destroy the very essence of what makes these cultures attractive to begin with. Controlling tourism behaviour can take many forms. Occasionally, as in the case of the Hopi Indians, it requires clear, explicit and enforceable behavioural policies and regulations.

The Hopi Indian Tribe was never conquered by the Spanish conquistadores or American cavalry forces. In the early 1900s, the Santa Fe Railroad Company began to promote the Hopi Mesa villages in northern Arizona, along with other Indian tribes, as a tourist attraction to boost travel along its new rail lines through the American Southwest (Lew and Kennedy 2002). Famous painters and illustrators were brought from the East Coast to produce posters and calendar art that presented the Indian as a romanticized 'Noble Savage' – an image that was widely distributed throughout the United States and Europe. Today, the Hopi Indians are among the most traditional tribes in the United States. Part of this comes from their isolation from both Spanish and American population centres in North America. Their villages and traditional dance ceremonies continue to be a popular tourist destination. However, to maintain the dignity of their traditions, and the secrecy of their religion, they have adopted a standard set of rules that apply to all non-Hopi visitors to their 11 villages on First Mesa, Second Mesa and Third Mesa. The rules vary from one village to the next, but generally include:

- Visitors are welcome, but should remember that they are guests of the Hopi, and should act accordingly.
- Possession of alcohol or drugs anywhere on the reservation is prohibited by Tribal Law.
- Archaeological resources and ruin sites are off-limits to all non-tribal members – removal of artefacts is a criminal offence.
- Photographing, recording or sketching of villages, religious ceremonies or individuals is strictly prohibited on the reservation, unless permission is granted by the village chief or tribal governor.
- If spending an unusually lengthy period of time in a village, permission must be obtained from the village chief or tribal governor.
- Drivers are cautioned to obey posted speed limits on the reservation and to watch for livestock on roads and highways, especially at night.
- Violations of these rules can lead to a visitor being expelled from the reservation or even arrested by reservation police.

Most of the villages are open to visitors and quite a few Hopi sell arts and crafts from their homes. This includes Oraibi (also known as Old Oraibi and Orayvi), which is considered the oldest site of continuous habitation in the United States. Villages, however, may also be closed during some dances, although most dance ceremonies are open for public viewing. Because their religion is secret, ceremonial dates are usually not known until about a week before they take place. The small village of Walpi sits on a dramatic precipice atop First Mesa and has intentionally chosen to forgo electricity and running water. It is closed to unguided tourists, though the First Mesa Consolidated Villages government provides free or low-cost guides on most days, when there are no ceremonies.

The restriction on photography is probably the most difficult for tourists, for whom photographs have become a major way of personalizing and remembering their visits. The Hopi ban on photography was imposed by religious leaders soon after tourists started arriving in the 1920s. They rightly saw that photography was highly disruptive to the practice of their religion, which also includes traditional dry land agriculture, as well as to their privacy. Some of the pueblo Indian tribes in the state of New Mexico also forbid photography, though others, most notably the Taos and Acoma tribes, charge fees for non-commercial photography and videos by tourists.

Source: Lew and Kennedy 2002.

Summary and conclusions

Because tourism is a complex industry, comprised of many different segments that do not always see their mutual relationships, it is difficult to address the development and impacts of tourism from one point of authority. Instead, many different organizations, representing different interest groups and different political orientations address different aspects of travel and tourism. This chapter has identified some of the most important of these at the global and national level. Many more exist that were not covered, though some have been mentioned in other chapters. As presented in Figure 6.1, the broader, more global and long-term issues are mostly addressed by international organizations that adopt general policy statements, which they hope will influence corporate practices and government laws. Thus the focus was on the planning and implementation of those practices and laws.

The list of options provided in this chapter can be thought of as a cookbook. Ingredients include matching the right mix of approaches and techniques with the fixed ingredients of the social, political, cultural and environmental site conditions, as well as their relationship to the broader situational context in which they exist. Each situation will provide a different complexity of problems, which can be assessed from a *tame* or *wicked* perspective, as discussed above. Other important considerations include the degree to which a private–public conflict resolution must be achieved, and whether the planning is proactive (before the problem) or reactive (fixing an existing problem). And there is the degree to which tourism is central or peripheral to the larger planning process and goals of an organization, enterprise or community. Fortunately, the cookbook provided in this chapter can be adjusted to ingredients beyond those of travel and tourism.

Self-review questions

1　What organizations are involved in planning and managing tourism-related impacts at the international, regional, national and local scales? What forms of authority do they have?
2　What is the basic rational planning process and how is it modified to meet the needs of different organizations and communities?
3　What are the basic tools of tourism planning and management at the local community level?
4　What political challenges might a planner encounter when a large, new tourism project is proposed?
5　How does the concept of sustainable tourism differ at the global, regional and local scales?
6　How is tourism planning different between a developed country setting and a developing country setting?

Recommended further reading

Campbell, S. and Fainstein, S. (eds) (2003) *Readings in Planning Theory*, 2nd edn, London: Blackwell.
Provides one of the best introductions to urban and regional planning theory.

Eagles, P.F.J., McCool, S.F. and Haynes, C.D. (2002) *Sustainable Tourism in Protected Areas: Guidelines for Planning and Management*, Gland, Switzerland: International Union for the Conservation of Nature and Natural Resources (IUCN).
A good introduction to some of the issues and methods involved in natural area planning

Hall, C.M. and Lew, A.A. (eds) (1998) *Sustainable Tourism: A Geographical Perspective*, London: Addison Wesley Longman.
Provides a range of different perspectives on the development of sustainable tourism.

Deery, M., Fredline, L. and Jago, L. (2005) 'A framework for the development of social and socio-economic indicators for sustainable tourism in communities', *Tourism Review International*, 9(1): 69–78.
Provides one approach towards indicator development for tourism.

Hall, C.M. (2008) *Tourism Planning*, 2nd edn, Harlow: Pearson Education.
Provides an outline of the tourism planning process at various scales.

Web resources

UNWTO Global Code of Ethics for Tourism: http://www.unwto.org/code_ethics/eng/principles.htm and: http://www.unwto.org/code_ethics/eng/brochure.htm
International Air Transport Association (IATA) Building a Greener Future, 2nd edn: http://www.iata.org/whatwedo/environment/index.htm
Kangaroo Island Tourism Optimization Management Model: TOMM Process: http://www.tomm.info/Background/tomm_process.aspx
Office of Travel and Tourism Industries (OTTI): http://www.tinet.ita.doc.gov/about/index.html#TD

Tourism Concern: http://www.tourismconcern.org.uk/

US Department of Transportation (USDOT) Aviation Consumer Protection Division: http://airconsumer.ost.dot.gov/org.htm

World Conservation Union (IUCN): http://cms.iucn.org/

Gross, Julian (2005) *Community Benefits Agreements: Making Development Projects Accountable*, Washington, DC: Good Jobs First: http://www.goodjobsfirst.org/pdf/cba2005final.pdf

Ecosystem Touring Good Practice Guides and Checklists: http://tourism.jot.com/WikiHome/MainstreamBiodiv

US Environmental Protection Agency, Environmental Management System: http://www.epa.gov/ems/

Key concepts and definitions

Governance The pattern or structure that emerges in a socio-political system as a 'common' result or outcome of the interacting intervention efforts of all involved actors in the management of a policy issue.

Whole of government approach Public service agencies working across portfolio boundaries to achieve a shared goal and an integrated government response to particular issues. Approaches can be formal and informal. They can focus on policy development, programme management and service delivery.

Recreation Opportunity Spectrum (ROS) A land management approach that zones different areas of the resource for different levels of use, from the most pristine and natural to the most urban. Zones may require reclamation to achieve their required level of use, or policies to manage use levels. The approach works better in natural areas, and has been applied to tourism and ecotourism.

Rational planning The fundamental planning approach that starts with problem identification and goal setting, and includes scenario building and the selection and implementation of a preferred plan. Many other models build on the rational planning approach, which assumes that objective analysis can ensure the best decision-making, though this is subject to debate.

Public participation The process of involving all stakeholders (those affected by the issue) in the planning and decision-making process. Involves a wide range of techniques, including surveys, committees and meetings, and is crucial for gaining public support in goal setting and selecting a preferred plan.

7 Futures of tourism

Learning objectives

By the end of this chapter, students will be able to:

- Define the concept of foresighting and its main approaches.
- Understand the difficulties of forecasting.
- Define the concept of a wildcard event.
- Appreciate the positive and negative factors that will affect tourism at a global and local scale in the future, including technology advances, climate change and energy supply.

This chapter provides a greater understanding of the potential futures of tourism impacts and their assessment. It first provides an introduction to forecasting and foresight studies, followed by a discussion of some of the positive and negative assessments of the future of tourism. The chapter concludes by reinforcing the importance to the future of tourism of understanding impacts and change.

Introduction

This book has examined the ways in which tourism is interrelated with economic, socio-cultural and environmental change and the ways in which tourism's contribution to change can be managed. As the previous chapter discussed, strategic planning is a good means to try and ensure the sustainability of tourism destinations. However, the future, by definition, is unknown and even the most comprehensive business or destination strategy may be affected by *wildcard events* – high impact, very low probability events. Examples of such wildcard events that have had a global impact include the terrorist attacks in the United States on September 11, 2001 (9/11) and pandemic diseases, such as the Asian SARS outbreak from 2002 to 2003 (see Chapter 5). The 9/11 attacks clearly show how events in a specific location can affect the global political and economic system. The impacts of that day and the subsequent 'war against terror' are still reverberating through the tourism system to the present in terms of airport and transport security and the perception of destinations (see Photo 7.1).

Photo 7.1 A Ground Zero memorial in New York. This sculpture, titled *The Sphere*, originally stood in the plaza of the World Trade Center as a symbol of world peace. It survived the terrorist attack and now sits in Battery Park, a couple of blocks from the World Trade Center site. The eternal flame in the foreground was lit on September 11, 2002, one year after the attack in honour of those who died. International tourism to the United States dropped precipitously following the attack and, for a variety of reasons, had not yet fully recovered as of 2007. (Photo by Alan A. Lew)

Wildcard events can occur at all scales and can have a dramatic effect on tourism. Examples of such events are shown in Table 7.1. Generally, the larger the scale of the wildcard event, in terms of effects on the tourism system, the longer such effects will last.

Some events, such as the devastation of New Orleans by Hurricane Katrina in 2005 and the subsequent effects on tourism and the creation of environmental refugees, although high magnitude low probability events, are not wildcards in the strict sense. This is because the probabilities of a hurricane at that location are well recognized. This is also a reason why the attacks of 9/11 are a wildcard event because, although the likelihood of a terrorist attack was arguably predictable and preventable, the precise location of such an attack was more problematic. Furthermore, in the case of Hurricane Katrina, because of the particular urban and physical geography of New Orleans and the Gulf region of the United States, the economic and social impact of Katrina was regarded by some researchers as a disaster waiting to happen (Colten 2000; Brinkley 2006; Squires and Hartman 2006; Bergal et al. 2007). As Colten (2005: 2) stated in a book on New Orleans, which came out just prior to Hurricane Katrina, 'Wrestling the site from its watery excesses and the associated problems was, and remains, a central issue in this city's existence.. . . Keeping the city dry, or separating the human-made

Table 7.1 Wildcard events

Scale	Event	Effects on tourism
Global	Terrorist attacks of September 11, 2001, in New York and Washington	Dramatic sudden decrease in international travel around the world, but particularly on international routes to and from the USA.
Supranational (Indian Ocean)	Tsunami, 6 December 2004, following an earthquake off the coast of Sumatra, Indonesia	Impacted coastal tourism destinations throughout the Indian Ocean through damage to infrastructure. Led to a decline in tourism arrivals in a number of Indian Ocean countries.
National	Coups d'état in Fiji: May–October 1987; May 2000; December 2005	Although no tourists were killed in the coups each coup led to a loss of investor and tourist confidence in Fiji for a number of years before visitor numbers again started to increase.
Regional	Terrorist attack of 12 October 2002 in the tourist district of Kuta, Bali, Indonesia	Eighty-eight Australian tourists were killed with the attack leading to a longer-term loss of confidence in visitation to the island resort by Australians and, to a lesser extent, New Zealanders and other nationalities.

environment from its natural endowment, has been the perpetual battle for New Orleans'.

However, the importance of an event such as Hurricane Katrina is not just with respect to its immediate, albeit highly significant, impact (see Case 7.1) but it is also with respect to the lessons it can provide regarding preparedness for change. Hurricane Katrina is often held up as an example of what can happen as a result of environmental change, particularly climate change, and is, therefore, regarded by some as forewarning of what might happen around the world if current predictions on climate change and sea level rise come true (e.g. O'Brien 2006; Rayner 2006; Baker and Refsgaard 2007; Gupta et al. 2007). Just as significantly, the Katrina experience reinforced the possibility that climate events might evolve into environmental, social and economic catastrophes as a result of pre-impact failures to address legitimate scientific concerns (Ackerman and Stanton 2006; Baker and Refsgaard 2007).

Case 7.1: The impacts of Hurricane Katrina on tourism in New Orleans

On the morning of 29 August 2005, the Gulf Coast region of the United States was hit by one of the most destructive hurricanes in modern times. Hurricane Katrina, a Category 4 storm, made landfall on the Louisiana coast and travelled northeast through Mississippi and Alabama before being downgraded to a tropical storm the following day. A Category 4 storm implies:

- Winds 210–249 km/hr (131–155 mph)
- Storm surge waves that are generally 4 to 5.5 metres (13 to 18 feet) above normal
- More extensive curtain wall failures with some complete roof structure failures on small residences
- Shrubs, trees and all signs are blown down; complete destruction of mobile homes; extensive damage to doors and windows
- Low-lying escape routes may be cut by rising water three to five hours before arrival of the centre of the hurricane
- Major damage to lower floors of structures near the shore
- Terrain lower than 3 metres (10 feet) above sea level may be flooded requiring massive evacuation of residential areas as far inland as 10 kilometres (6 miles).

This final characteristic greatly increased the severity of the impact of the hurricane in New Orleans, Louisiana, because when the city's flood-control levees were breached approximately 80 per cent of the city was covered by water. Hurricane Katrina therefore became the city's ultimate 'worst-case scenario' and the chain of events that followed indicated a general lack of preparedness, even though the likelihood of such a disaster occurring had been predicted by many scientists and had also been the subject of several television documentaries (Cooper and Hall 2008).

The impact of Hurricane Katrina on tourism in Louisiana and New Orleans was dramatic. By mid-2006, the Louisiana Department of Culture, Recreation and Tourism (DCRT 2006a: 1) reported that 'negative images of affected areas portrayed by the media . . . had resulted in a significant loss of interest in tourism'. Louisiana was expected to have lost 20 per cent of its visitor spending in one year (approximately US$2.2 billion) and New Orleans was losing an average of $15.2 million per day, with a consequent loss in state and local tax revenues of approximately US$125 million for the year.

The impact on the state and city economy was magnified as tourism, along with port operations and education, was regarded as one of the cornerstones of the economy and an area of employment creation at a time when jobs in other sectors were being lost. In 2000, the tourism industry represented 16 per cent of employment and 8 per cent of the total wages generated in the city (Dolfman et al. 2007). Between 1990 and 2004, jobs in the tourism sector grew by 33 per cent and were continuing to grow in the months preceding Katrina. According to Dolfman et al. (2007: 6), New Orleans was doubly hit as tourism experienced the largest job loss among all sectors and 'would have shown further gains in employment had the hurricane not struck the city'. Over the ten months following the hurricane, the city's tourism industry lost approximately 22,900 jobs along with a loss in wages of about $382.7 million.

Total enquiries to the Louisiana Office of Tourism via the Internet, telephone and mail dropped by over a third after Hurricane Katrina and Hurricane Rita (which hit Louisiana shortly after Katrina) from over 3 million in 2003–5 to just under 2 million in 2005–6 (DCRT 2006b). However, the effects with respect to the state's and city's image was wider than just potential visitors, as the state also acknowledged that the hurricanes had 'weakened investor confidence with regards to spending money in Louisiana' (DCRT 2006a: 3). The impact on the region's image was demonstrated in a study conducted by Cunningham Research (2005), which indicated that travellers were most concerned over their personal safety and their finances when considering a trip to New Orleans. Of travellers who stated they were interested in visiting New Orleans, 50 per cent were somewhat or extremely concerned over a hurricane being predicted before they leave home, and an equal number were just as concerned over losing money from pre-paid travel expenses and additional costs from delays. An overall negative impact on confidence with leadership at all levels of government was also regarded as influencing the attractiveness of New Orleans as a destination (CBS 2005).

Nevertheless, even though negative news of the impact of Katrina were affecting the city's and the state's image and economy, the media were also seen as part of the solution with respect to redeveloping tourism. In response to the damage wrought by Hurricane Katrina on the physical and image dimensions of New Orleans and Louisiana tourism, the State of Louisiana launched a 'Rebirth plan' (DCRT 2005). Employment and the rebuilding of physical and social infrastructure were seen as important but so too were the provision of information and the development of a public relations campaign. According to the plan, 'People need new images of Louisiana, to replace the weeks of negative images on television' (DCRT 2005: 3). Five strategies were identified with respect to tourism:

1   *Public relations campaign* – 'Immediately and aggressively promote those areas of the state currently open for business. An aggressive public relations campaign will be implemented'
2   *Business assistance* – 'Rapidly develop and implement a statewide tourism small business assistance program, utilizing any and all available federal, private and state funds'
3   *Infrastructure* – 'Facilitate the rapid rebuilding and improvement of the state's infrastructure'
4   *Image* – 'Rebuild and enhance Louisiana's national and international image as an attractive, compelling, unique tourism destination'
5   *Lead agency* – 'The Office of Lieutenant Governor to serve as the lead agency with respect to policy setting, intergovernmental relations, and as an information clearing house' (DCRT 2005: 7).

In 2006, the state amended its tourism marketing plan to increase the number of visitors in the areas impacted by Katrina and invested $28.5 million in a national campaign and other initiatives designed to bring out-of-state travellers back to the region. The primary targets were leisure visitors, though convention and business travellers were also major markets. The initiative included:

1   Convention and interactive marketing, promotions to travel agents, and related activities.
2   A local awareness campaign focusing on New Orleans, Southeast Louisiana, and Southwest Louisiana.
3   Niche marketing programmes designed to promote family activities and festivals and cultural attractions unique to the areas most affected by the storms.
4   Coordinated marketing efforts between the state's Department of Culture, Recreation and Tourism and the Louisiana Department of Economic Development to regain investor confidence for spending money and creating jobs in Louisiana (DCRT 2006a: 4).

By September 2006, significant progress had been made with respect to restoring the tourism resources of New Orleans and Louisiana. Prior to Hurricanes Katrina and Rita, 21 per cent of US travellers and 53 per cent of Louisiana's regional market expressed an intention to travel to Louisiana in the next 24 months. Immediately after the hurricanes, the national number dropped to 13 per cent and the regional market dropped to 37 per cent. As of May 2006, the regional number had increased to 47 per cent (DCRT 2006b: 4). Nevertheless, the long-term future of tourism in New Orleans will likely be affected by the degree of redevelopment of the city (Gladstone and Préau 2008), including new flood protection measures, as well as the impact of any future hurricanes on the city (Cooper and Hall 2008).

Sources: Louisiana Office of Tourism: www.crt.state.la.us/tourism

New Orleans Convention and Visitors Bureau: www.neworleanscvb.com

Severe and undesirable change can also occur without being associated with a single large impact event, or even with just a relatively small event, that may trigger a series of unexpected consequences. In such cases, it is important to recognize the significance of incremental change that usually comes with only a gradual realization that there is a problem at hand. A 'bit-by-bit' slow incremental change may not be recognized as being undesirable or even actually occurring until the system reaches a *tipping point*, beyond which the nature of the system changes markedly and change becomes irreversible.

Tipping points have been recognized in both natural and social systems. For example, in the case of climate change, there is concern that this point has already been reached with respect to global warming and sea level rise as a result of the amount of greenhouse gases that are now in the atmosphere, whether or not rates of emissions are reduced in the near future (e.g. Hansen 2006). In the social sciences, the term has been made popular by Gladwell's (2000) book on the diffusion of ideas and products, although the concept is now widely used in sociology, geography and health studies to describe the way in which the momentum for change or adoption becomes unstoppable and, therefore, irreversible.

In this book, we have addressed the ways that tourism provides momentum for change in economic, socio-cultural and physical environmental settings, as well as how the environment and the agents within it affect tourism. In many situations where the impacts of tourism are concerns, the fear of many communities is that the destinations will become irreversibly changed, thereby affecting the desirability and attractiveness of such places for locals and visitors alike. As the previous chapter suggested, however, strategic planning and strategy may help avert such dramatic change and at least restrict change to a pace at which it becomes acceptable by providing sufficient time for adaptation to occur.

Good tourism planning and impact management is as much grounded in the development of capacities to recognize factors that may lead to change and, therefore, anticipate potential responses to them to respond effectively as in responding to impacts that have already occurred (Hall 2008c). However, the scenarios for the future of tourism, and hence understanding the potential dimensions for planning and managing for change, currently seem almost schizophrenic with respect to forecasts for substantial growth on the one hand, and fears over the effects of environmental change on the other. Furthermore, the future, by definition, is difficult to predict accurately (see Case 7.2).

This chapter examines some of the different alternative futures for tourism, as well as, where possible, the means by which they are forecast. The first section discusses some of the positive forecasts for growth in tourism as well as the development of new tourism opportunities. The second section discusses potential limits to tourism growth in the form of climate change and the increasing costs of energy. The final section addresses the obvious issue of the extent to which the two future visions can be reconciled.

Case 7.2: Forecasting the future: a 1974 forecast of future leisure environments

> As an aid to policy- and decision-making about future environmental problems, a panel of experts was asked to predict the probabilities of future events associated with natural-resource management, wildland-recreation management, environmental pollution, population-workforce-leisure, and urban environments. Though some of the predictions projected to the year 2050 may sound fantastic now, the authors think that some of the events predicted may occur sooner than forecast.
>
> (Shafer et al. 1974: ii)

In 1973 and 1974 a Delphi study of a panel of 904 experts was undertaken by US Forest Service to identify those developments likely to occur in the United States that would influence the leisure environment (Shafer et al. 1974). The panel concluded a range of situations that would be in place by the year 2000; some of these are outlined below. The actual situation in 2000 is noted in [brackets], and shows the challenge in forecasting into the future even just a couple of decades.

1   500 miles (800 km) is considered a reasonable one-way distance for weekend pleasure travel [this is only applicable to the small minority who undertake weekend pleasure travel by plane].
2   Average retirement age is 50 [average retirement age has remained at over 60 and is increasing].
3   Middle-class American families vacation on other continents as commonly as they vacation in the USA in the 1970s [middle-class Americans still vacation domestically far more than internationally].
4   Electric power or other non-polluting propulsion methods replace internal combustion engines in recreational vehicles [very few recreational vehicles use electric power or non-polluting engines].
5   Travel in large parks is limited to minimal impact mass transit, e.g. tramways, air transport and underground rapid transit [despite discussions on adopting alternative transit, travel in large parks is still mainly conducted by cars on roads. It is also interesting to see air transport being described as minimal impact given present-day concerns over noise and pollution].
6   Most marine and estuarine areas are managed for fish and wildlife habitat [while conservation and management has improved since 1974, most marine and estuarine areas, if managed at all, are managed for a far wider range of multiple uses].
7   Coordinated environmental planning between government and private enterprise [although coordination has improved in a number of areas, there are still substantial differences between public and private approaches towards environmental planning].

8   Airstrips and helicopter pads established at most popular recreation areas [most public natural areas are restricting air activities due to noise and visual pollution].

9   Small private recreational submarines used as commonly as snowmobiles are used today [very few of these exist].

10   International agency established to halt further air and water pollution [some international agencies and conventions exist, but national legislative and regulatory power still dominates].

11   Consumers accept major costs of pollution control [although attitudes towards pollution control are increasingly positive, widespread behavioural changes and support for higher costs is extremely limited and problematic].

12   'Weekends' distributed throughout the week [four-day working weeks have increased for some, but only accompanied by more hours for each day worked; Americans are working more hours a year than ever].

13   Leisure is an accepted lifestyle [apparently only among some segments of the younger generations, which may not be long-lasting].

14   Attempt to control population through tax incentives [tax incentives are commonly used to encourage more children, rather than less, which was more of a goal in 1974 than in 2000].

15   Simulated indoor environments available that provide recreational opportunities now available only in the outdoors [climbing walls, wave pools and indoor ski slopes are popular examples of this prediction].

16   Green space preserved between most metropolitan areas [very limited].

17   Some city parks or parts of parks enclosed in all-weather protective bubbles [mostly used for tennis courts, very limited for other recreation].

18   Computers used to direct and control movements of individual transportation units [limited use, mostly for mass transit rail and subway systems in some cities].

Sources: Shafer et al. 1974; Cooper and Hall 2008.

Foresighting

Foresighting is the effort to assess future conditions. Although the popular perception is that the future will become more predictable as foresighting methods improve, experts involved in futures studies and foresighting have developed a perspective of the future as inherently *contingent* (i.e. not able to be directly determined by current conditions regardless of how much more detailed and rigorous any foresighting method becomes) (Skumanich and Silbernagel 1997). Case 7.2 is an example of this problem. Some researchers have even argued that there may not be much point in trying to forecast the future, given its contingent nature, particularly with respect to potential low frequency high impact events. For example, in the tourism context Leiper (2000: 808) argued, 'If nobody foresaw the severe downturn in tourism across Asia that began in mid-1997, when all the official forecasting agencies were predicting strong growth, there seems little point in trying to predict, using existing research approaches, what will happen to tourism in the next century'. However, Hall (2005c) noted that such situations do not mean that tourism forecasting is completely without value. Qualitative methods, such as environmental scanning and key force identification, as well as the results of quantitative techniques, are an important part of scenario building for government and industry, which allows *what if?* type questions to be asked with appropriate preparations in response to potential future events and issues (see also Yeoman and McMahon-Beattie 2005; Yeoman et al. 2006).

> What does not occur very often in tourism, however, is *backcasting:* the selection of desired futures and how they may be reached. Elements of the process are to be found in strategic planning with the formulation of visions, missions, goals and action plans, but they are generally not sophisticatedly applied to places. Indeed, such a change in direction may be significant, as it also gives focus to the question of what sort of future do we actually want.
>
> (Hall 2005c: 224)

A number of different techniques are available for looking at the future. Skumanich and Silbernagel (1997) identified six different categories of foresighting:

* environmental scanning and monitoring
* trend extrapolation
* opinion surveys
* scenario-building
* modelling
* morphological analysis.

Environmental scanning and monitoring are primarily data-gathering activities that provide much of the basic empirical data required to understand the current environment to provide a basis for the identification and analysis of environmental trends. An example is a state of the environment report. It is important to note that the environment, in this sense, is not just the physical environment, but can also include the socio-cultural, economic and political environments for particular organizations and locations, such as firms, destinations and governments.

Trend extrapolation is the extension of past and present trends into the future and, together with environmental scanning and monitoring, is the simplest form of foresighting. Much of the UNWTO's tourism forecasts appear to be substantially based on monitoring and trend extrapolation. However, according to Skumanich and Silbernagel (1997):

> This method has two major weaknesses. First, it is often a fallacy to assume that the future will follow the pattern of the past. While people often make such assumptions due to a lack of better information, any picture of the future that is developed on this basis can be inaccurate. The second weakness of this method is that it typically provides information on only a single variable. Especially in current world conditions, it is rare for any variable to act

independently. More often, the influence of outside forces can dramatically alter the future of any one event or condition.

Opinion surveys are usually undertaken to gather both expert opinion as well as broader public opinion. However, evidence suggests that forecasts based on expert judgement are seldom accurate, with a substantial body of research indicating that experts in various areas do not generate better predictions than non-experts who have received some training (as seen in Case 7.2). Even more, the qualitative predictions of experts are completely outperformed by those made by simple statistical models (Hall 2005a, c).

Qualitative scenario-building is regarded as being best suited to preparing for surprise events that cannot be easily incorporated within more quantitative forecasts. The approach involves the development of different scenarios to explore a range of possible future outcomes. The resulting scenarios, based on different combinations or changes in drivers, could be a business-as-usual scenario, a worst-case scenario and a best-case scenario. The events of September 11, 2001, for example, would only have been anticipated under a worst-case scenario (Commission for Environmental Cooperation (CEC) 2003).

Scenario-building (see Case 7.3) is often supplemented by *modelling* and *morphological analysis*, the latter being modelling with less reliance on quantitative data. Models are also used to understand linkages between the economy and the environment and the implications for the future (Yeoman and McMahon-Beattie 2005). Economic modelling has become an integral component of public policy analysis and, more recently, bio-economic models have been developed that utilize both economic and scientific modelling of the environment. Visit Scotland, the Scottish NTO, has been an extensive user of scenarios for tourism planning (Yeoman et al. 2006). Perhaps better known are the various climate change models of the Intergovernmental Panel on Climate Change (IPCC), which have demonstrated that such models can result in quantitative predictions around different assumptions, which can lead to a focused policy debate. However, the IPCC forecasts have also illustrated the difficulties in managing expectations of what forecast data can provide with respect to the desire for certainty. We just do not know enough to predict the future with a high degree of confidence, which considerably complicates the political debates over climate change.

The use of such forecasts with respect to tourism have similarly shown the difficulties of assuming that reports produced by experts are deterministic in predicting actual behaviour. Table 7.2 outlines some of the major weaknesses with respect to current models of different climate change scenarios in predicting passenger travel flows. Indeed, the conclusions of the CEC are salutary with respect to the overall value of different techniques of foresighting:

> The general public and many decision-makers derive comfort from the apparent authority of highly quantitative approaches, such as modeling and trend extrapolation, to future work. For this reason, such methods represent a powerful tool for persuading the public and advancing decision-making.

Table 7.2 Major weaknesses of current models in predicting passenger travel flows

- Validity and structure of statistical databases
- Temperature assumed to be the most important weather parameter
- Importance of other weather parameters largely unknown (rain, storms, humidity, hours of sunshine, air pollution)
- Role of weather extremes unknown
- Role of information in decision-making unclear
- Role of non-climatic parameters unclear (e.g. social unrest, political instability, risk perceptions, destination perception)
- Existence of fuzzy variables problematic (terrorism, war, epidemics, natural disasters)
- Assumed linearity of change in behaviour unrealistic
- Future costs of transport and availability of tourism infrastructure uncertain
- Future levels of personal disposable income (economic budget) and availability of leisure time (time budget) that are allocated to travel uncertain

Source: after Gössling and Hall 2006a, b.

However, such techniques can rarely predict the unexpected events that can have such a powerful influence on future environmental conditions. For such purposes, more imaginative techniques such as scenario-building can be highly useful. The appropriateness of using any one approach will depend on the goal and circumstances of the analysis, the kind of data available and the nature of the problem requiring analysis.

(CEC 2003: 49)

Case 7.3: The UK Foresight on Intelligent Infrastructure Systems

The UK Foresight Directorate undertakes futures research designed to assist the development of government policy and strategy (Foresight 2007). In 2006 the Directorate undertook a project on Intelligent Infrastructure Systems, the main objectives of which were:

- To focus on transportation of goods and people, and the alternatives to mass movement;
- To look at the future of transportation systems, and the application of information technologies and infrastructure, and to provide:
  - a view of the future technologies and an exploration in the scenarios of how we might deploy those technologies,
  - an understanding of how best to use those technologies to deliver our objectives,
  - a view of the opportunities and challenges that intelligent infrastructure will deliver,
  - an understanding of the presumptions that underpin the decisions we make, and
  - the strategic choices that the UK faces, along with most other societies.

The project was of relevance to tourism because of the focus on transport and mobility issues. This was regarded as being important because of the varied pressures being placed on UK transport policy and infrastructure with respect to energy costs, climate change, pollution and opposition to some major infra-structure development projects. As the UK Foresight Directorate (2006a) report on Intelligent Infrastructure Systems observed:

> Energy is not cheap, and is most unlikely to be cheaper 50 years hence. Indeed, most people would anticipate significantly higher prices. The idea that the UK could build new roads at the same pace as it did during the past half-century is simply untenable – 'road protests' did not exist 50 years ago. As to market forces, the new presumptions of future circumstances – that we have to anticipate and ameliorate the likely impacts of climate change, and that sustainability now deserves as much attention as economic growth – make it hard to see how the private sector alone can make the difficult choices.

The Foresight Directorate outlined four scenarios set in a post-oil world of personal transport for the future of human mobility, including tourism (2006b). The main uncertainties that drove the scenarios were whether or not low environmental impact transport systems will be developed and whether or not people will accept intelligent infrastructure. Four scenarios were developed which were labelled 'perpetual motion', 'urban colonies', 'tribal trading' and 'good intentions' (Foresight Directorate 2006a: 43–4).

- *Good intentions scenario*
  - The need to reduce carbon emissions constrains personal mobility.
  - Traffic volumes have fallen and mass transportation is used more widely.
  - Businesses have adopted energy-efficient practices: they use wireless identification and tracking systems to optimize logistics and distribution.
  - Some rural areas pool community carbon credits for local transport provision, but many are struggling.
  - Airlines continue to exploit loopholes in the carbon enforcement framework.

- *Perpetual motion scenario*
  - Society is driven by constant information, consumption and competition. In this world, instant communication and continuing globalization has fuelled growth: demand for travel remains strong.
  - New, cleaner, fuel technologies are increasingly popular. Road use is causing less environmental damage, although the volume and speed of traffic remains high. Aviation still relies on carbon fuels – it remains expensive and is increasingly replaced by 'telepresencing' for business, and rapid trains for travel.

- *Tribal trading scenario*
  - The world has been through a sharp and savage energy shock. The global economic system is severely damaged and infrastructure is falling into disrepair.

- Long-distance travel is a luxury that few can afford and for most people the world has shrunk to their own community.
- Cities have declined and local food production and services have increased.
- There are still some cars, but local transport is typically by bike and by horse.
- There are local conflicts over resources: lawlessness and mistrust are high.

- *Urban colonies scenario*
  - Investment in technology primarily focuses on minimizing environmental impact.
  - Good environmental practice is at the heart of the UK's economic and social policies: sustainable buildings, distributed power generation and new urban planning policies have created compact, dense cities.
  - Transport is permitted only if green and clean – car use is energy-expensive and restricted.
  - Public transport – electric and low energy – is efficient and widely used.

Each of these scenarios provides a different picture of society in the developed world in relation to transport and mobility in the year 2055. The role of the scenarios is to allow people to see how certain combinations of events, innovations and social changes could change the future. As the Directorate noted, the real world in 2055 will likely contain some elements of all these scenarios: 'The scenarios allow us to see what we might need to prepare for and the opportunities that await us if we set the right path ahead' (Foresight Directorate 2006a: 43).

Nevertheless, the different scenarios raise fundamental questions about the future of tourism and the world we will be living in (Cooper and Hall 2008), and can provide guidance in developing policies that address current needs and concerns. People are more mobile than ever before, particularly in terms of long-distance travel. Many believe that they have a right to mobility. But at the same time the opportunities for mobility are likely going to become increasingly constrained.

See: Foresight (2007) Intelligent Infrastructure Systems, online at: http://www. foresight.gov.uk/Drumbeat/OurWork/CompletedProjects/IIS/

Positive futures for tourism

Probably the most widely cited source of international tourism futures is the UN World Tourism Organization's (1997) *Tourism 2020 Vision* long-term forecast that international arrivals are expected to reach nearly 1.6 billion by the year 2020. According to UNWTO:

By the year 2020, tourists will have conquered every part of the globe as well as engaging in low orbit space tours, and maybe moon tours. The Tourism 2020 Vision study forecasts that the number of international arrivals worldwide will increase to almost 1.6 billion in 2020. This is 2.5 times the volume recorded in the late 1990s.. . . Although the pace of growth will slow down to a forecast average 4 percent a year – which signifies a doubling in 18 years, there are no signs at all of an end to the rapid expansion of tourism.. . . Despite the great volumes of tourism forecast for 2020, it is important to recognise that international tourism still has much potential to exploit . . . the proportion of the world's population engaged in international tourism is calculated at just 3.5 percent.

(UNWTO 2001: 9, 10)

Of these worldwide arrivals in 2020, 1.2 billion are forecast to be intra-regional and 378 million long-haul inter-regional travellers. According to UNWTO the top three international tourist-receiving regions by 2020 will be Europe (717 million tourists), East Asia and the Pacific (397 million) and the Americas (282 million), followed by Africa, the Middle East and South Asia. The world average international tourism growth rate over the period is estimated at 4.1 per cent with East Asia and the Pacific, South Asia, the Middle East and Africa forecast to record growth at rates of over 5 per cent, and Europe and Americas expected to show lower than average growth rates. Europe is also expected to maintain the highest share of world arrivals, although there will be a decline from 60 per cent in 1995 to 46 per cent in 2020. Long-haul travel worldwide is also expected to grow faster, at 5.4 per cent per year over the period 1995–2020, than intra-regional travel (within the major regions identified above), at 3.8 per cent per year. As a result the ratio between intra-regional and long-haul travel is forecast by UNWTO to shift from around 82:18 in 1995 to close to 76:24 in 2020.

Despite a number of events since 2000 that have affected international tourism, such as the 9/11 terrorist attacks, the second invasion of Iraq, the bombing of Madrid railways and the increasing costs of aviation fuel, UNWTO has maintained a positive outlook with respect to their forecasts. As of June 2008 the UNWTO website stated:

Although the evolution of tourism in the last few years has been irregular, UNWTO maintains its long-term forecast for the moment. The underlying structural trends of the forecast are believed not to have significantly changed. Experience shows that in the short term, periods of faster growth (1995, 1996, 2000) alternate with periods of slow growth (2001 to 2003). While the pace of growth till 2000 actually exceeded the Tourism 2020 Vision forecast, it is generally expected that the current slowdown will be compensated in the medium to long term.

(UNWTO 2008a)

Similar optimism was also expressed late in 2008, with respect to the impacts of economic uncertainty following the subprime mortgage crises and economic prospects in the United States, alongside 'global imbalances' and 'high oil prices'. Nevertheless, UNWTO concluded their release of preliminary international tourist figures for 2007 by stating, 'International tourism might be affected by this global context. But based on past experience, the sector's proven resilience and given the current parameters, UNWTO does not expect that growth will come to a halt' (UNWTO 2008b).

The World Travel and Tourism Council is also bullish with respect to future growth for tourism. In March 2008, it announced 'Looking past this present cyclical downturn, the long-term forecasts point to a mature but steady phase of growth for world Travel & Tourism between 2009 and 2018, averaging a growth rate of 4.4 percent per annum, supporting 297 million jobs and 10.5 percent of global GDP by 2018'. According to WTTC President Jean-Claude Baumgarten

> Challenges come from the US slowdown and the weak dollar, higher fuel costs and concerns about climate change. However, the continued strong expansion in emerging countries – both as tourism destinations and as an increasing source of international visitors – means that the industry's prospects remain bright into the medium term.. . . Moreover, even in countries where economic growth slows, there is likely to be a switch from international to domestic travel rather than a contraction in demand for Travel & Tourism.
>
> (WTTC 2008a)

Interestingly, an earlier WTTC press release did express concerns for the tourism industry with respect to the economic downturn and higher energy costs, although with respect to the latter the WTTC noted:

> Higher energy prices are a two pronged challenge as they squeeze household budgets globally and raise the cost of a key input for the Travel & Tourism industry. Baumgarten stated that even this challenge has a positive angle, explaining how 'higher revenues are boosting oil producers' incomes and raising available funds for investment in diversification projects often focussing on tourism's undoubted potential'.
>
> (WTTC 2008b)

Other influential forecasts of future travel includes aircraft companies, such as Boeing and Airbus. Aviation company forecasts are also extremely important because of the potential future contribution of emissions from aircraft. In the period 2006 to 2026 Boeing predicts that the number of airline passengers will increase 4.5 per cent per annum, while airline traffic expressed as revenue passenger-kilometres (RPKs) will increase 5.0 per cent. This means that world passenger traffic will grow from 4.2 trillion RPKs in 2006 to 11.4 trillion in 2026. In order to cater for such predicted demand Boeing forecast that there will be 36,420 airplanes in service in 2026 as compared with 18,230 in 2006, although 28,600 airplanes will have been delivered new after 2006 (Boeing 2007).

Airbus SAS has a similarly positive outlook on the future of aviation. Airbus (2008) foresees a demand of 24,300 new passenger and freighter aircraft between 2008 and 2026. Passenger traffic is expected to grow at an average rate of 4.9 per cent per year, leading to a near threefold increase in the forecast period and the figure is also expected to remain resilient to the cyclical effects of the aviation industry. The Airbus forecast means that they expect the world's airlines to more than double their passenger aircraft fleets of 100 seats or more, from approximately 13,300 at the beginning of 2008 to some 28,550 in 2026. In terms of markets

> The greatest demand for passenger aircraft will be from the Asia-Pacific region, which will account for 31 percent of the total world demand for aircraft. It is followed by North America (27 percent) and Europe (24 percent). Emerging markets are also driving traffic demand. Whilst China and India will remain the largest, Airbus forecasts that some 30 additional emerging economies, including Argentina, Brazil, South Africa and Vietnam, with a combined population of almost three billion people, will grow increasingly prominent by 2026.
>
> (Airbus 2008)

These forecasts for growth in international travel and tourism all point to a seemingly bright future for tourism. However, as the next section discusses, there is substantial concern as to the impacts of a number of factors such as environmental change, and particularly climate change, on tourism (also discussed in Chapter 5), as well as the increasing costs of oil in the longer term.

Changes in communications technology

In 1965, the co-founder of Intel, Gordon Moore, observed that digital technology, as measured by capacity and speed, was growing exponentially, resulting in a doubling every one to two years (Moore 1965; Hutcheson 2005). Although *Moore's law* was originally conceived in relation to the density of transistors on a circuit board, the concept has been recognized in a host of consumer and industrial technology, including computer processor speed, digital memory capacity, the physical size of digital devices, and the amount of information available in devices and accessible on the Internet. While there are theoretical limits to Moore's law, based on the size of atomic particles, technological innovations have, so far, managed to advance at this remarkable rate. Innovations that are projected to allow technology to further advance in the future include new materials and new manufacturing techniques that will allow even greater digital capacity and speeds, and advances in nanotechnology, biology and chemistry that will play new roles in digital devices (Kanellos 2005).

These technological advances have contributed to the phenomenal growth in travel and tourism over the past several decades by reducing costs for the industry, and thereby for the tourists. Global positioning system units, for example, have made flying safer for airplane pilots, and are increasingly commonplace as consumer devices that enable location and way-finding, along with location-based information on traveller services and attractions (Lew 2009) (see Photo 7.2). The Internet, which is now accessible via handheld telephones, provides consumers with access to trip planning and booking services that were inconceivable a couple of decades ago. In the future, location-based services, through cell phones and other devices that are always connected to the Internet, will offer new opportunities for travel and tourism innovation.

These advances in communication technology give businesses new ways to communicate with and provide services to consumers, and new ways for consumers to learn about destinations and different options available to them as tourists (Fessenmaier and Gretzel 2004). Niche tourism destinations and attractions are, thereby, able to create their own speciality markets of highly motivated consumers, through the use of Web 2.0 *social media* tools, such as blogs, discussion forums, podcasts and collaborative websites (sharing stories and photos, for example) (Gillin 2007; Lew 2009). Audio and video podcasts, for example, are gradually expanding in use by destination convention and visitors bureaus to provide alternative forms of information on attractions and services available to visitors (Xie and Lew 2009). These can take the form of one-off introductions to various attractions at a destination, or as audio walking or driving tours, or as weekly updates on special events. As this technology becomes more widely used by destinations, and more common among consumers, other innovative applications are likely to arise in the future.

Photo 7.2 GoCars in San Francisco. Advertising themselves as 'computer guided tours', these two passenger vehicles use GPS systems and location-based audio tracks to guide tourists through the city of San Francisco. Attraction and service providers can customize the audio so special messages play when the car is near their establishment, and companies can design special tours that only focus on the interests of particular market clientele. GoCars are also found in a few other US and European cities. (Photo by Alan A. Lew)

Virtual travel is a longer-term technological innovation for the future of travel and tourism. Virtual worlds and destinations are currently highly contrived and animated experiences. However,

> In addition to location-based technologies, the future holds considerable promise for virtual experiences of real places, which may alleviate the need for everyone to physically travel to every corner of the globe. Such experiences are likely to be very social, as most virtual worlds are today, and provide real out of body and in place experiences.
>
> (Lew 2009)

Three-dimensional digital rooms, in which people are physically surrounded on all sides by a moving display of real and imaginary views, are currently in the experimental stage, and only a few such rooms exist in the world today. Explorations into simulating out-of-body experiences are barely starting, but hold the ultimate promise of being able to experience a different place without physically leaving your home. It is also likely that as the technology of virtual experiences advances, opportunities to experience these will become leisure and recreation attractions in themselves.

Advances in transportation technology will also alleviate some of the greenhouse gas emission and oil consumption issues of travel and tourism, discussed below and in Chapter 5. For example, a United Airlines flight from Sydney, Australia, to San Francisco, California, was able to save 1,564 gallons and 32,656 pounds of CO_2, in comparison to a standard flight, 'by using up-to-the-minute fuel data, priority takeoff clearance, normally restricted airspace around Sydney's airport, and new arrival procedures – all of which are possible with new technology' (United Airlines 2008). The emission reductions were in comparison to average flights for the year on the same route and required special training for the pilots, and new fuel monitoring gauges in the cockpit. These kinds of technologies, along with alternative fuel sources, hold much promise for the future of travel and tourism.

Negative futures for tourism

Cost of oil

The capacity to engage in tourism is subject to a number of constraints, including cost, time, regulation, accessibility and motivation (see Chapter 4). Several of these constraints will be directly and indirectly affected by the challenges to the future growth of tourism, such as the cost of energy, and especially aviation fuel. According to the Energy Committee at the Royal Swedish Academy of Sciences *Statement on Oil* (ECRSRS 2005: 1): 'It is very likely that the world is now entering a challenging period for energy supply, due to the limited resources and production problems now facing conventional (easily accessible) oil. Nearly 40 percent of the world's energy is provided by oil, and over 50 percent of the latter is used in the transport sector.'

The cost of fuel is extremely important for tourism as it affects travel costs and, therefore, the relative accessibility of destinations. Oil is also central to tourism because it is not easily replaceable as a source of energy for many of the main modes of tourism transport, such as aviation, shipping (cruise ships and ferries) and, to a great extent, the use of cars and buses, especially over longer distances (see Case 7.4). The only mode of transport in which oil can be substituted as a fuel with relative ease is railways, where lines can be electrified. However, while this is significant for continental destinations where railway connections can still carry international travellers, this is obviously not a relevant scenario for trans-oceanic and island destinations.

Case 7.4: Aviation – 'oil is everything'

In June 2008 the International Air Transport Association (IATA), representing 230 airlines and accounting for 94 per cent of scheduled air traffic, revised its industry financial forecast for 2008 significantly downwards to a loss of US$2.3 billion. The June 2008 forecast used a consensus oil price of US$106.5 per barrel crude (Brent) and represented a drop of US$6.8 billion from the previously forecasted industry profit of US$4.5 billion that had been announced in March and was based on an average oil price of US$86 per barrel. The change in the air transport industry's total fuel bill has been dramatic:

- In 2002 the bill was US$40 billion, equal to 13 per cent of costs
- In 2006 the bill was US$136 billion, equal to 29 per cent of operating costs
- In 2008 the bill was expected to be US$176 billion (based on oil at US$106.5 per barrel) accounting for 34 per cent of operating costs.

(IATA 2008a)

According to Giovanni Bisignani, IATA Director General and CEO,

We also need to take a reality check. Despite the consensus of experts on the oil price, today's oil prices make the US$2.3 billion loss look optimistic. For every dollar that the oil price increases, we add US$1.6 billion to costs. If we see US$135 oil for the rest of the year, losses could be US$6.1 billion.. . . Oil is changing everything. There are no easy answers. In the last six years, airlines improved fuel efficiency by 19 percent and reduced non-fuel unit costs by 18 percent. There is no fat left. To survive this crisis, even more massive changes will be needed quickly.

(IATA 2008a)

Even though the price for oil had dropped substantially by November 2008 to under $60 per barrel, the economic damage had been done, with IATA (2008d) forecasting that airline industry net losses would total $5.2 billion in 2008 and weak financial performance will continue into 2009, with further net losses of $4.1 billion. The cost of oil and the downturn in the global economy had a substantial impact on airlines, even those from oil-rich states. For example, Emirates Airline reported an 88 per cent drop in its net profit for the six months to 30 September 2008, with the reason largely due to higher oil prices. The Dubai government-owned airline's fuel costs were $469 million higher than it had planned in its budget. Emirates chairman Sheikh Ahmed bin Saeed al-Maktoum said in a statement, 'The first half of the year has been very tough for the airline industry, with record fuel prices forcing many carriers to shut shops or consolidate' (BBC News 2008f). It has also been hard on destinations, with more isolated communities and distant islands seeing a dramatic drop in the number of arrivals as airlines cut back on their flights, or even declare bankruptcy (see Photo 7.3).

The impact of high oil process can also be seen in Milmo's (2008) report that British Airways, which was one of only 14 airlines in the world with a profit margin of more than 10 per cent at the beginning of 2008, had admitted it could move from a profit of £883m to a loss by 2010, despite its robust balance sheet, because of fuel cost pressures. The rest of the aviation industry is much less resilient, operating on an average profit margin of just over 1 per cent, thus underlining the threat posed by a sudden spike in fuel costs. In fact, between December 2007 and April 2008 three UK-based carriers (Maxjet, Eos and Silverjet) and three US-based airlines (Aloha, ATA and Frontier) had grounded their fleets and declared bankruptcy primarily due to rising oil prices. Worldwide, some 30 airlines had failed through the middle of 2008, and another 30 were expected to declare bankruptcy by mid-2009, despite a return to lower fuel prices, due to the decline in consumer spending resulting from the global credit crisis.

In his State of the Industry address at IATA's 64th Annual General Meeting and World Air Transport Summit in Istanbul, Turkey, Bisignani argued the 'fuel crisis' should be a catalyst for change in the bilateral system that provides for the various Freedoms of the Air in four different areas: restrictions on trade, security, regulation of monopolies and the environment (IATA 2008b). According to Bisignani,

> To fight crises effectively, brands not flags must define our business.. . . We must communicate clearly to governments the dimension of the oil crisis, the potential impact on the global economy if the air transport industry fails, the measures that airlines are taking to survive and the action we need from them. . . . It's time to tear-up the 3,500 bilateral agreements and replace them with a clean sheet of paper without any reference to commercial regulation. Airlines would be free to innovate, compete, grow, become financially healthy or even disappear. Governments also have an important role: to ensure a level playing field and regulate safety, security and environmental performance.
>
> (IATA 2008b)

Nevertheless, with respect to the environment Bisignani was also critical of some forms of government intervention, especially EU actions on carbon emissions trading:

The current fuel crisis must be a catalyst for governments to deliver results on the environment that reduce fuel burn. Our vision for carbon neutral growth leading to a carbon-free future sets the benchmark.. . . But governments remain fixated on punitive economic measures such as the EU Emissions Trading Scheme. These are reckless decisions when the oil price could re-shape the industry. Governments must drive progress by taking politics out of air traffic management, acting globally on emissions trading and supporting positive economic measures to drive innovation.

(IATA 2008b)

A number of key issues associated with oil supply can be identified (ECRSAS 2005; International Energy Agency (IEA) 2007).

Shortage of oil

The global demand for oil has been growing by almost 2 per cent per year, with consumption for 2008 averaging 85.89 million barrels per day (1 barrel = 159 litres

Photo 7.3 Puerto Rico's Condado Beach. This is the tourist core of the island of Puerto Rico, a popular sun, sand and surf destination for Americans and Canadians from colder northern climates. Airline industry moves to drop flights and lay employees off in 2008, in response to high fuel prices, resulted in a significant drop in tourists to Puerto Rico and other Caribbean island destinations. Puerto Rico responded with marketing campaigns and incentives for airlines to keep their routes to this commonwealth territory of the United States. (Photo by Alan A. Lew)

or 42 gallons) or over 31 billion barrels for the year; and consumption was expected to increase in 2009, despite the global economic slowdown (Schenk 2008). Finding additional supplies is increasingly problematic since most major oil fields have already 'matured', 54 of the 65 most important oil-producing countries have declining productions, and the rate of discoveries of new reserves is less than a third of the rate of increased consumption, as of the end of 2005. Although the IEA (2007: 5) forecast that 'World oil resources are judged to be sufficient to meet the projected growth in demand to 2030, with output becoming more concentrated in OPEC countries – on the assumption that the necessary investment is forthcoming'. They also add a significant cautionary noted:

> Although new oil-production capacity additions from greenfield projects are expected to increase over the next five years, it is very uncertain whether they will be sufficient to compensate for the decline in output at existing fields and keep pace with the projected increase in demand. A supply-side crunch in the period to 2015, involving an abrupt escalation in oil prices, cannot be ruled out.
>
> (IEA 2007: 5)

Concerns over a shortage of oil have also led to significant public debate over the concept of *peak oil* and its impact on tourism and travel. IATA (2008e: 1), for example, argues that the rise in oil prices in 2008 had 'not been because of "peak oil", the idea that we are fast running out of oil', but instead was driven by a futures market 'bubble'. In the UK, the British government endorses the IEA forecast, yet in October 2008 eight leading UK companies (Arup, FirstGroup, Foster + Partners, Scottish and Southern Energy, Solarcentury, Stagecoach Group, Virgin Group, Yahoo), known as the UK Industry Taskforce on Peak Oil and Energy Security (ITPOES) or 'The Peak Oil Group', launched a report warning that a peak in cheap, easily available oil production is likely to hit by 2013, posing a major risk to the UK and world economy. Clearly, peak oil and oil availability are a major policy issue.

Peak oil refers to the point in time when the maximum rate of petroleum production is reached in any area under consideration (e.g. from a specific site to the global scale), after which the rate of production goes into decline, often becoming progressively more expensive to produce. (Table 7.3 outlines some of the projections of the peaking of world oil production.) Peak oil, therefore, does not mean the same as 'no oil' or 'running out' of oil. Oil will always be available but the key issue is its accessibility in terms of cost. Peak oil is, therefore, concerned with the flow of oil. As Andrews and Udall (2008) note,

> Why does the obfuscation of peak oil deniers matter? The coming end of the 'supply growth' world will require an enormous paradigm shift: there will be a little less oil to divvy up among more people. We will need to conserve with a vengeance, and substitute ingenuity, intelligence, and efficiency where we can. Treating this [imminent] event as a non-problem could end up being enormously painful.

Table 7.3 Projections of the peaking of world oil production

Projected date	Source of projection	Background and reference
2005	Pickens, T. Boone	Oil & gas investor; Boone Pickens warns of petroleum production peak, *EV World*. May 4, 2005
December 2005	Deffeyes, K.	Retired Princeton professor & retired Shell geologist; http://www.defense-and-society.org/fcs/crisis_unfolding.htm. Feb. 11, 2006
2006–2007	Bakhitari, A.M.S.	Iranian oil executive; World oil production capacity model suggests output peak by 2006–7. *Oil & Gas Journal*. April 26, 2004
Close or past	Herrera, R.	Retired BP geologist; Bailey, A., Has oil production peaked? *Petroleum News*. May 4, 2006
2007–2009	Simmons, M.R.	Investment banker; Simmons, M.R. ASPO Workshop. May 26, 2003; CFA Society of St Louis, Brentwood, MO, slide 23, 'World should assume we are at peak for oil and gas', May 24, 2006
After 2007	Skrebowski, C.	Petroleum journal editor; Oil field mega projects – 2004. *Petroleum Review*. January 2004
At hand	Westervelt, E.T. et al.	US Army Corps of Engineers; Energy trends and implications for US Army installations. ERDC/CERL TN-05–1, Sept. 2005
Before 2009	Deffeyes, K.S.	Oil company geologist (ret.); *Hubbert's Peak: The Impending World Oil Shortage*. Princeton University Press, 2003
Very soon	Groppe, H.	Oil/gas expert & businessman; Peak oil: myth vs. reality. Denver world oil conference. Nov. 10, 2005
Before 2010	Goodstein, D.	Vice Provost, Cal Tech; *Out of Gas – The End of the Age of Oil*. W.W. Norton, 2004
Around 2010	Bentley, R.	University energy analyst; The case for peak oil, DOE/EPA Modelling the Oil Transition, April 21, 2006
Around 2010	Campbell, C.J.	Oil company geologist (ret.); Industry urged to watch for regular oil production peaks, depletion signals. *Oil & Gas Journal*. July 14, 2003; An updated depletion model, *The Association for the Study of Peak Oil and Gas, ASPO Newsletter* no. 64 April, 2006
2010 +/–1 year	Skrebowski, C.	Editor of *Petroleum Review*; Peak oil: the emerging reality, ASPO-5 Conference, Pisa, Italy, July 18, 2006
After 2010	World Energy Council	Non-government organization; *Drivers of the Energy Scene*. World Energy Council. 2003.
A challenge around 2011	Meling, L.M.	Statoil oil company geologist; Statoil ASA. Oil supply, Is the peak near? Centre for Global Energy Studies. Sept. 29–30, 2005

Table 7.3 Continued

Projected date	Source of projection	Background and reference
Around 2012	Pang, X. et al.	China University of Petroleum; Pang, X., Meng, Q., Zhang, J. and M. Natori, The challenges brought by the shortage of oil and gas in China and their countermeasures, ASPO IV International Workshop on Oil and Gas Depletion, May 19–20, 2005
Around 2012	Koppelaar, R.H.E.M.	Dutch oil analyst; *World Oil Production & Peaking Outlook*. Stichting Peakoil-Nederland, 2005
2010–2020	Laherrere, J.	Oil company geologist; Seminar Center of Energy Conversion. Zurich. May 7, 2003
Within a decade	Volvo Trucks	Swedish company; Volvo website
Within a decade	De Margerie, C.	Oil company executive; *ASPO Newsletter* no. 65, May 2006
Around 2015	al Husseini, S.	Retired exec. VP of Saudi Aramco; End of an era, *Cosmos*, April 2006
2015–2020	ECRSAS	*Statements on Oil*, October 14, Stockholm: Energy Committee at the Royal Swedish Academy of Sciences, 2005
2015–2020	West, J.R., PFG Energy	Consultants; Energy Insecurity, Testimony before the Senate Committee on Commerce, Science & Transportation. Sept. 21, 2005
2016	Energy Information Administration (EIA) nominal case	US Department of Energy (DOE) analysis/information; DOE EIA. *Long Term World Oil Supply*. April 18, 2000
Around 2020 or earlier	Maxwell, C.T., Weeden & Co.	Brokerage/financial; The gathering storm, *Barron's*. Nov. 14, 2004
Within 15 years	Amarach Consulting	Ireland; *A Baseline Study on Oil Dependence in Ireland*. Amarach Consulting, Ireland. Dec. 2005
Tight balance by 2020	Wood Mackenzie	Energy consulting; *MacroEnergy Long Term Outlook – March 2006*. Wood Mackenzie
Around 2020	Total	French oil company; Bergin, T., Total sees 2020 oil output peak, urges less demand, Reuters, June 7, 2006
After 2020	CERA	Energy consultants; Jackson, P. et al. Triple witching hour for oil arrives early in 2004 – but, as yet, no real witches. *CERA Alert*. April 7, 2004
2025 or later	Shell	Major oil company; Davis, G. *Meeting Future Energy Needs*. The Bridge. National Academies Press. Summer 2003

Table 7.3 Continued

Projected date	Source of projection	Background and reference
Mid- to late 2020s	UBS	Brokerage/financial; Oil output set to peak, but no fuel shortage-UBS. Reuters. August 24, 2006
After 2030	EIA	US DOE analysis; Morehouse, D., Private communication. June 1, 2006 in Hirsch 2007
After 2030	CERA	Energy consultants; 'Barring unforeseen events there is no reason to believe capacity couldn't meet demand well after 2030, CERA researchers said.' Strahan, A. Global Petroleum Capacity to Rise 25 Percent by 2015. Bloomberg. 2006–08–08
Now–2040	Congressional General Accountability Office	*Uncertainty about Future Oil Supply Makes It Important to Develop a Strategy for Addressing a Peak and Decline in Oil Production.* GAO-07–283. February 2007
No sign of peaking	Exxon Mobil	Major oil company; www.exxonmobil.com. http://www.exxonmobil.com/Corporate/Files/Corporate/OpEd_peakoil.pdf (as of March 2007)
No visible peak	Lynch, M.C.	Energy economist; Petroleum resources pessimism debunked in Hubbert model and Hubbert modelers' assessment. *Oil and Gas Journal*, July 14, 2003
Impossible to predict	BP CEO	Major oil company CEO; As profit rises, BP chief seeks to allay anger. *Washington Post*. July 27, 2006
Do not subscribe to peak oil theory	OPEC	Neil Chatterjee, OPEC needs clear demand signals for spare capacity, *Mail & Guardian* Online. July 11, 2006

Sources: derived from ECRSAS 2005; Hirsch et al. 2005; Hirsch 2007a, b.

Reserves of conventional oil

Since the mid-1990s, two-thirds of the increases in reserves of conventional oil have been based on increased estimates of recovery from existing fields and only one-third on the discovery of new fields. According to the US Geological Survey, a conservative estimate of discovered oil reserves and undiscovered recoverable oil resources is about 1,200 billion barrels, including 300 billion barrels in the world's, as yet unexplored, sedimentary basins (ECRSAS 2005). IATA (2008e: 1) argue that, 'If no more oil was discovered and we continued to consume at a rate of 85mb/d, known oil reserves would still last over 40 years until 2052'.

The factors determining present and future oil flow rates are:

- 85 per cent of the world's oil is produced by the 21 largest producer countries.
- Production declines are occurring in six of the large producers: the USA, Indonesia, the UK, Norway, Mexico and Venezuela.

- Flat or volatile production is occurring in Russia, Iraq, Iran, Nigeria and Algeria.
- Production is increasing in the rest. 'But China is nearing peak. Saudi Arabia, Kuwait, Qatar and the United Arab Emirates are not planning much more expansion. Canada and Libya can continue growing, within limits . . . only Brazil, Kazakhstan and Angola are likely to grow production sufficiently to make a difference past 2010' (Andrews and Udall 2008).

The key role of the Middle East

According to the Energy Committee (ECRSAS 2005) only in the Middle East, and possibly in the countries of the former Soviet Union, is there potential (proven reserves of 130 billion barrels) to significantly increase production rates to compensate for decreasing rates in other countries. As of the end of 2005, Saudi Arabia provided 9.5 million barrels per day (11 per cent of the global production rate).

Unconventional oil resources

There are very large hydrocarbon resources, so-called unconventional oil, including natural gas (about 1,000 billion barrels of oil equivalent, much of which could be converted to liquid fuels), heavy oil and tar sands (about 800 billion barrels), oil shales (about 2,700 billion barrels) and coal (ECRSAS 2005). Problems with unconventional resources include long lead times in development, significant environmental impacts, and the availability of water and natural gas for the production and refining processes.

Immediate action on supplies

Improvements in the search for, and recovery of, conventional oil, as well as the production rate of unconventional oil, are required to avoid price spikes. According to the Energy Committee (ECRSAS 2005), unstable oil price fluctuations will lead to instability of the world economy over the next few decades. The IEA (2007: 13) also note that there is a significant global energy challenge that will affect the world economy, but argue 'the primary scarcity facing the planet is not of natural resources nor money, but time. Investment now being made in energy-supply infrastructure will lock in technology for decades, especially in power generation'.

Liquid fuels and the transport system

Oil supply is a severe liquid fuels problem and less of a general energy supply problem, as 57 per cent of the world's oil is consumed in the transportation sector (ECRSAS 2005). Alternatives need to be developed to oil in the transport sector, otherwise not only will there be increased oil prices, but also increased competition between transport and other oil users. The lack of transferability (substitution) of energy sources for many forms of transport means that transport demand for oil is relatively price-inelastic although there is also a degree of elasticity and substitution with respect to consumer demand for different modes of transport.

Economic considerations

In the long run, the price of crude oil will be determined by the price of substitutes. The Energy Committee (ECRSAS 2005) anticipate continued high oil prices as long as the pressure from the expanding Asian economies is maintained. The IEA (2007: 9) also note that 'growing exports from China and India also increase competitive pressures on other countries, leading to structural adjustments.. . . Rising commodity needs risk driving up international prices for commodities, including energy – especially if supply-side investment is constrained'.

Environmental concerns

Unconventional oil will significantly extend the length of the hydrocarbon era and its subsequent contributions to greenhouse gas emissions. Constraints similar to those imposed on other fossil fuels (e.g. emission controls and CO_2 sequestration) will be necessary and provide major challenges for industry. The IEA (2007) also report that rising CO_2 and other greenhouse gas concentrations in the atmosphere, resulting largely from fossil-energy combustion, are contributing to higher global temperatures and to changes in climate. Noting in their forecasts for the period 2005–30, 'Growing fossil-fuel use will continue to drive up global energy-related CO_2 emissions over the projection period' (IEA 2007: 11). In the reference scenario used by the IEA for their forecasts, 'emissions jump by 57 percent between 2005 and 2030. The United States, China, Russia and India contribute two-thirds of this increase. China is by far the biggest contributor to incremental emissions, overtaking the United States as the world's biggest emitter in 2007. India becomes the third-largest emitter by around 2015' (IEA 2007: 11).

Significantly, the IEA argue that 'Urgent action is needed if greenhouse-gas concentrations are to be stabilised at a level that would prevent dangerous interference with the climate system' (2007: 12) with the curbing of emissions from coal-fired power stations being a major focal point for IEA activities, along with emissions savings from improvements in the efficiency of fossil-fuel use, switching to nuclear power and renewables, and the widespread deployment of CO_2 capture and storage (CCS) mechanisms. 'Exceptionally quick and vigorous policy action by all countries, and unprecedented technological advances, entailing substantial costs, would be needed to make this case a reality' (IEA 2007: 12). Perhaps more positively, they also note that 'Many of the policies available to alleviate energy insecurity can also help to mitigate local pollution and climate change, and vice-versa. As the Alternative Policy Scenario demonstrates, in many cases, those policies bring economic benefits too, by lowering energy costs – a "triple-win" outcome' (IEA 2007: 13).

Tourism and greenhouse gas emissions: adaptation and mitigation

As noted in Chapter 5, one of the major focal points for the relationship between the environment and tourism is that of climate change. Even if some of the forecasts for tourism growth already discussed are cut by half, tourism is still almost certainly going to contribute to increasing emissions. The UNWTO et al.'s (2008) estimate for late 2007 was that emissions of CO_2 from tourism activities will grow by a factor 2.5 in the period 2005–35, primarily as a result of increasing air travel. However, such growth stands in contrast to the global emission reduction needs, as agreed upon in the Kyoto Agreement and IPCC documents. In order to limit the average increase in global temperatures to a maximum of 2.4°C, the smallest increase in any of the IPCC scenarios, the concentration of greenhouse gases in the atmosphere would need to be stabilized at around 450 ppm. To achieve this, CO_2 emissions would need to peak by 2015 at the latest, and to fall between 50 per cent and 85 per cent below 2000 levels by 2050. The IPCC (2007) suggests the goal of cutting emissions to about 1 ton of CO_2 per person per year by the end of the century, and countries like Sweden and the UK are already discussing emission cuts by 60 to 80 per cent by 2050.

However, very few governments have taken any action to address emissions from aviation or tourism more generally, although it is an area in which much debate is occurring (Gössling et al. 2009). This comes, however, as the European Council, against protests by the IATA and the US government, adopted a directive that will include aeroplanes in its carbon Emissions Trading Scheme (ETS) (Stone 2008). This will not take effect until 2012, but it will include all flights within the EU, as well as inter-regional flights that originate or depart from an EU airport. Under the ETS, activities that emit carbon dioxide are granted an allowance for their emissions that is slightly below what their actual emissions are. They must either adopt methods or technologies to reduce their output to the lower levels, or purchase (trade) additional allowances from users who are better able to reduce their emissions. For the airlines, the initial emission caps in 2012 will be 97 per cent of their estimated 2004 to 2006 emission level, which will drop to 95 per cent in 2013.

The reasons for this highly controversial environment policy primarily relates to the costs of any mitigation and adaptation strategies (Monbiot 2006). For example, applying the 'polluter pays' principle to an area such as aviation would have significant impacts on the costs for airline passengers and freight, and their levels of demand. Furthermore, climate change is now part of wider political debates and appears to be a factor in electoral campaigns. For example, Australia's non-ratification of the Kyoto Agreement under the conservative coalition government of Prime Minister John Howard (1996 to 2007) was an issue that emerged a number of times during Australia's 2007 federal election (e.g. Murphy 2007). Tourism and climate change relationships were directly linked in the election campaign with Opposition Labour Party leader, Kevin Rudd, announcing an A$200m (£90m, US$185.5m) plan to help save the Great Barrier Reef, one of Australia's leading tourist attractions, which is being damaged by water pollution, primarily from land run-off, and rising sea temperatures (BBC News 2007f).

Similarly, climate change and environmental issues were seen as playing a significant part in the election of the Democrat President, Barack Obama, in the United States in 2008, while the political parties in the UK were also seeking to position themselves with respect to climate change in the run-up to their next election, with the Labour government announcing a 'stronger' climate change bill to help lower Britain's carbon emissions and strengthen the country's international position in response to climate change (Ryan and Stewart 2007). Following a period for public consultation and scrutiny, suggested amendments to that bill included adding emissions from the aviation and shipping industry in the UK's GHG emission targets. The measure built on the British government's change in position in October 2007, when it announced that starting in 2009 air taxes would be levied on the flight, rather than the passenger, a measure that was regarded as better targeted at cutting carbon emissions (Wintour and Elliott 2007), and which was also a policy of the then opposition Conservative Party. Interestingly, the policy position was also publicly supported by the budget airline EasyJet, which argued that air passenger duty should be dropped in favour of a scheme that grades aircraft according to their carbon dioxide emissions and the length of their journey (Milmo 2007). However, given the potential lifespan of aircraft this may also have the effect of moving older aircraft with higher emissions to regions with weaker emissions rules such as Africa, Asia or Latin America.

As Chapter 5 indicated, the impacts of climate change will be substantial and will profoundly affect not only tourism but also the communities and environments with which tourism is related. Tourism may see changes in travel flows and patterns in both the short and long term as a result of climate change, such as reductions in ski seasons, loss of some ski areas and coastal resources, and changes in the competitiveness of tourist destinations because of changes in climate and extreme events (see Case 7.1). Although the impacts of climate change on tourism may be difficult to predict at a local scale, Table 7.4 outlines the distribution of major climate change impacts affecting tourism destinations by regions. In addition, it also identifies six destinations that were most at risk from the cumulative effects of climate and environmental change: South Africa, the Mediterranean, Queensland (Australia), South-Western United States and Polynesia/Micronesia.

Such forecasts are extremely difficult because there are substantial gaps in our knowledge of the impacts of climate and environmental change on tourism (Table 7.5), as well as on the adaptive and mitigative practices and behaviours of destinations and tourists. Indeed, there will be winners as well as losers from environmental change, with respect to tourism, at least in the short term. Ecotourism, for example, has been cited as a potential substitute for the ski industry in Asia (Fukushima et al. 2002). Similarly, Anisimov et al. (2007) noted the potential significance of ecotourism as an opportunity for adaptation of indigenous peoples in the Arctic and sub-Arctic regions, even though the biodiversity and landscape of that region is experiencing some of the most rapid climate-related environmental changes on the planet today, with average Arctic temperatures having increased at almost twice the rate of the rest of the world in

Table 7.4 Distribution of major climate and environmental change impacts affecting tourism destinations by region

Impact	Africa	Australia/ New Zealand	Caribbean	Indian Ocean Small Island Nations	Mediter-ranean	Middle East	North America	Northern Europe	Pacific Ocean Small Island Nations	South America	South/ East Asia
Increase in disease outbreaks	o+	o	+		o+		o	o+			+
Increase in extreme events			+	+			+		+	+	+
Land biodiversity loss	o+			+	o+		o+	+	o+	+	o
Marine biodiversity loss	+		o+	o+	o+			+	o+	+	o+
Political instability	o+		+			+					+
Sea level rise		o	+	+	o		o+		o+		o+
Travel cost increase from main markets as a result of mitigation policy	o+	o	+	+					o+	+	
Urbanization					o		o				o
Warmer summers	o+	o	+		o+	+	o+	o+			
Warmer winters		o					o+	o+			
Water security	o+	o	+	+	o+	+	o+		+		

Key: + identified in UNWTO et al. 2008, UNWTO and UNEP 2008; o identified in Gössling and Hall 2006d.

Table 7.5 Relative level of tourism specific climate change knowledge and estimated impact of climate change on tourism by region

Region	Estimated impact of climate change on tourism	Relative level of tourism specific climate change knowledge
Africa	Moderately–strongly negative	Extremely poor
Asia	Weakly–moderately negative	Extremely poor
Australia & New Zealand	Moderately–strongly negative	Poor–moderate (high in Great Barrier Reef)
Europe	Weakly–moderately negative	Moderate (high in alpine areas)
Latin America	Weakly–moderately negative	Poor
North America	Weakly negative	Moderate (high in coastal and ski areas)
Polar regions	Weakly negative–weakly positive	Poor
Small islands	Strongly negative	Moderate

Sources: derived from Gössling and Hall 2006a; Parry et al. 2007; UNWTO et al. 2008; Hall 2008b.

the past 100 years (Arctic Climate Impacts Assessment (ACIA) 2005; Solomon et al. 2007).

Even one of the most high-profile dimensions of climate change in recent years – the opening of the Northwest Passage through the Arctic Ocean, as the result of the loss of Summer ice in 2007 – is seen as potentially beneficial for tourism. Anisimov et al. (2007: 676) state 'the Northern Sea Route will create new opportunities for cruise shipping. Projections suggest that by 2050, the Northern Sea Route will have 125 days/yr with less than 75 percent sea-ice cover'. Similarly, Instanes et al. (2005) noted that increased possibilities for marine navigation and the extension of the warm-weather season will improve conditions for tourism.

The fact that there are potential 'winners', at least in the short term, as well as 'losers' from climate change also reflects the adaptation capacities of different regions, sectors, actors and firms. '*Adaptive capacity* is the ability or potential of a system to respond successfully to climate variability and change, and includes adjustments in both behaviour and in resources and technologies' (Adger et al. 2007: 727). Because of the explicit focus on real-world behaviour, assessments of adaptation practices differ from the more theoretical assessments of potential response (see Photo 7.4). However, at this stage, an understanding of the adaptive capacities and practices of the various elements of tourism in relation to climate change are quite limited (Becken 2005; Hall and Higham 2005; Gössling & Hall 2006a; UNWTO et al. 2008), a situation that arguably reflects the large knowledge gaps that surround a number of areas of tourism and change related research.

Photo 7.4 Singapore's Changi International Airport. About half the airport, one of the busiest in Asia, can be seen here, along with a large part of the eastern part of the modern island city-state. As a global transportation hub, many of the planes coming in and out of Singapore are long-haul flights to and from Europe, Australia and northeast Asia (China and Japan). Living in one of the smallest countries in the world, in land area, Singaporeans are avid travellers, which contributes to their having carbon footprints similar to those of the United States and Australia. Changing our human desire to travel will be one of the biggest challenges facing the world in the coming decades. (Photo by Alan A. Lew)

Which way for tourism? Responding to change

Tourism appears to be at a significant crossroads in its development. One path, as epitomized by the UNWTO and WTTC rosy tourism growth forecasts, is suggesting that tourism will experience almost unlimited growth in the future, or at least to 2020 or 2030, provided that further deregulation occurs and greater economic integration is encouraged. This is also regarded as potentially assisting in the implementation of the United Nations Millennium Goals to raise people from poverty in the less developed countries. For the UNWTO, increased focus on tourism by these countries will be integral for such a strategy.

Another path is far more cautious and, while not anti-tourism, suggests that the impacts of environmental change, and global warming in particular, threats to oil supplies and growing concern over the environmental impacts of tourism may lead to mitigative government actions that seek to place greater limits and controls on tourism. In addition, this more cautious approach also suggests that the general economic turbulence of the new energy and emissions environment may affect people's consumer behaviour too, thereby influencing travel and tourism (e.g. Monbiot 2006). Perhaps rather unfortunately for the future of the tourism industry, the reality seems to be that

> Much of today's tourism infrastructure is shaped by almost half a century's assumptions that we would have cheap oil and energy, that it was possible to respond to market demand by building more capacity, and that 'predict and provide' assumptions of infrastructure supply would meet transport needs. Not only has infrastructure been shaped by such assumptions but also successive generations of people in [developed] countries who take long-distance mobility as a norm.
>
> (Cooper and Hall 2008: 367–8)

Nevertheless, one of the clear messages from research on environmental change, as well as other forms of tourism-related change, is that people, places and

Photo 7.5 Climate change as an election issue: campaign billboard from the 2008 New Zealand election campaign. As part of their post-election negotiations with the ACT Party the new National Party government suspended the previous government's emissions trading scheme and removed the moratorium on coal-powered electricity generation. (Photo by C. Michael Hall)

firms can adapt (Gössling et al. 2009). Hall (2009a) notes that the capacity of tourism businesses to survive and grow is a measure of their individual resilience, as well as a potential indicator of system resilience at a destination level. For example, firm survival is recognized as being linked to the innovative capacity of those firms (Hall and Williams 2008), which often involves various adaptations to their changing physical, social and economic environment. The capacity of tourism businesses to adapt and innovate is, therefore, a product of both the characteristics of the firm and their embeddedness within tourism and innovation systems (see Case 2.4). From a system-wide perspective, the resilience of a social–ecological system, such as a tourism destination, will be affected by the innovative diversity of the tourism firms within it. This means that the different sensitivities of firms to specific disturbances to the system, such as economic and social change and restructuring, will decrease the vulnerability of that destination to such disturbances. As noted several times in this book, change is a normal process. The fundamental issue, in terms of the sustainability of firms and destinations, is not so much change per se, but the magnitude of change in relation to a system's capacity to adapt and develop without the loss of the natural capital essential to social, economic and political well-being.

Case 7.5: A model of tourism firm adaptation

In order to understand how tourism firms may adapt to the challenge of change, and to environmental change in particular, it is helpful to understand how a firm adds value and organizes itself. This can be done via its business model, which is a conceptual tool that contains a set of elements and their relationships that allows the expression of the business logic of a specific firm (Hall and Williams 2008). 'It is a description of the value a company offers to one or several segments of customers and of the architecture of the firm and its network of partners for creating, marketing, and delivering this value and relationship capital, to generate profitable and sustainable revenue streams' (Osterwalder et al. 2005: 17–18).

Figure 7.1 presents a business model of the *innovation value creation points* in a tourism firm (Hall and Williams 2008). The main idea of creating a reference model is that it identifies the domains, concepts and relationships shared among tourism firms, no matter where the firm is located. Nevertheless, the service aspects of tourism allow for the consideration of a number of common elements or dimensions. Eleven dimensions are identified: (1) the business model (referring to the overall approach), (2) networks and alliances, (3) enabling processes, (4) service design and development, (5) service value, (6) distribution, (7) brand, (8) servicescape, (9) customer service experiences, (10) customer satisfaction, and (11) customer loyalty. It should be noted that many of the elements of the model are closely related to one another. In addition, a further dimension exists, which is the way in which all the elements of the business model are put together – what would commonly be described as the *business strategy*. Significantly, each of these dimensions is an opportunity for innovation and adaptation at the firm level.

INCREASING CUSTOMER FOCUS →

Innovation points in value chain	Business model	Networks and alliances	Enabling process	Service design and development	Service value	Distribution	Brand	Servicescape	Customer's service experience	Customer satisfaction	Customer loyalty
Intellectual capital	How the firm makes its money and how it is structured	Economic and knowledge relationships with other firms and actors	Support firm's core processes and employees	Service development architecture and systems. Matching of service to market segments	Provision of value to customers and consumers with respect to products	How the product offerings reach the market	The symbolic embodiment of all the information connected to the product, values and expectations	The physical evidence of service and manifestation of brand	The experiences that customers have at point of co-creation / production	Levels of customer satisfaction with experience over time	Long-term relationships with customers
Innovative capacities and potential	New business models, including ownership structure, payment and cash flow system, and location	New sets of actor relations create new economic and social capital. New supply chains also affect product offerings	Improving employee satisfaction, retention and productivity	New service design, including service blueprints, assessment of internal service quality processes, and market segmentation	New services and means of adding value for customers	Utilization of new channels, combinations of channel and intermediaries	Utilization of new ways to communicate offerings and create brand value	Development and design of new servicescapes, including the virtual world	Providing new experiences	Tracking customer satisfaction	Customer relationship management strategies
Examples of potential sustainability-related firm innovations	Place-based ownership reinforces local supply chains	Developing local supply chain minimizes emission costs of supply	Improvements in productivity through reducing energy consumption and waste	Services can be designed so as to reduce emissions and be more energy efficient	New green services can attract new customers as well as generate new firm product	Developing distribution channels with low emissions and energy demands	Utilizing the brand to emphasize positive environmental values at a firm and destination scale	Design can incorporate measures to minimize emissions and energy use as well as enhance experiences	Knowledge of experience can help make environmental measures acceptable to customers	Better knowledge can lead to increased yield per customer at lower environmental cost	Loyal customers become advocates for firm's green business activities

Source: Hall 2009a, after Hall and Williams 2008.

Figure 7.1 Innovation value creation points in the tourism firm in relation to sustainability.

The model indicates the need to understand innovation for sustainable tourism as being primarily based in business organization and behaviour – the social dimensions of innovation – rather than in the utilization of technology. The model also suggests that as actors in the tourism system, sustainable firms should be understood in terms of survivability and profit, as clearly they are not sustainable if they cease to exist. However, the notion of survivability also suggests that the conceptualization of firm behaviour needs to be extended beyond that of being solely responsible to shareholders, with a focus on short-term returns, and instead indicates that return needs to be understood over extended time periods. This approach, in a sense, approximates the concept of sustained yield, rather than immediate gratification. Sustainable tourism, like innovation, is therefore socially constructed because it requires new attitudes, measurements and ways of thinking about profit, success and shareholder value.

Source: Hall 2009a.

Summary and conclusions

This book has provided an overview of the impacts of tourism and their relationship to economic, social and environmental change at a critical time for the

Photo 7.6 Sidmouth, UK. In order to protect the beaches of the holiday resort of Sidmouth large groynes have had to be built so as to reduce the impacts of storms on the coast. Increased storms and higher sea levels as a result of climate change may mean that many other coastal resorts will have to utilize expensive engineering solutions for coastal protection. (Photo by C. Michael Hall)

future of the tourism industry. Tourism can no longer be regarded as a 'smokeless' industry and may even be behind a number of other industries or fields of consumption with respect to the overall level of environmental activity and the provision of environmentally sound alternatives (Figure 7.2) (Martens and Spaargaren 2005). Increasingly, we are coming to realize that tourism is like any other form of development: it has its costs and benefits. But for some reason, tourism often seems to be regarded as different, and seems to regard itself as different. Maybe this is because, in much of the developed world at least, the majority of the population, and certainly nearly all of this book's readers, have been tourists at one time or another.

Because of this, tourism is subject to something akin to Hardin's (1968) 'tragedy of the commons', whereby free access and unrestricted demand for a finite resource ultimately structurally dooms the resource through over-exploitation. In Hardin's essay, as well as in the world of tourism, the 'solution' has been to manage the problems of the commons via various means, such as regulation, applying the polluter pays principle, and even corporatization or privatization of what was previously an unregulated public resource. In the case of some of the commons properties, such as the air or the sea, regulation and polluter pays policies are often substantially resisted by the tourism industry. The industry typically refers to the wider 'public good' that their economic activity generates as a justification to minimize regulations. Of course, such arguments are

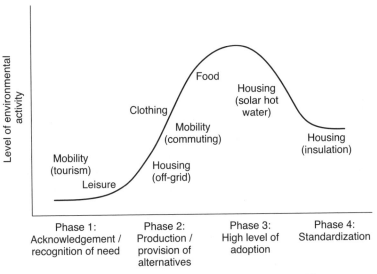

Stages in the greening of an industry via the provision
and adoption of environmentally sound
consumption alternatives

Figure 7.2 The relative level of environmental activities and the provision and adoption of environmentally sound alternatives for different consumption domains.

nonsense, as all businesses contribute to a common economic good via their purchasing and employment capacities, and yet common public goods (such as air and roads) still need to be managed to avoid degradation from overuse.

In the case of the individual tourist, it is also extremely likely that there are very few of them who deliberately aim to abuse the destination they are visiting and the social and environmental resources they are consuming. Increasingly, people are even seeking to make environmentally and socially friendly tourism purchasing decisions. As in the tragedy of the commons, individual travellers generally do not believe that their consumption is going to lead to negative impacts. In fact, most will likely believe that their being a tourist is good for a destination because of the employment it must bring. Nevertheless, it is the totality of all the individual tourist consumption decisions that is causing economic, social and environmental challenges.

Unfortunately, there are no easy solutions to these challenges, as preventing people from travelling will only create political problems, while all that consumption (and subsequent pollution) will just be shifted elsewhere. Instead, solutions lie in the realm of the subtitle of Hardin's famous essay: 'it requires a fundamental extension in morality'. The issue is not just one for tourism, significant as it is, but is grounded with respect to morality in terms of self-reflexive, conscious considerations of what and why we consume, as well as how cooperation can be encouraged for mutual benefit (Ostrom et al. 1999). Undoubtedly, some steps in the tourism field have been taken with respect to better understanding how we can do this, but apart from isolated instances, tourism has neither become a 'force for peace' nor fulfilled the initial hope of sustainability and ecotourism. Instead, much of tourism has become, and is becoming, a *greenwashing*.

What then is to be done? If the business of tourism is to make a contribution to the sustainability of this planet, as well as the industry, then it becomes vital that there is greater public acknowledgement of what the positive and negative impacts of tourism are, and thereby to see tourism as part of the larger socio-economic and biophysical systems. To modify Goodland and Daly (1996: 1002), sustainable tourism development is tourism development without growth in throughput of matter and energy beyond regenerative and absorptive capacities. Although usually regarded as a service industry, tourism still consumes tangible items. In order to reduce its environmental footprint, tourism needs to become part of a *circular economy* rather than a linear one, so that inputs of virgin raw material and energy and outputs in the form of emissions and waste requiring disposal are reduced. Such a change is often categorized as *sustainable consumption*, and is also similar to the concept of a steady-state economy (Hall 2009c) (see Case 7.6).

Within tourism there are two main forces for change in encouraging this approach (see Figure 7.3). First is the efficiency approach, which seeks to reduce the rate of consumption by using materials more productively. This approach places more focus on recycling, using energy more efficiently, and reducing emissions, but otherwise operating in a 'business-as-usual' manner. Examples of this include the efforts of Boeing and Airbus to produce more fuel-efficient aircraft. The second approach may be referred to as 'slow consumption', which is primarily consumer led, but which is also influenced by policy initiatives at various scales of governance. This includes:

- the development of environmental standards at the regional, national and international scales, e.g. such as the Nordic Swan label (Bohdanowicz 2006);
- localization schemes such as farmers' markets and the 100-mile diet which reinforce the potential economic, social and environmental benefits of purchasing locally and which may receive substantial support from local authorities (Hall et al. 2008; Hall and Sharples 2008);
- ethical consumption, through ethical and responsible tourism (Goodwin and Francis 2003); and
- the so-called 'new politics of consumption' such as culture jamming and anti-consumerism (Ropke and Godsaken 2007).

Significant examples of this second approach include the Slow Food Movement (Hall and Sharples 2008; Miele 2008), and the Fair Trade Movement (Clevedon and Kalisch 2000; Goodwin and Francis 2003). Arguably, there are also attempts to meld various elements of the two approaches together as with the Industry Taskforce on Peak Oil and Energy Security's (ITPOES 2008) position on the response to the problems of peak oil. Increasing product lifespans and decreased energy use may also enable both efficiency and sufficiency (Cooper 2005; Jackson 2005). This includes means by which materials are used more productively (i.e. the same quantity providing a longer service) and throughput is slowed (i.e. products are replaced less frequently, plus, in the case of tourism, distance of travel is less – the so-called 'staycation' approach). Such *life-cycle* thinking has been brought to other sectors, such as manufacturing, but it would be timely to start to consider this approach with respect to tourism.

The increasing costs of energy, and oil for transport in particular, is providing an external factor for change. As Lord Ron Oxburgh, former chairman of Shell, wrote in the foreword to the ITPOES report on peak oil:

Today's high prices are sending a message to the world that words alone have failed to convey, namely that not only are we leaving the era of cheap energy but that we have to wean ourselves off fossil fuels. For once what is right is also what is expedient – we know that we have to stop burning fossil fuels because of the irreversible environmental damage they cause, and now it may be cheaper to do so as well! The problem is that in the developed world our power and transport infrastructure is based almost entirely on fossil fuels. With the best will and the best technology in the world this will take decades to change.

(ITPOES 2008: 3)

Sustainable tourism consumption does not necessarily mean people travelling less, although in the case of long-distance air travel there may be a decline, but it

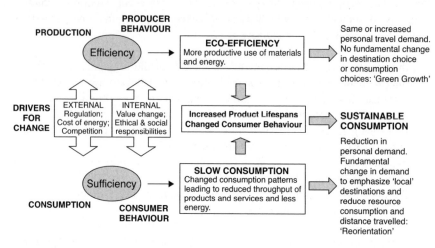

Figure 7.3 Towards sustainable tourism consumption (Hall 2007f).

will mean people travelling more locally and, when people do travel long-distance, potentially staying longer so as to reduce the overall environmental effects of their trip in terms of emissions and energy consumption. Although some destinations distant from their origin markets will be affected many will benefit.

Case 7.6: Towards steady-state tourism?

Building on the notion that sustainable tourism development is tourism development without growth in throughput of matter and energy beyond regenerative and absorptive capacities, Hall (2009c) has argued that there needs to be a focus on the development of tourism within a steady-state economy, what can be described as steady-state tourism. Much tourism growth, as with much economic growth in general, is already uneconomic at the present margin as we currently measure it. As Daly (2008: 2) commented,

> The growth economy is failing. In other words, the quantitative expansion of the economic subsystem increases environmental and social costs faster than production benefits, making us poorer not richer, at least in high-consumption countries. Given the laws of diminishing marginal utility and increasing marginal costs, this should not have been unexpected.. . . It is hard to know for sure that growth now increases costs faster than benefits since we do not bother to separate costs from benefits in our national accounts. Instead we lump them together as 'activity' in the calculation of GDP.

Steady-state tourism is therefore a tourism system that encourages qualitative development but not aggregate quantitative growth to the detriment of natural and human capital (Hall 2009c). A steady-state economy, including at the destination level, can be defined in terms of 'a constant flow of throughput at a sustainable (low) level, with population and capital stock free to adjust to whatever size can be maintained by the constant throughput beginning with depletion and ending with pollution' (Daly 2008: 3). Such an approach consciously seeks to make the costs of growth transparent in any set of national or regional accounts, including satellite accounts (see Chapter 3). In addition, visitor numbers would be included in assessments of population stock for locations.

Hall (2009c) has argued that sustainable tourism policies should be geared to stop tourism growing where marginal costs equal marginal benefits. Yet much of the thinking about tourism, even if it does acknowledge economic costs, does not fully consider the extent to which the marginal benefits of economic growth relates to those costs. Instead, 'any consumption of capital, manmade or natural, must be subtracted in the calculation of income' (Daly 2008: 10). Hall (2009c) agrees that climate change and meeting the challenge of global poverty are two major issues for tourism but that the 'solutions' of organizations such as UNWTO and the World Bank are wrong. Instead, promoting medium and long-haul tourism without consideration of environmental change only continues to use scarce energy resources and increase emissions that contribute to global warming and ultimately deplete natural capital, particularly as the rate of population growth and per capita tourism mobility is exceeding the environmental efficiencies presently being gained by the tourism industry. To be truly sustainable human capital cannot expand at the expense of natural capital. Under the dominant doctrine of 'economism' in which growth is the ultimate good, developed countries as well as the newly industrialized countries should continue to grow as rapidly as possible to provide markets for the poor and accumulate capital to invest in the less developed world (assuming that GDP growth is actually reasonably distributed in those countries). In contrast, a steady-state approach which focuses on 'right-sizing' the economy in relation to human and natural capital constraints indicates that the developed world 'should reduce their throughput growth to free up resources and ecological space for use by the poor, while focusing their domestic efforts on development, technical and social improvements, that can be freely shared with poor countries' (Daly 2008: 2).

This, therefore, takes us back to the start of the book and the need to understand tourism as a system. The last chapter highlighted the notion of integrated approaches to tourism impacts with respect to organizational roles and seeking to find common ground between different value positions. This chapter, if not the book as a whole, highlights the importance of integrating our understanding of sustainability into our entire approach to planning, managing and, hence, critically for students of the subject, thinking about tourism. Research and thinking on tourism and its impacts needs to broaden interest in visitor consumption beyond the points of purchase to all phases in the life of the tourism product and experience, from start to finish and from conception to final disposal. Understanding the local impacts of tourism will remain important, but increasingly we realize that impacts occur at more than just the attraction or destination, but instead occur through the entire course of tourist consumption.

Tourism is a major industry. But it is time to go beyond the promotional phrases of it being the world's 'fastest-growing' or 'largest industry' to develop a much more balanced, systematic and integrated assessment of its economic, social and environmental relationships. Until this is done, the long-term potential of tourism contributing to sustainable change cannot be fulfilled.

Self-review questions

1　What are the main approaches to foresight research?
2　How are advances in technology likely to impact travel and tourism in the future?
3　Why is oil supply such a significant issue for international tourism?
4　Why is it hard to predict the impacts of climate change on travel flows to destinations?
5　In November 2008 the first democratically elected President of the Maldives, Mohamed Nasheed, announced that the country will begin to divert a portion of the country's annual billion-dollar tourist revenue into buying a new homeland in another country – as an insurance policy against climate change that threatens to turn the 300,000 islanders into environmental refugees (Ramesh 2008). To what extent is it ethically appropriate to use international tourism, which has significant emissions, to do this?

Recommended further reading

Gössling, S., Hall, C.M. and Weaver, D. (eds) (2009) *Sustainable Tourism Futures: Perspectives On Systems, Restructuring and Innovations*, New York: Routledge.
Provides a range of chapters on some of the various dimensions of achieving sustainable tourism.

Gössling, S. and Hall, C.M. (eds) (2006) *Tourism and Global Environmental Change*, London: Routledge.
The most comprehensive account of the relationships between tourism and global environmental change.

Hall, C.M. (2005) *Tourism: Rethinking the Social Science of Mobility*, Harlow: Prentice-Hall.
The last four chapters deal with the future of tourism as well as the future of tourism studies.

Lew, A.A. (2009) 'Long tail tourism: new geographies for marketing niche tourism products', *Journal of Travel and Tourism Marketing* 25(3/4): 409–19.
Provides an overview of Web 2.0 technology and social media, with applications for marketing tourism on the Internet.

Monbiot, G. (2006). *Heat: How to Stop the Planet Burning*, London: Allen Lane.
Provides a challenging read on the issues associated with climate change and the possibility of a sustainable future, including the issue of aviation and tourism which are described as 'love miles'.

Web resources

Association for the Study of Peak Oil and Gas (ASPO): http://www.peakoil.net/

Energy Committee at the Royal Swedish Academy of Sciences: http://www.kva.se/KVA_Root/eng/society/energy/index.asp?br=ns&ver=6up

Energy Information Administration (US official energy statistics): http://www.eia.doe.gov/

International Air Transport Association (IATA): http://www.iata.org/index.htm

International Energy Agency: http://www.iea.org/

Life after the Oil Crash: Deal with Reality or Reality Will Deal with You: http://www.lifeaftertheoilcrash.net/

Peak Oil News and Message Boards (a community peak oil portal): http://www.peakoil.com/

UK Industry Taskforce on Peak Oil and Energy Security: http://peakoil.solarcentury.com/

Air Transport Action Group: http://www.environ.aero

Tourism Place in Second Life (Virtual Worlds Connected): http://snipr.com/5k33e

Geographical Magazine, Future of Tourism issue. The December 2008 issue of the Royal Geographical Society's magazine was devoted to the future of tourism: http://www.geographical.co.uk/Home/index.html

Key concepts and definitions

Adaptive capacity The ability or potential of a system to respond successfully to climate variability and change, and includes adjustments in both behaviour and in resources and technologies.

Peak oil Refers to the point in time when the maximum rate of petroleum production is reached in any area under consideration (e.g. from a specific site to the global scale), after which the rate of production goes into decline, often becoming progressively more expensive to produce.

Steady-state tourism A tourism system that encourages qualitative development but not aggregate quantitative growth to the detriment of natural and human capital.

Steady-state destination economy A constant flow of throughput at a sustainable (low) level, with permanent population, visitor numbers and natural and human capital stock free to adjust to whatever size can be maintained by the constant throughput beginning with depletion and ending with pollution.

Sustainable tourism development Development of tourism without growth in throughput of matter and energy beyond regenerative and absorptive capacities.

Tipping point A point beyond which the nature of a natural or other system changes markedly and change becomes irreversible.

Wildcard events High impact, very low probability events.

Social media Internet-based tools that enable the creation of social communities, including discussion forums, audio and video podcasts and blogs with commenting space, photo and story sharing sites, and product review websites.

References

Aa, B.J.M. van der, Groote, P.D. and Huigen, P.P.P. (2004) 'World heritage as NIMBY: the case of the Dutch part of the Wadden Sea', *Current Issues in Tourism*, 7(4–5): 291–302.

Ackerman, F. and Stanton, E. (2006) *Climate Change – The Costs of Inaction*. Report to Friends of the Earth England, Wales and Northern Ireland, Global Development and Environment Institute, Tufts University.

Adger, W.N., Agrawala, S., Mirza, M.M.Q., Conde, C., O'Brien, K., Pulhin, J. et al. (2007) 'Assessment of adaptation practices, options, constraints and capacity', in M.L. Parry, O.F. Canziani, J.P. Palutikof, P.J. van der Linden and C.E. Hanson (eds) *Climate Change 2007: Impacts, Adaptation and Vulnerability*, Cambridge: Cambridge University Press.

Agrawala, S. (2007) *Climate Change in the European Alps: Adapting Winter Tourism and Natural Hazard Management*, Paris: OECD.

Airbus (2008) Airbus foresees demand for some 24,300 aircraft in the next 20 years, Press Release, 7 February, ttp://www.airbus.com/en/presscentre/pressreleases/press releases_items/08_02_07_airbus_forecast_2008.html

Alcamo, J., Moreno, J.M., Nováky, B., Bindi, M., Corobov, R., Devoy, R.J.N. et al. (2007) 'Europe', in M.L. Parry, O.F. Canziani, J.P. Palutikof, P.J. van der Linden and C.E. Hanson (eds) *Climate Change 2007: Impacts, Adaptation and Vulnerability*, Cambridge: Cambridge University Press.

Amelung, B. and Viner, D. (2006) 'Mediterranean tourism: exploring the future with the tourism climatic index', *Journal of Sustainable Tourism*, 14: 349–66.

Anderson, M.R. (1996) 'Human rights approaches to environmental protection: an overview', in A.E. Boyle and M.R. Anderson (eds) *Human Rights Approaches to Environmental Protection*, pp. 21–2, New York: Oxford University Press.

Andrews, S. and Udall, R. (2008) Peak Oil: 'It's the Flows, Stupid!', 12 May. Association for the Study of Peak Oil and Gas – USA. Energy Actions for a Healthy Economy and a Clean Environment, http://www.aspo-usa.com/index.php?option=com_content&task=view&id=370&Itemid=91

Ang, A. (2008) 'Tibet tourism plunges in the wake of protests', *USA Today*, 19 May, http://www.usatoday.com/travel/destinations/2008-05-19-tibet-protests_N.htm

Anisimov, O.A., Vaughan, D.G., Callaghan, T.V., Furgal, C., Marchant, H., Prowse, T.D. et al. (2007) 'Polar regions (Arctic and Antarctic)', in M.L. Parry, O.F. Canziani, J.P. Palutikof, P.J. van der Linden and C.E. Hanson (eds) *Climate Change 2007: Impacts, Adaptation and Vulnerability*, Cambridge: Cambridge University Press.

Archer, B.H. (1982) 'The value of multipliers and their policy implications', *Tourism Management*, 3: 236–41.

Arctic Climate Impacts Assessment (ACIA) (2005) *Impacts of a Warming Arctic: Arctic Climate Impacts Assessment*, Cambridge: Cambridge University Press.

Arkell, J. (2003) Background paper on GATS issues. Presented at Commonwealth Business Council, Symposium on Trade Policy Challenges in East Asia: The New Regionalism and the WTO, 20 February.

Ashworth, G.J. and Kuipers, M.J. (2001) 'Conservation and identity: a new vision of pasts and futures in the Netherlands', *European Spatial Research Policy,* 8(2): 55–65.

Ashworth, G.J. and Tunbridge (2004) 'Whose tourist-historic city? Localizing the global and globalizing the local', in A.A. Lew, C.M. Hall and A.M. Williams (eds) *Companion to Tourism*, London: Blackwell.

Ateljevic, I. and Doorne, S. (2000) 'Staying within the fence: lifestyle entrepreneurship in tourism', *Journal of Sustainable Tourism,* 8(5): 378–92.

Australian Bureau of Statistics (2006) *Innovation in Australian Business 2005, Catalogue no. 8158.0,* Canberra: Australian Bureau of Statistics.

—— (2007) *ABS Catalogue no. 8158.0*, Canberra: Australian Bureau of Statistics.

Ausubel, J.H., Marchetti, C. and Meyer, P. (1998) 'Toward green mobility: the evolution of transport', *European Review,* 6(2): 137–56.

Baade, R. and Matheson, V. (2002) 'Bidding for the Olympics: fool's gold?' in C. Barros, M. Ibrahimo and S. Szymanski (eds) *Transatlantic Sport*, London: Edward Elgar.

Badalamenti, F., Ramos, A.A., Voultsiadou, E., Sánchez Lizaso, J.L., D'anna, G., Pipitone, C. et al. (2000) 'Cultural and socio-economic impacts of Mediterranean marine protected areas', *Environmental Conservation,* 27: 110–25.

Baggio, R. (2007) Symptoms of complexity in a tourism system, arXiv:physics/0701063v2 [physics.soc-ph]

Baker, D. and Refsgaard, K. (2007) 'Institutional development and scale matching in disaster response management', *Ecological Economics,* 63(2–3): 331–43.

Bartlett, R.V. and Kurian, P.A. (1999) 'The theory of environmental impact assessment: implicit models of policy making', *Policy and Politics*, 27(4): 415–33.

Baum, T. (2006) *Human Resource Management for the Tourism, Hospitality and Leisure Industries: An International Perspective*, London: Thompson.

BBC News (2004) 'What makes Heritage List and why?' *BBC News*, 1 July, http://news.bbc.co.uk/1/hi/world/asia-pacific/3858079.stm (accessed 18 February 2008).

—— (2006) 'Trump "in talks" over development', *BBC News*, 12 January, http://news.bbc.co.uk/2/hi/uk_news/scotland/4607062.stm (accessed 25 January 2006).

—— (2007a) 'Tibet tourism "hits record high"', *BBC News, Asia-Pacific*, 17 December, http://news.bbc.co.uk/2/hi/asia-pacific/7147366.stm

—— (2007b) 'Scheme to boost beaver population', *BBC News*, 1 October, http://news.bbc.co.uk/2/hi/uk_news/scotland/glasgow_and_west/7021983.stm

—— (2007c) 'Trump's £1bn golf plan rejected', 29 November, http://news.bbc.co.uk/2/hi/uk_news/scotland/north_east/7118105.stm (accessed 3 December 2007).

—— (2007d) 'Trump's £1bn resort plan "dead"', 2 December, http://news.bbc.co.uk/2/hi/uk_news/scotland/north_east/7123779.stm (accessed 3 December 2007).

—— (2007e) 'Ministers to decide Trump plans', 5 December, http://news.bbc.co.uk/2/hi/uk_news/scotland/north_east/7127760.stm (accessed 22 December 2007).

—— (2007f) 'Rudd unveils Barrier Reef plan', 29 October, http://news.bbc.co.uk/2/hi/asia-pacific/7067158.stm (accessed 29 October 2007).

—— (2008a) 'Tourism crash threatens big cats', *BBC News*, 29 April, http://news.bbc.co.uk/go/pr/fr/-/2/hi/science/nature/7372298.stm

—— (2008b) 'Senegal city is "most threatened"', *BBC News*, 13 June, http://news.bbc.co.uk/2/hi/africa/7452352.stm

—— (2008c) 'Trump "signals a new local era"', 5 November, http://news.bbc.co.uk/2/hi/uk_news/scotland/north_east/7710375.stm (accessed 5 November 2008).

—— (2008d) 'Trump's £1bn golf resort approved', 3 November, http://news.bbc.co.uk/2/hi/uk_news/scotland/north_east/7700074.stm (accessed 4 November 2008).

—— (2008e) 'Trump golf inquiry in full swing', 10 June, http://news.bbc.co.uk/2/hi/uk_news/scotland/north_east/7444123.stm (accessed 14 June 2008).

—— (2008f) 'Emirates profit sinks on oil cost', *BBC News*, 10 November, http://news.bbc.co.uk/2/hi/business/7719620.stm (accessed 10 November, 2008).

Becken, S. (2005) 'Harmonizing climate change adaptation and mitigation. The case of tourist resorts in Fiji', *Global Environmental Change – Part A*, 15(4): 381–93.

Becker, H.A. (2000) 'Social impact assessment', *European Journal of Operational Research*, 128(2): 311–21.

Becker, H.A. and Vanclay, F. (eds) (2003) *The International Handbook of Social Impact Assessment: Conceptual and Methodological Advances*, Cheltenham: Edward Elgar.

Beeton, S. (2005) *Film-Induced Tourism*, Clevedon: Channel View Publications.

Beeton, S. and Benfield, R. (2002) 'Demand control: the case for demarketing as a visitor and environmental management tool', *Journal of Sustainable Tourism*, 10(6): 497–513.

Beldona, S. (2005) 'Cohort analysis of online travel information search behavior: 1995–2000', *Journal of Travel Research*, 44(2): 135–42.

Bell, C. and Lewis, M. (2005) 'The economic implications of epidemics old and new', Centre for Global Development Working Paper No. 54, Washington, DC: Centre for Global Development.

Bendor, J. (1995) 'A model of muddling through', *American Political Science Review*, 89(4): 819–40.

Bergal, J., Hiles, S.S. and Koughan, F. (2007) *City Adrift: New Orleans Before and After Katrina*, Baton Rouge: Louisiana State University Press.

Berger, J. (2007) 'Fear, human shields and the redistribution of prey and predators in protected areas', *Biology Letters*, 3(6): 620–3.

BEUC (2008) Airline tickets on the internet: fair practices are still to take off/ Billets d'avion sur Internet: ça vole bas! BEUC Press Release, PR 023/2008, 8 May.

Biosecurity Strategy Development Team (2001) *A Biosecurity Strategy for New Zealand, Strategy Vision Framework Background Paper for Stakeholder Working Groups*, Wellington: Biosecurity Strategy Development Team.

Blair, J. (1995) *Local Economic Development: Analysis and Practice*, Thousand Oaks, CA: Sage.

Boden, M. and Miles, I. (2000) *Services and the Knowledge-based Economy*, London: Continuum.

Boeing (2007) *Current Market Outlook 2007: How Will You Travel Through Life*, Seattle: Boeing Commercial Airplanes, Market Analysis.

Bohdanowicz, P. (2006) 'Environmental awareness and initiatives in the Swedish and Polish hotel industries—survey results', *International Journal of Hospitality Management,* 25(4): 662–82.

—— (2009) 'Theory and practice of environmental management and monitoring in hotel chains', in S. Gössling, C.M. Hall and D. Weaver (eds) *Sustainable Tourism Futures: Perspectives on Systems, Restructuring and Innovations*, New York: Routledge.

Boko, M., Niang, I., Nyong, A., Vogel, C., Githeko, A., Medany, M. et al. (2007) 'Africa', in M.L. Parry, O.F. Canziani, J.P. Palutikof, P.J. van der Linden and C.E. Hanson (eds) *Climate Change 2007: Impacts, Adaptation and Vulnerability*, Cambridge: Cambridge University Press.

Boyd, S.W. and Butler, R.W. (1996) 'Managing ecotourism: an opportunity spectrum approach', *Tourism Management*, 17(8), 557–66.

Boyle, P., Halfacree, K. and Robinson, V. (1998) *Exploring Contemporary Migration*, Harlow: Addison Wesley Longman.

Bradley, D.J. (1989) 'The scope of travel medicine: an introduction to the conference on international travel medicine', in R. Steffen, H.O. Lobel and J. Bradley (eds) *Travel Medicine: Proceedings of the First Conference on International Travel Medicine, Zurich, Switzerland, April 1988*, Berlin: Springer-Verlag.

Brandon, K. (1996) *Ecotourism and Conservation: A Review of Key Issues*, Environment Department Paper no.33. Washington DC: World Bank.

Bread Not Circuses (1998a) *Bread Alert! (E-mail edition)* 2(2) February 20.

—— (1998b) *Bread Alert! (E-mail edition)* 2(3) February 26.

Brimblecombe, P. (1987) *The Big Smoke: A History of Air Pollution in London Since Medieval Times*, London: Routledge.

Brinkley, D. (2006) *The Great Deluge: Hurricane Katrina, New Orleans, and the Mississippi Gulf Coast*, New York: Morrow.

Britton S. (1991) 'Tourism, capital, and place: towards a critical geography of tourism', *Environment and Planning D: Society and Space,* 9(4) 451–78.

Brown, D. and Reeder, R. (2005) 'Rural areas benefit from recreation and tourism development', *Amber Waves* (September), http://ers.usda.gov/AmberWaves/September05/Features/RuralAreasBenefit.htm (accessed 28 February 2008).

Brown, L. (1981) *Building a Sustainable Society*, New York: W.W. Norton.

Bryman, A., Bytheway, B., Allatt, P. and Keil, T. (1987) 'Introduction', in A. Bryman, B. Bytheway, P. Allatt and T. Keil (eds) *Rethinking the Life Cycle*, Basingstoke: Macmillan.

Budowski, G. (1976) 'Tourism and conservation: conflict, coexistence or symbiosis', *Environmental Conservation*, 3(1): 27–31.

Bull, A. (1994) *The Economics of Travel and Tourism*, 2nd edn, South Melbourne: Longman Australia.

Burdge, R.J. (2002) 'Why is social impact assessment the orphan of the assessment process?' *Impact Assessment and Project Appraisal*, 20(1): 3–9.

Butler, R.W. (ed.) (2006a) *The Tourism Life Cycle. Vol.1, Applications and Modifications*, Clevedon: Channel View Publications.

—— (2006b) *The Tourism Life Cycle, Vol.2, Conceptual and Theoretical Issues*, Clevedon: Channel View Publications.

Butler, R.W. and Waldbrook, L.A. (2003) 'A new planning tool: the Tourism Opportunity Spectrum', *Journal of Tourism Studies*, 14(1): 21–32, http://www.jcu.edu.au/business/idc/groups/public/documents/journal_article/jcudev_012854.pdf (accessed 8 May 2008).

Buultjens, J., Ratnayake, I., Gnanapala, A. and Aslam, M. (2005) 'Tourism and its implications for management in Ruhuna National Park (Yala), Sri Lanka', *Tourism Management*, 26(5): 733–42.

Candela, G., Figini, P. and Scorcu, A.E. (2005) *The Economics of Local Tourist Systems*, Nota di Lavoro 185.2005. Milan: Fondazione Eni Enrico Mattei.

Carlsen, J. (1999) 'A systems approach to island tourism destination management', *Systems Research and Behavioral Science,* 16(4): 321–7.

Carlton, J.T. (1999) 'The scale and ecological consequences of biological invasion in the World's oceans', in O.T. Sandlund, P.J. Schei and Á. Viken (eds) *Invasive Species and Biodiversity Management*, Dordrecht: Kluwer.

Carroll, R. (2008) 'Tourism curbed in bid to save Galapagos haven', The Observer, 12 October, http://www.guardian.co.uk/travel/2008/oct/12/galapagosislands-travelnews

Carson, R. (1962) *Silent Spring*, New York: Houghton Mifflin.

Caton, K. and Santos, C.A. (2007) 'Heritage tourism on Route 66: deconstructing nostalgia', *Journal of Travel Research,* 45(4): 371–86.

CBS (2005) *Poll: Katrina Response Inadequate.* http://www.cbsnews.com/stories/2005/09/08/opinion/polls/main824591.shtml (accessed August 23, 2006)

Ceballos-Luscarain, H. (1990) 'La Ruta Maya: a transfrontier ecocultural tourist circuit in the Yucatan region', in J. Thorsell (ed.) *Parks on the Borderline: Experiences in Transfrontier Conservation*, Gland, Switzerland: IUCN.

Centre on Housing Rights and Evictions (COHRE) (2007) *Fair Play for Housing Rights. Mega-Events, Olympic Games and Housing Rights. Opportunities for the Olympic Movement and Others*, Geneva: COHRE.

Ceron, J-P. and Dubois, G. (2003) 'Tourism and sustainable development indicators: the gap between theoretical demands and practical achievements', *Current Issues in Tourism,* 6(1): 54–75.

Chan, W.W. and Lam, J.S. (2000) 'The lodging industry's contribution to Hong Kong's gross domestic product', *International Journal of Contemporary Hospitality Management,* 12(2): 86–98.

Chang, T.C. (2000) 'Singapore's Little India: a tourist attraction as a contested landscape', *Urban Studies,* 37(2): 343–66.

Chang, T.C. and Huang, S. (2004) 'Urban tourism: between the global and the local', in A.A. Lew, C.M. Hall and A.M. Williams (eds) *Companion to Tourism*, London: Blackwell.

Chape, S., Blyth, S., Fish, L., Fox, P. and Spalding, M. (compilers) (2003) *2003 United Nations List of Protected Areas*, Gland and Cambridge: IUCN and UNEP-WCMC.

Chen, A. (2007) 'For Olympics in China, a dizzying hotel boom', *International Herald Tribune*, 22 August 2007, http://www.iht.com/articles/2007/08/22/travel/trbeijing.php (accessed 28 February 2008).

Chernushenko, D. (1994) *Greening our Games: Running Sports Events and Facilities that Won't Cost the Earth*, Ottawa: Centurion.

China National Tourism Organization (CNTO) (2007) China Tourism Statistics, http://www.cnto.org/chinastats.asp#transport (accessed 27 February 2008).

Christ, C., Hilel, O., Matus, S. and Sweeting, J. (2003) *Tourism and Biodiversity: Mapping Tourism's Global Footprint*, Washington, DC: Conservation International.

Clark, W. and Munn, R.E. (eds) (1986) *Ecologically Sustainable Development of the Biosphere*, New York: Cambridge University Press.

Clarke, J.N. and McCool, D. (1985) *Staking Out the Terrain: Power Differentials Among Natural Resource Management Agencies*, Albany: State University of New York Press.

Clevedon, R. and Kalisch, A. (2000) 'Fair trade in tourism', *International Journal of Tourism Research,* 2(3): 171–87.

Cliff, A. and Haggett, P. (2004) 'Time, travel and infection', *British Medical Bulletin,* 69: 87–99.

Cohen, E. (1979) 'A phenomenology of tourist experiences', *Sociology*, 13: 179–201.

Cohen, M.P. (1984) *The Pathless Way: John Muir and American Wilderness*, Madison: University of Wisconsin Press.

Cole, S. (2007) 'Beyond authenticity and commodification', *Annals of Tourism Research*, 34(4): 943–60.

Coles, T.E. (2008) 'Introduction: tourism and international citizenship and the state: hidden features in the internationalization of tourism', in T.E. Coles and C.M. Hall (eds) *International Business and Tourism: Global Issues, Contemporary Interactions*, London: Routledge.

Coles, T.E. and Hall, C.M. (eds) (2008) *International Business and Tourism: Global Issues, Contemporary Interactions*, London: Routledge.

Coles, T.E. and Timothy, D. (eds) (2004) *Tourism, Diasporas and Space*, London: Routledge.

Coles, T.E., Duval, D. and Hall, C.M. (2004) 'Tourism, mobility and global communities: new approaches to theorising tourism and tourist spaces', in W. Theobold (ed.) *Global Tourism*, 3rd edn, Oxford: Heinemann.

Coles, T.E., Hall, C.M. and Duval, D. (2005) 'Mobilising tourism: a post-disciplinary critique', *Tourism Recreation Research*, 30(2), 31–41.

Colten, C.E. (ed.) (2000) *Transforming New Orleans and Its Environs: Centuries of Change*, Pittsburgh, PA: University of Pittsburgh.

Colten, C.E. (2005) *An Unnatural Metropolis: Wrestling New Orleans from Nature*, Baton Rouge: Louisiana State University Press.

Commission for Environmental Cooperation (CEC) (2003) *Understanding and Anticipating Environmental Change in North America: Building Blocks for Better Public Policy*, Montréal: Commission for Environmental Cooperation.

Committee for the Activities of the Council of Europe in the Field of Biological and Landscape Diversity (CO-DBP) (1999) *European Code of Conduct for Coastal Zones*, CO-DBP (99)11, Secretariat General Directorate of Environment and Local Authorities, Strasbourg.

Community Affairs. http://www.msu.edu/course/prr/840/econimpact/pdf/ecimpvol1.pdf (accessed 28 February 2008).

Convery, I. and Dutson, T. (2008) 'Rural communities and landscape change: a case study of wild Ennerdale', *Journal of Rural Community Development*, 3: 104–18.

Cooper, C. and Hall, C.M. (2008) *Contemporary Tourism: An International Approach*, Oxford: Butterworth Heinemann.

Cooper, T. (2005) 'Sustainable consumption: reflections on product lifespans and "throwaway society"', *Journal of Industrial Ecology*, 9(1–2): 51–67.

Cornelissen, S. (2005) *The Global Tourism System: Governance, Development and Lessons from South Africa*, Aldershot: Ashgate.

Council of the European Communities (CEC) (1985) EU European Council Directive of 27 June 1985, The assessment of the effects of certain public and private projects on the environment, 85/337/EEC, http://eur-lex.europa.eu/LexUriServ/LexUriServ.do?uri= CELEX:31985L0337:EN:HTML

Cox, G., Darcy, M. and Bounds, M. (1994) *The Olympics and Housing: A Study of Six International Events and Analysis of Potential Impacts*, Sydney: University of Western Sydney.

Crick, M. (1989) 'Representations of international tourism in the social sciences: sun, sand, sex, sights, savings and servility', *Annual Review of Anthropology*, 18: 307–44.

Crumbo, K. and George, R. (2005) *Protecting and Restoring the Greater Grand Canyon Ecoregion: Finding Solutions for an Ecoregion at Risk*, Flagstaff: Sierra Club Grand Canyon Chapter.

Cunningham Research (2005) *Study of 2005 Hurricane Impacts on Tourism*, http://www. cunninghamresearch.com/release.html (accessed 17 July 2006).

Daly, H.E. (2008) *A Steady-State Economy*, London: Sustainable Development Commission.

Daniels, P.W. and Bryson, J.R. (2002) 'Manufacturing services and servicing manufacturing: knowledge-based cities and changing forms of production', *Urban Studies,* 39(5/6): 977–91.

Davidoff, P. (2003) 'Advocacy and pluralism in planning', in S. Campbell and S.S. Fainstein (eds) *Readings in Planning Theory*, 2nd edn, London: Blackwell.

Davies, C. (2008) 'Great apes face threat from germs carried by eco-tourists: jungle holidays raise funds to protect wildlife, but humans harbour viruses that have killed chimps and could be fatal for gorillas and orangutans', *The Observer*, 3 February.

De Lollis, B. (2008) 'Economic woes cause travelers to postpone, cancel trips', *USA Today*, 9 October, http://www.usatoday.com/travel/news/2008-10-09-travel-mood_N.htm (accessed 31 October 2008).

De Stefano, L. (2004) *Freshwater and Tourism in the Mediterranean*, WWF Mediterranean Programme, Rome.

Debbage, K. and Ioannides, D. (2004) 'The Cultural Turn? Towards a more critical economic geography of tourism', in A.A. Lew, C.M. Hall and A.M. Williams (eds) *Companion to Tourism*, London: Blackwell.

Department of Culture, Recreation and Tourism (DCRT) (2005) *Louisiana Rebirth: Restoring the Soul of America*, Baton Rouge, LA: Department of Culture, Recreation and Tourism.

—— (2006a) *Tourism Action Plan Amendment: Louisiana Tourism Marketing Program*, Baton Rouge, LA: Department of Culture, Recreation and Tourism.

—— (2006b) *Louisiana Rebirth Scorecard*, Baton Rouge, LA: Department of Culture, Recreation and Tourism.

Department of Industry, Tourism and Resources (DITR) (2002) *Research Report Number 2: Tourism Productivity and Profitability*, Canberra: DITR.

d'Hautserre, A-M. (1999) 'The French mode of social regulation and sustainable tourism development: the case of Disneyland Paris', *Tourism Geographies,* 1(1): 86–107.

Dockerty, T., Lovett, A. and Watkinson, A. (2003) 'Climate change and nature reserves: examining the potential impacts, with examples from Great Britain', *Global Environmental Change,* 13: 125–35.

Dolfman, M.L., Wasser, S.F. and Bergman, B. (2007) 'The effects of Hurricane Katrina on the New Orleans economy', *Monthly Labor Review*, June: 3–18.

Dooley, K. (1997) 'A complex adaptive systems model of organization change', *Nonlinear Dynamics, Psychology, and Life Science*, 1(1); 69–97.

Doxey, G.V. (1972) 'Administration in the tourist industry: a comment', *Canadian Public Administration*, 15(3): 487–9.

—— (1975) 'A causation theory of visitor–resident irritants, methodology and research inferences', in *Conference Proceedings: Sixth Annual Conference of Travel Research Association*, San Diego, CA.

Duffield, B.S. and Walker, S.E. (1984) 'The assessment of tourism impacts', in B.D. Clark, A. Gilad, R. Bisset and P. Tomlinson (eds) *Perspectives on Environmental Impact Assessment*, Dordrecht: D. Reidel.

Duggan, I.C., van Overdijk, C.D.A., Bailey, S.A., Jenkins, P.T., Limén, H. and MacIsaac, H.J. (2005) 'Invertebrates associated with residual ballast water and sediments of cargo-carrying ships entering the Great Lakes', *Canadian Journal of Fisheries and Aquatic Sciences,* 62: 2463–74.

Dunlop, A. (2003) *Tourism Services Negotiation Issues: Implications for Cariform Countries*, Barbados: Caribbean Regional Negotiating Machinery.

Dye, T. (1992) *Understanding Public Policy*, 7th edn, Englewood Cliffs, NJ: Prentice Hall.

Eagles, P.F.J., McCool, S.F. and Haynes, C.D. (2002) *Sustainable Tourism in Protected Areas: Guidelines for Planning and Management*, Gland, Switzerland: IUCN.

Eden, S. (2000) 'Environmental issues. Sustainable progress?' *Progress in Human Geography,* 24: 111–18.

Edquist, C. (ed.) (1997) *Systems of Innovation: Technology, Institutions and Organizations*, London: Pinter.

Elegant, S. (2006) 'China's hotel boom', *Time*, 26 June, http://www.time.com/time/magazine/article/0,9171,1207866,00.html (accessed 28 February 2008).

Elsasser, H. and Bürki, R. (2002) 'Climate change as a threat to tourism in the Alps', *Climate Research*, 20: 253–7.

Energy Committee at the Royal Swedish Academy of Sciences (ECRSAS) (2005) *Statements on Oil*, 14 October, Stockholm: ECRSAS.

Environmental Protection Agency (EPA) (1997) *Terms of Environment: Glossary, Abbreviations, and Synonyms*, EPA 175–B-97–001, Washington, DC: EPA.

eTurboNews (ETN) (2008) 'US travel industry lost $26 billion in 2007 due to "air security measures"', eTurboNews: Global Travel Industry News (29 June), http://www.eturbonews.com/3403/us-travel-industry-lost-26-billion-2007-due-air-security-measures (accessed 1 November).

European Communities (2007) *Inbound and Outbound Tourism in Europe*. Statistics in Focus. Industry, Trade and Services: Population and Social Conditions 52/2007. Luxembourg: European Communities.

European Union (EU) (2006) Assessment of the environmental impact of projects, Europa Activities of the European Union, Summaries of Legislation, http://europa.eu/scadplus/leg/en/lvb/l28163.htm

Evans, G. (2001) *Cultural Planning: Towards an Urban Renaissance*, London: Routledge.

Evans-Pritchard, D. (1989) 'How "they" see "us": Native American images of tourists', *Annals of Tourism Research,* 16: 89–105.

Farrell, B.H. and Twinning-Ward, L. (2004) 'Reconceptualizing tourism', *Annals of Tourism Research,* 31(2): 274–95.

Faulkner, B. and Russell, R. (2003) 'Chaos and complexity in tourism: in search of a new perspective', in H.W. Faulkner, L. Fredline, L. Jago and C.P. Cooper (eds) *Progressing Tourism Research – Bill Faulkner*, Clevedon: Channel View Publications.

Fennell, D. (2001) 'A content analysis of ecotourism definitions', *Current Issues in Tourism,* 4: 403–21.

Fesenmaier, D.R. and Gretzel, U. (2004) 'Searching for experience: technology-related trends shaping the future of tourism', in K. Weimair and C. Mathies (eds) *The Tourism and Leisure Industry: Shaping the Future*, Binghamton, NY: Haworth Press.

Fidler, D.P. (1999) *International Law and Infectious Diseases*, Oxford: Oxford University Press.

Field, C.B., Mortsch, L.D., Brklacich, M., Forbes, D.L., Kovacs, P., Patz, J.A. et al. (2007) 'North America', in M.L. Parry, O.F. Canziani, J.P. Palutikof, P.J. van der Linden and C.E. Hanson (eds) *Climate Change 2007: Impacts, Adaptation and Vulnerability*, Cambridge: Cambridge University Press.

Flack, W. (1997) 'American microbreweries and neolocalism: "ale-ing" for a sense of place', *Journal of Cultural Geography,* 16(2): 37–53.

Flyvbjerg, B., Bruzelius, N. and Rothengatter, W. (2003) *Megaprojects and Risk*, Cambridge: Cambridge University Press.

Foresight (2007) Intelligent Infrastructure Systems, online at: http://www.foresight.gov.uk/Drumbeat/OurWork/CompletedProjects/IIS/Index.asp.

Foresight Directorate (2006a) *Intelligent Infrastructure Futures: Project Overview*, London: Foresight Directorate.

—— (2006b) *Intelligent Infrastructure Futures: The Scenarios – Towards 2055*, London: Foresight Directorate.

Fotsch, P.M. (2004) 'Tourism's uneven impact – history on Cannery Row', *Annals of Tourism Research*, 31(4): 779–800.

Frändberg, L. and Vilhelmson, B. (2003) 'Personal mobility – a corporeal dimension of transnationalisation. The case of long-distance travel from Sweden', *Environment and Planning A,* 35, 1751–68.

Franklin, A. and Crang, M. (2001) 'The trouble with tourism and travel theory?' *Tourist Studies,* 1: 5–22.

Fridgen, J.D. (1984) 'Environmental psychology and tourism', *Annals of Tourism Research*, 11: 19–40.

Frost, W. and Hall, C.M. (eds) (2009) *Tourism and National Parks*, London: Routledge.

Fukushima, T., Kureha, M., Ozaki, N., Fujimori, Y. and Harasawa, H. (2002) 'Influences of air temperature change on leisure industries: case study on ski activities', *Mitigation and Adaptation Strategies for Global Change*, 7, 173–89.

Garcia, O. (2008) 'Weak US dollar can't entice foreign tourists', *ABC News* (8 July), http://abcnews.go.com/Travel/BusinessTravel/wireStory?id=5321573 (accessed 1 November 2008).

Gaston, K.J. (2003) *The Structure and Dynamics of Geographic Ranges*, Oxford: Oxford University Press.

Geertz, C. (1973) *The Interpretation of Cultures*, New York: Basic Books.

German Federal Agency for Nature Conservation (1997) *Biodiversity and Tourism: Conflicts on the World's Seacoasts and Strategies for Their Solution*, Berlin: Springer Verlag.

Getz, D. and Carlsen, J. (2000) 'Characteristics and goals of family and owner-operated businesses in the rural tourism and hospitality sectors', *Tourism Management*, 21: 547–60.

Gielen, D.J., Kurihara, R. and Moriguchi, Y. (2002) 'The environmental impacts of Japanese tourism and leisure', *Journal of Environmental Assessment Policy and Management,* 4(4): 397–424.

Gillin, P. (2007) *The New Influencers: A Marketer's Guide to Social Media*, Sanger, CA: Quill Driver Books-World Dancer Press.

Gilpin, A. (1995) *Environmental Impact Assessment (EIA): Cutting Edge for the Twenty-first Century*, Cambridge: Cambridge University Press.

Gladstone, D. and Préau, J. (2008) 'Gentrification in tourist cities: evidence from New Orleans before and after Hurricane Katrina', *Housing Policy Debate* ,19(1): 137–75.

Gladwell, M. (2000) *The Tipping Point: How Little Things Can Make A Big Difference*, New York: Little Brown.

Glasson, J., Therivel, R. and Chadwick, A. (2005) *Introduction to Environmental Impact Assessment: Principles and Procedures, Process, Practice and Prospects*, 3rd edn, London: Routledge.

Goodland, R. (1995) 'The concept of environmental sustainability', *Annual Review of Ecology and Systematics*, 26: 1–24.

Goodland, R. and Daly, H. (1996) 'Environmental sustainability: universal and non-negotiable', *Ecological Applications*, 6(4): 1002–17.

Goodwin, H. and Francis, J. (2003) 'Ethical and responsible tourism: consumer trends in the UK', *Journal of Vacation Marketing*, 9(3): 271–84.

Gordon, I. and Goodall, B. (2000) 'Localities and tourism', *Tourism Geographies*, 2(3): 290–311.

Gössling, S. (2002) 'Global environmental consequences of tourism', *Global Environmental Change*, 12: 283–302.

Gössling, S. and Hall, C.M. (eds) (2006a) *Tourism and Global Environmental Change, Ecological, Social, Economic and Political Interrelationships*, London: Routledge.

Gössling, S. and Hall, C.M. (2006b) 'Uncertainties in predicting tourist flows under scenarios of climate change', *Climatic Change*, 79(3–4): 163–73.

Gössling, S. and Hall, C.M. (2006c) 'An introduction to tourism and global environmental change', in S. Gössling and C.M. Hall (eds) *Tourism and Global Environmental Change*, London: Routledge.

Gössling, S. and Hall, C.M. (2006d) 'Conclusion: wake up . . . this is serious', in S. Gössling and C.M. Hall (eds) *Tourism and Global Environmental Change*, London: Routledge.

Gössling, S. and Hall, C.M. (2008) 'Swedish tourism and climate change mitigation: an emerging conflict?' *Scandinavian Journal of Hospitality and Tourism*, 8(2): 141–58.

Gössling, S., Hall, C.M. and Weaver, D. (eds) (2009) *Sustainable Tourism Futures: Perspectives on Systems, Restructuring and Innovations*, New York: Routledge.

Gössling, S., Hansson, C.B., Hörstmeier, O. and Saggel, S. (2002) 'Ecological footprint analysis as a tool to assess tourism sustainability', *Ecological Economics*, 43: 199–211.

Gössling, S., Lindén, O., Helmersson, J., Liljenberg, J. and Quarm, S. (2007) 'Diving and global environmental change: a Mauritius case study', in B. Garrod and S. Gössling (eds) *New Frontiers in Marine Tourism: Diving Experiences, Management and Sustainability*, Oxford: Elsevier.

Goudie, A. (2005) *The Human Impact on the Natural Environment: Past, Present, and Future*, 6th edn, Oxford: Blackwell.

Graburn, N.H. (ed.) (1976) *Ethnic and Tourist Arts. Cultural Expressions from the Fourth World*, London: University of California Press.

Grant, A. (1995) 'Human impacts on terrestrial ecosystems', in T. O'Riordan (ed.) *Environmental Science for Environmental Management*, Harlow: Longman.

Gross, Julian (2005) *Community Benefits Agreements: Making Development Projects Accountable*, Washington, DC: Good Jobs First, http://www.goodjobsfirst.org/pdf/cba2005final.pdf

Guidetti, P., Fanellib, G., Fraschettia, S., Terlizzia, A. and Boero, F. (2002) 'Coastal fish indicate human-induced changes in the Mediterranean littoral', *Marine Environmental Research*, 53(1), 77–94.

Gupta, J., van der Leeuw, K. and de Moel, H. (2007) 'Climate change: a "glocal" problem requiring "glocal" action', *Environmental Sciences*, 4(3): 139–48.

Hall, C.M. (1994) *Tourism and Politics: Policy, Power and Place*, Chichester: John Wiley.

—— (1997) 'Geography, marketing and the selling of places', *Journal of Travel and Tourism Marketing*, 6(3/4): 61–84.

—— (1998) 'Historical antecedents of sustainable development and ecotourism: new labels on old bottles?' in C.M. Hall and A.A. Lew (eds) *Sustainable Tourism: A Geographical Perspective*, London: Addison Wesley Longman.

—— (2000) *Tourism Planning. Policies, Processes and Relationships*, Harlow: Prentice-Hall.

—— (2001) 'Imaging, tourism and sports event fever: the Sydney Olympics and the need for a social charter for mega-events', in C. Gratton and I.P. Henry (eds) *Sport in the City: The Role of Sport in Economic and Social Regeneration*, London: Routledge.

—— (2003) 'Tourism and temporary mobility: circulation, diaspora, migration, nomadism, sojourning, travel, transport and home', International Academy for the Study of Tourism (IAST) Conference, 30 June–5 July, Savonlinna, Finland.

—— (2004a) 'Tourism and mobility', in *Creating Tourism Knowledge, 14th International Research Conference of the Council for Australian University Tourism and Hospitality Education, 10–13 February*, St Lucia: School of Tourism and Leisure Management, University of Queensland.

—— (2004b) 'Scale and the problems of assessing mobility in time and space', paper presented at the Swedish National Doctoral Student Course on Tourism, Mobility and Migration, hosted by Department of Social and Economic Geography, University of Umeå, Umeå, Sweden, October.

—— (2005a) *Tourism: Rethinking the Social Science of Mobility*, Harlow: Prentice-Hall.

—— (2005b) 'Reconsidering the geography of tourism and contemporary mobility', *Geographical Research*, 43(2): 125–39.

—— (2005c) 'The future of tourism research', in B. Ritchie, P. Burns and C. Palmer (eds) *Tourism Research Methods*, Wallingford: CAB International.

—— (2006a) 'Tourism urbanisation and global environmental change', in S. Gössling and C.M. Hall (eds) *Tourism and Global Environmental Change,* London: Routledge.

—— (2006b) 'Tourism, disease and global environmental change: the fourth transition', in S. Gössling and C.M. Hall (eds) *Tourism and Global Environmental Change*, London: Routledge.

—— (2006c) 'Tourism, biodiversity and global environmental change', in S. Gössling and C.M. Hall (eds) *Tourism and Global Environmental Change,* London: Routledge.

—— (2007a) *Introduction to Tourism in Australia: Development, Issues and Change*, 5th edn, Melbourne: Pearson Education Australia.

—— (2007b) 'Tourism and regional competitiveness', in J. Tribe and D. Airey (eds) *Advances in Tourism Research, Tourism Research, New Directions, Challenges and Applications*, Oxford: Elsevier.

—— (ed.) (2007c) *Pro-Poor Tourism: Who Benefits? Perspectives on Tourism and Poverty Reduction*, Clevedon: Channel View Publications.

—— (2007d) 'Scaling ecotourism: the role of scale in understanding the impacts of ecotourism', in J. Higham (ed.) *Critical Issues in Ecotourism*, Oxford: Elsevier.

—— (2007e) 'Response to Yeoman et al: the fakery of "The authentic tourist"', *Tourism Management*, 28(4): 1139–40.

—— (2007f) 'The possibilities of slow tourism: can the slow movement help develop sustainable forms of tourism consumption?' paper presented at Achieving Sustainable Tourism, Helsingborg, 11–14 September.

—— (2008a) 'Regulating the international trade in tourism services', in T.E. Coles and C.M. Hall (eds) *International Business and Tourism: Global Issues, Contemporary Interactions*, London: Routledge.

—— (2008b) 'Tourism and climate change: knowledge gaps and issues', *Tourism Recreation Research,* 33(3): 339–50.

—— (2008c) *Tourism Planning: Policies, Processes and Relationships,* 2nd edn, Harlow: Pearson Education.

—— (2009a) 'Tourism firm innovation and sustainability', in S. Gössling, C.M. Hall and D. Weaver (eds) *Sustainable Tourism Futures*, New York: Routledge.

—— (2009b) 'Innovation and tourism policy in Australia and New Zealand: never the twain shall meet?' *Journal of Policy Research in Tourism, Leisure and Events*, 1(1): 2–18.

—— (2009c) 'Degrowing tourism: décroissance, sustainable consumption and steady-state tourism', *Anatolia: An International Journal of Tourism and Hospitality Research*, 20(1), in press.

Hall, C.M. and Boyd, S. (2005) 'Nature-based tourism and regional development in peripheral areas: Introduction', in C.M. Hall and S. Boyd (eds) *Tourism and Nature-based Tourism in Peripheral Areas: Development or Disaster*, Clevedon: Channel View Publications.

Hall, C.M. and Coles, T.E. (2008a) 'Introduction: tourism and international business – tourism as international business', in T.E. Coles and C.M. Hall (eds) *International Business and Tourism*, London: Routledge.

Hall, C.M. and Coles, T.E. (2008b) 'Conclusion: mobilities of commerce', in T.E. Coles and C.M. Hall (eds) *International Business and Tourism,* London: Routledge.

Hall, C.M. and Frost, W. (2009) 'The future of the national park concept', in W. Frost and C.M. Hall (eds) *Tourism and National Parks*, London: Routledge.

Hall, C.M. and Higham, J. (eds) (2005) *Tourism, Recreation and Climate Change*, Clevedon: Channel View Publications.

Hall, C.M. and Jenkins, J. (1995) *Tourism and Public Policy*, London: Routledge.

Hall, C.M. and Lew, A.A. (eds) (1998) *Sustainable Tourism: A Geographical Perspective*, London: Addison Wesley Longman.

Hall, C.M. and McArthur, S. (1996) 'The human dimension of heritage management: different values, different interests . . . different issues', in C.M. Hall and S. McArthur (eds) *Heritage Management in Australia and New Zealand: The Human Dimension*, Sydney: Oxford University Press.

Hall, C.M. and McArthur, S. (1998) *Integrated Heritage Management*, London: The Stationery Office.

Hall, C.M. and Müller, D. (eds) (2004) *Tourism, Mobility and Second Homes: Between Elite Landscape and Common Ground*, Clevedon: Channel View Publications.

Hall, C.M. and Page, S.J. (2006) *The Geography of Tourism and Recreation: Environment, Place and space*, 3rd edn, London: Routledge.

Hall, C.M. and Rusher, K. (2004) 'Risky lifestyles? Entrepreneurial characteristics of the New Zealand bed and breakfast sector', in R. Thomas (ed.) *Small Firms in Tourism: International Perspectives*, Oxford: Elsevier.

Hall, C.M. and Rusher, K. (2005) 'Entrepreneurial characteristics and issues in the small-scale accommodation sector in New Zealand', in E. Jones and C. Haven (eds) *Tourism SMEs, Service Quality and Destination Competitiveness: International Perspectives*, Wallingford: CAB International.

Hall, C.M. and Sharples, L. (2008) 'Food events and the local food system: marketing, management and planning issues', in C.M. Hall and L. Sharples (eds) *Food and Wine Festivals and Events Around the World*, Oxford: Butterworth Heinemann.

Hall, C.M. and Williams, A. (2008) *Tourism and Innovation*, London: Routledge.

Hall, C.M., Mitchell. R., Scott, D. and Sharples, L. (2008) 'The authentic market experiences of farmers' markets', in C.M. Hall and L. Sharples (eds) *Food and Wine Festivals and Events Around the World,* Oxford: Butterworth Heinemann.

Hall, C.M., Müller, D. and Saarinen, J. (2009) *Nordic Tourism*, Clevedon: Channel View Publications.

Hall, P. (2002) *Urban and Regional Planning*, 4th edn, London: Routledge.

Hannigan, J. (1998). *Fantasy City: Pleasure and Profit in the Postmodern Metropolis*, London: Routledge.

Hansen, J.E. (2006) 'Can we still avoid dangerous human-made climate change?' *Social Research: An International Quarterly of Social Sciences*, 73(3): 949–74.

Hanson, A. and Hanson, L. (eds) (1990) *Art and Identity in Oceania*, Bathurst: Crawford House Press.

Hardin, G. (1968) 'The tragedy of the commons: the population problem has no technical solution; it requires a fundamental extension in morality', *Science*, 162(3859): 1243–8.

Harper, D.A. (1996) *Entrepreneurship and the Market Process,* London: Routledge.

Harvey, D. (1969) *Explanation in Geography*, London: Edward Arnold.

—— (1989) *The Condition of Postmodernity: An Enquiry into the Origins of Cultural Change,* Oxford: Basil Blackwell.

Hassan, J.A. (2008) 'The growth and impact of the British water industry in the nineteenth century', *Economic History Review*, 38(4): 531–47.

Hauvner, E., Hill, C. and Milburn, R. (2008) *Here to Stay: Sustainability in the Travel and Leisure Sector*, Hospitality Directions Europe 17, London: PricewaterhouseCoopers.

Hawaii Visitors and Convention Bureau (HVCB) (2007) Annual Visitor Research Report – 2006, http://www.hawaii.gov/dbedt/info/visitor-stats/visitor-research/2006-annual-research-r.pdf (accessed 28 February 2008).

Hawkins, D.E. and Mann, S. (2007) 'The World Bank's role in tourism development' *Annals of Tourism Research*, 34(2): 348–63.

Hay, A.M. (2000) 'System', in R.J. Johnston, D. Gregory, G. Pratt and M. Watts (eds) *The Dictionary of Human Geography*, Oxford: Blackwell.

Hays, S.P. (1957) *The Response to Industrialism, 1885–1914*, Chicago, IL: University of Chicago Press.

—— (1959) *Conservation and the Gospel of Efficiency: The Progressive Conservation Movement 1890–1920*, Cambridge, MA: Harvard University Press.

Healy, P. (2003) 'The communicative turn in planning theory and its implications for spatial strategy formation', in S. Campbell and S.S. Fainstein (eds) *Readings in Planning Theory*, 2nd edn, London: Blackwell.

Hewison, R. (1991) 'Commerce and culture', in J. Corner and S. Harvey (eds) *Enterprise and Heritage: Crosscurrents of National Culture*, London: Routledge.

Hill, P. (1999) 'Tangibles, intangibles and services: a new taxonomy for the classification of output', *Canadian Journal of Economics*, 32(2), 426–46.

Hirsch, R.L. (2007a) Peaking of world oil production: Recent forecasts. DOE/NETL-2007/ 1263, 5 February. National Energy Technology Laboratory.

—— (2007b) Peaking of world oil production: Recent forecasts. *WorldOil.com The oilfield information source*, 228(4) April. This work (except for the GAO Addendum) was sponsored by the US Department of Energy, National Energy Technology Laboratory, under Contract No. DE-AM26–04NT41817, http://www.worldoil.com/ Magazine/MAGAZINE_DETAIL.asp?ART_ID=3163&MONTH_YEAR=Apr-2007

Hirsch, R.L., Bezdek, R. and Wendling, R. (2005) *Peaking of World Oil Production: Impacts, Mitigation, and Risk Management*, US report prepared for the US Department of Energy.

Hjalager, A. (1999) 'Tourism destinations and the concept of industrial districts', paper for ESRA conference, Dublin, August.

—— (2002) 'Repairing innovation defectiveness in tourism', *Tourism Management*, 23: 465–74.

Holbrook, J.A. and Wolfe, D.A. (eds) (2000) *Knowledge, Clusters and Regional Innovation: Economic Development in Canada*, Montreal and Kingston, McGill-Queen's University Press for the School of Policy Studies, Queen's University.

Holling, C.S. (ed.) (1978) *Adaptive Environmental Assessment and Management*, New York: John Wiley.

Horst, R. and Webber, M. (1973) 'Dilemmas in a general theory of planning', *Policy Sciences*, 4: 155–69.

Hudson, B.M. (1979) 'Comparison of current planning theories: counterparts and contradictions', *Journal of the American Planning Association,* 45(October): 387–98.

Hughes, G. (2004) 'Tourism, sustainability and social theory', in A.A. Lew, C.M. Hall and A.M. Williams (eds) *Companion to Tourism*, London: Blackwell.

Hulme, M., Conway, D. and Lu, X. (2003) *Climate Change: An Overview and its Impact on the Living Lakes.* A report prepared for the 8th Living Lakes Conference Climate Change and Governance: Managing Impacts on Lakes, Zuckerman Institute for Connective Environmental Research, University of East Anglia, Norwich, UK, 7–12 September. Norwich: Tyndall Centre for Climate Change Research, University of East Anglia.

Hutcheson, D.G. (2005) 'Moore's law: the history and economics of an observation that changed the world', *The Electrochemical Society INTERFACE,* 14(1): 17–21.

Industry Taskforce on Peak Oil and Energy Security (ITPOES) (2008) *The Oil Crunch Securing the UK's Energy Future*, London: ITPOES.

Instanes, A., Anisimov, O., Brigham, L., Goering, D., Ladanyi, B., Larsen, J.O. and Khrustalev, L.N. (2005) 'Infrastructure: buildings, support systems, and industrial facilities', in C. Symon, L. Arris and B. Heal (eds) *Impacts of a Warming Arctic: Arctic Climate Impacts Assessment*, Cambridge: Cambridge University Press.

Intergovernmental Panel on Climate Change (IPCC) (1999) *Special Report on Aviation and the Global Atmosphere*, Geneva: IPCC.

—— (2007) *Summary for Policymakers of the Synthesis Report of the IPCC Fourth Assessment Report*, draft copy, 16 November, 23:04, Subject to final copy edit. Geneva: IPCC.

International Air Transport Association (IATA) (2008a) Crisis again – deep losses projected, IATA Press Release No. 26, 2 June, http://www.iata.org/pressroom/pr/2008-06-02-01.htm

—— (2008b) Fuel crisis a catalyst for change, IATA Press Release No. 27, 2 June, http://www.iata.org/pressroom/pr/2008-06-02-02.htm

—— (2008c) *Building a Greener Future*, 2nd edn, Geneva: IATA.

—— (2008d) *Financial Forecast: Significant Losses Continue into 2009* (September), Geneva: IATA.

—— (2008e) *IATA Economic Briefing: Medium-term Outlook for Oil and Jet Fuel Prices* (October), Geneva: IATA.

International Association for Impact Assessment (IAIA) (1999) *Principles of Environmental Impact Assessment Best Practice*, Fargo, ND: IAIA and Lincoln: Institute of Environmental Assessment.

—— (2003) *Social Impact Assessment: International Principles*, Fargo, ND: IAIA and Lincoln: Institute of Environmental Assessment.

International Energy Agency (IEA) (2007) *World Energy Outlook 2007 Executive Summary: China and India Insights*, Paris: OECD/IEA.

International Union for the Conservation of Nature (IUCN) (1994) *Guidelines for Protected Area Management Categories*, Gland, Switzerland: IUCN.

International Union for the Conservation of Nature and Natural Resources (IUCN) (1980) *World Conservation Strategy*, IUCN with the advice, cooperation and financial assistance of UNEP and WWF and in collaboration with the UN FAO and UNESCO, Morges: IUCN.

International Union of Tourism Organizations (IUOTO) (1974) 'The role of the state in tourism', *Annals of Tourism Research*, 1: 66–72.

Jackson, T. (2005) 'Live better by consuming less? Is there a "double dividend" in sustainable consumption?' *Journal of Industrial Ecology,* 9(1–2): 19–36.

Jacques, P. and Ostergen, D.M. (2006) 'The end of wilderness: conflict and defeat of wilderness in the Grand Canyon', *Review of Policy Research,* 23(2): 573–88.

Jones, D. and Smith, K. (2005) 'Middle-earth meets New Zealand: authenticity and location in the making of The Lord of the Rings', *Journal of Management Studies,* 42(5): 923–45.

Jutla, R.S. (2000) 'Visual image of the city: tourists versus resident perceptions of Simla, a hill state in Northern India', *Tourism Geographies,* 2(4): 403–20.

Kanellos, M. (2005) 'New life for Moore's Law', CNET News (14 April), http://news.cnet.com/New-life-for-Moores-Law/2009–1006_3–5672485.html (accessed 15 November 2008).

Kaplan, M. (2007) 'Moose use roads as a defence against bears', *Nature,* published online 10 October, doi:10.1038/news.2007.155

Kennedy, C.B. and Lew, A.A. (2000) 'Living on the edge: defending American Indian reservation lands and culture', in J.R. Gold and G. Revill (eds) *Landscapes of Defence,* London: Addison Wesley Longman.

Kent, M., Newnham, R. and Essex, S. (2002) 'Tourism and sustainable water supply in Mallorca: a geographical analysis', *Applied Geography,* 22(4): 351–74.

Kerr, B., Barron, G. and Wood, R.C. (2001) 'Politics, policy and regional tourism administration: a case examination of Scottish area tourist board funding', *Tourism Management,* 22(6), 649–57.

Köndgen, S., Kühl, H., N'Goran, P.K., Walsh, P.D., Schenk, S., Ernst, N. et al. (2008) 'Pandemic human viruses cause decline of endangered great apes', *Current Biology,* article in press doi:10.1016/j.cub.2008.01.012

Kooiman, J. (1993) 'Findings, speculations and recommendations', in J. Kooiman (ed.) *Modern Governance: New Governmment–Society Interactions,* London: Sage.

Ksaibati, K., Wilson, E.M.,Warder, D.S. and Bryan, G. (1994) Determining Economic Effects of Wyoming's Loop Tours. Mountain-Plains Consortium, MPC Report No. 94–29A, http://www.mountain-plains.org/pubs/html/mpc-94–29A/ (accessed 28 February 2008).

Kuneva, M. (2008) 'European consumer commissioner, airline ticket sweep investigation', press conference speaking points, Brussels, 8 May, Reference SPEECH/08/235, http://europa.eu/rapid/pressReleasesAction.do?reference=SPEECH/08/235&format=HTML&aged=0&language=EN

Lafferty, G. and van Fossen, A. (2001) 'Integrating the tourism industry: problems and strategies', *Tourism Management,* 22(1): 11–19.

Lamers, M. and Amelung, B. (2007) 'The environmental impacts of tourism to Antarctica. A global perspective', in P.M. Peeters (ed.) *Tourism and Climate Change Mitigation. Methods, Greenhouse Gas Reductions and Policies,* Breda: NHTV.

Lazanski, T.J. and Kljajic, M. (2006) 'Systems approach to complex systems modeling with special regards to tourism', *Kybernetes: International Journal of Systems and Cybernetics,* 35(7/8): 1048–58.

Lee, J.-W. and McKibbin, W.J. (2004) 'Globalization and disease: the case of SARS', Brookings Discussion Papers in International Economics No.156, Washington, DC: Brookings Institute.

Leffel, T. (2006) 'US passport holders up to 27 percent', Cheapest Destinations (blog). http://travel.booklocker.com/index.php?p=132 (accessed 27 February 2008).

Lehtonen, M. (2004) 'The environment–social interface of sustainable development: capabilities, social and capital institutions', *Ecological Economics,* 49(2): 199–214.

Leiper, N. (1983) 'An etymology of "tourism"', *Annals of Tourism Research*, 10: 277–81.
—— (1989) *Tourism and Tourism Systems*, Occasional Paper No. 1, Palmerston North: Department of Management Systems, Massey University.
—— (1990) 'Partial industrialization of tourism systems', *Annals of Tourism Research*, 17: 600–5.
—— (1999) 'A conceptual analysis of tourism-supported employment which reduces the incidence of exaggerated, misleading statistics about jobs', *Tourism Management*, 20(5): 605–13.
—— (2000) 'An emerging discipline', *Annals of Tourism Research*, 27(3): 805–9.
Lennon, J.J. (ed.) (2003) *Tourism Statistics: International Perspectives and Current Issues*, London: Continuum.
Lew, A.A. (1987) 'A framework of tourist attraction research', *Annals of Tourism Research*, 14(4): 553–75.
—— (1988) 'Tourism and place studies: an example of Oregon's older retail districts', *Journal of Geography*, 87(4): 122–6.
—— (1989) 'Authenticity and sense of place in the tourism development experience of older retail districts', *Journal of Travel Research*, 27(4): 15–22.
—— (1991) 'Place representation in tourist guidebooks: an example from Singapore', *Singapore Journal of Tropical Geography*, 12(2): 123–37.
—— (1992) 'Perceptions of tourists and tour guides in Singapore', *Journal of Cultural Geography*, 12(2): 45–52.
—— (1998a) 'The Asia Pacific ecotourism industry: putting sustainable tourism into practice', in C.M. Hall and A.A. Lew (eds) *Sustainable Tourism: A Geographical Perspective*, London: Addison Wesley Longman.
—— (1998b) 'American Indians in state tourism promotional literature', in A.A. Lew and G. Van Otten (eds) *Tourism and Gaming on American Indian Lands*, Elmsford, NY: Cognizant Communications.
—— (1999) 'Managing tourist space in Pueblo Villages of the American Southwest', in T.V. Singh (ed.) *Tourism Development in Critical Environments*, Elmsford, NY: Cognizant Communications.
—— (2007a) 'China's growing wander lust', *Far Eastern Economic Review*, (October): 60–3.
—— (2007b) 'Tourism planning and traditional urban planning theory: planners as agents of social change', *Leisure/Loisir: Journal of the Canadian Association of Leisure Studies*, 31(2): 383–92.
—— (2007c) 'Pedestrian shopping streets in the restructuring of the Chinese city', in T. Coles and A. Church (eds) *Tourism, Power and Place*, London: Routledge.
—— (2008) 'On the use and abuse of Tourism Satellite Accounts', Tourism Place blog (5 May), http://tourismplace.blogspot.com/2008/05/on-use-and-abuse-of-tourism-satellite.html (accessed 11 June 2008).
—— (2009) 'Long tail tourism: new geographies for marketing niche tourism products', *Journal of Travel and Tourism Marketing*, 25(3/4): 409–19.
Lew, A.A. and Hall, C.M. (1998) 'The geography of sustainable tourism: lessons and prospects', in C.M. Hall and A.A. Lew (eds) *Sustainable Tourism: A Geographical Perspective*, London: Addison Wesley Longman.
Lew, A.A. and Kennedy, C.L. (2002) 'Tourism and culture clash in American Indian Country', in S. Krakover and Y. Gradus (eds) *Tourism in Frontier Areas*, Lexington, MA: Lexington Books.
Lew, A.A. and McKercher, B. (2002) 'Trip destinations, gateways and itineraries: the example of Hong Kong', *Tourism Management*, 23(6): 609–21.

Lew, A.A. and Van Otten, G. (eds) (1998) *Tourism and Gaming on American Indian Lands*, Elmsford, NY: Cognizant Communications.

Lew, A.A. and Wong, A. (2003) 'News from the Motherland: a content analysis of existential tourism magazines in China', *Tourism Culture and Communication*, 4(2): 83–94.

Lew, A.A. and Wong, A. (2004) 'Sojourners, Gangxi and clan associations: social capital and overseas Chinese tourism to China', in D. Timothy and T. Coles (eds) *Tourism, Diasporas and Space*, London: Routledge.

Lew, A.A. and Wong, A. (2005) 'Existential tourism and the homeland seduction: the overseas Chinese experience', in C.L. Cartier and A.A. Lew (eds) *The Seduction of Place: Geographical Perspectives on Globalization and Touristed Landscapes*, London: Routledge.

Lew, A.A., Hall, C.M. and Timothy, D. (2008) *World Regional Tourism Geography*, Oxford: Elsevier.

Lew, A.A., Hall, C.M. and Williams, A.M. (eds) (2004) *A Companion to Tourism*, Oxford: Blackwell.

Lew, A.A., Wee, Tan, T.T. and Zafar, A.U. (1999) 'Tourism 21: keeping Singapore on top in the next millennium', in Hooi Den Huan (ed.) *Cases in Singapore Hospitality and Tourism Management*, Singapore: Prentice-Hall.

Ley, D. and Olds, K. (1988) 'Landscape as spectacle: World's Fairs and the culture of heroic consumption', *Environment and Planning D*, 6: 191–212.

Lindblom, C.E. (2003) 'The science of "muddling through"', in S. Campbell and S.S. Fainstein (eds) *Readings in Planning Theory*, 2nd edn, London: Blackwell.

Loizidou, X. (2003) 'Land use and coastal management in the Eastern Mediterranean: the Cyprus example', International Conference on the Sustainable Development of the Mediterranean and Black Sea Environment, May, Thessaloniki, Greece.

Lovelock, C. and Gummesson, E. (2004) 'In search of a new paradigm and fresh perspectives', *Journal of Service Research*, 7(1): 20–41.

Lowenthal, D. (1985) *The Past is a Foreign Country*, Cambridge: Cambridge University Press.

Lujan, C.C. (1998) 'A sociological view of tourism in an American Indian community: maintaining cultural integrity at Taos Pueblo', in A.A. Lew and G.A. Van Otten (eds) *Tourism and Gaming on American Indian Lands*, Elmsford, NY: Cognizant Communications. (Republished from *American Indian Culture and Research*, 17(3), (1993): 101–20.)

MacCannell, D. (1976) *The Tourist: A New Theory of the Leisure Class*, New York: Schocken Books.

McCool, S.F. (1994) 'Planning for sustainable nature dependent tourism development: the limits of acceptable change system', *Tourism Recreation Research*, 19(2), 51–5.

McCorran, R., Price, M.F. and Warren, C.R. (2008) 'The call of different wilds: the importance of definition and perception in protecting and managing Scottish wild landscapes', *Journal of Environmental Planning and Management*, 51(2): 177–99.

MacDonald, D., Crabtree, J.R., Wiesinger, G., Dax, T., Stamou, M., Fleury, P. et al. (2000) 'Agricultural abandonment in mountain areas of Europe: environmental consequences and policy response', *Journal of Environmental Management*, 59(1): 47–69.

Mcelroy, J. and De Albuquerque, K. (1986) 'The tourism demonstration effect in the Caribbean', *Journal of Travel Research*, 25(2): 31–4.

Macfarlane, R. (2007) 'Imaging a Britain running wild: bulldozers threaten the land around our cities, but elsewhere in Britain, untamed nature is being allowed to reassert itself', *The Observer*, 19 August, http://www.guardian.co.uk/environment/2007/aug/19/conservation

McGeoch, M.A., Chown, S.L. and Kalwij, J.M. (2006) 'A global indicator for biological invasion', *Conservation Biology,* 20(6): 1635–46.

McKercher, B. (1999) 'A chaos approach to tourism', *Tourism Management,* 20(4): 425–34.

McKercher, B. and Lau, G. (2007) 'Understanding the movements of tourists in a destination: testing the importance of markers in the tourist attraction system', *Asian Journal of Tourism and Hospitality Research,* 1(1): 39–53.

McKercher, B. and Lew, A.A. (2003) 'Distance decay and the impact of effective tourism exclusion zones on international travel flows', *Journal of Travel Research,* 42(2): 159–65.

Macleod, D.V.L. (2004) *Tourism, Globalisation and Cultural Change: An Island Community Perspective,* Clevedon: Channel View.

McMichael, A.J. (2001) *Human Frontiers, Environments and Disease: Past Patterns, Uncertain Futures,* Cambridge: Cambridge University Press.

MacNab, J. (1985) 'Carrying capacity and related slippery shibboleths', *Wildlife Society Bulletin,* 13(4): 403–10.

Malerba, F. (2001) 'Sectoral systems of innovation and production: concepts, analytical framework and empirical evidence', in *Conference 'The Future of Innovation Studies', Eindhoven University of Technology, the Netherlands, 20–23 September.* Eindhoven: Eindhoven Centre for Innovation Studies.

—— (2002) 'Sectoral systems of innovation and production', *Research Policy,* 31: 247–64.

—— (2005a) 'Sectoral systems of innovation: basic concepts', in F. Malerba (ed.) *Sectoral Systems of Innovation: Concepts, Issues and Analyses of Six Major Sectors in Europe,* Cambridge: Cambridge University Press.

—— (ed.) (2005b) *Sectoral Systems of Innovation: Concepts, Issues and Analyses of Six Major Sectors in Europe,* Cambridge: Cambridge University Press.

Malerba, F. and Orsenigo, L. (1996) 'The dynamics and evolution of industries', *Industrial and Corporate Change,* 5(1): 51–87.

Malerba, F. and Orsenigo, L. (1997) 'Technological regimes and sectoral patterns of innovation activities', *Industrial and Corporate Change,* 6(1): 83–118.

Malerba, F. and Orsenigo, L. (2000) 'Knowledge, innovative activities and industrial evolution', *Industrial and Corporate Change,* 9(2): 289–314.

Management Advisory Committee (2004) *Connecting Government: Whole of Government Responses to Australia's Priority Challenges,* Canberra: Commonwealth of Australia.

Mandelik, Y., Dayan, T. and Feitelson, E. (2005) 'Issues and dilemmas in ecological scoping: scientific, procedural and economic perspectives', *Impact Assessment and Project Appraisal,* 23(1): 55–63.

Mannion, A.M. (1991) *Global Environmental Change: A Natural and Cultural Environmental History,* New York: Longman.

Mark, S.R. and Hall, C.M. (2009) 'John Muir and William Gladstone Steel: activists and the establishment of Yosemite and Crater Lake National Parks', in W. Frost and C.M. Hall (eds) *Tourism and National Parks,* London: Routledge.

Marketwatch. (2008) 'Airlines on edge even as crude falls', *Wall Street Journal Digital Network* (15 October), http://www.eturbonews.com/5605/airlines-oil-proving-two-edged-sword (accessed 31 October 2008).

Markusen, A. (1999) 'Fuzzy concepts, scanty evidence, policy distance: the case for rigour and policy relevance in critical regional studies', *Regional Studie,s* 33(9): 869–84.

Marsh, G.P. (1965) *Man and Nature; Or, Physical Geography as Modified by Human Action*, orig. 1864, ed. D. Lowenthal, Cambridge, MA: The Belknap Press of Harvard University Press.

Martens, S . and Spaargaren, G. (2005) 'The politics of sustainable consumption: the case of the Netherlands', *Sustainability: Science, Practice, and Policy,* 1(1): 29–42.

Martin, C. (ed.) (1987) *The American Indian and the Problem of History*, New York: Oxford University Press.

Maslow, A.H. (1943) 'A theory of human motivation', *Psychological Review,* 50: 370–96.

Mason, P. (2003) *Tourism Impacts, Planning and Management*, Burlington, MA: Butterworth-Heinemann.

Mathews, G. (2000) *Global Culture/Individual Identity: Searching for Home in the Cultural Supermarket*, London: Routledge.

Mathieson, A. and Wall, G. (1982) *Tourism: Economic, Physical and Social Impacts*, Harlow: Longman.

Mattsson, J., Sundbo, J. and Fussing-Jensen, C. (2005) 'Innovation systems in tourism: the role of attractors and scene-takers', *Industry and Innovation*, 12(3): 357–81.

Max Planck Society (2008) Great Apes endangered by human viruses, Press Release, News B/2008(11), 25 January, http://www.mpg.de/english/illustrationsDocumentation/ documentation/pressReleases/2008/pressRelease20080125/index.html.

Max-Neef, M. (1991) *Human Scale Development*, New York: Apex Press.

Mayo, E.J. and Jarvis, L.P. (1981) *The Psychology of Leisure Travel, Effective Marketing and Selling of Travel Services*, Boston, MA: CBI Publishing.

Mbaiwa, J.E. (2005) 'Enclave tourism and its socio-economic impacts in the Okavango Delta, Botswana', *Tourism Management,* 26(2): 157–72.

Mendelovici, T. (2008) Kangaroo Island Tourism Optimisation Management Model: TOMM Process. http://www.tomm.info/Background/tomm_process.aspx (accessed 10 May 2008).

Meyer, W.B. and Turner II, B.L. (1995) 'The Earth transformed: trends, trajectories, and patterns', in R.J. Johnston, P.J., Taylor and M.J. Watts (eds) *Geographies of Global Change: Remapping the World in the Late Twentieth Century*, Oxford: Blackwell.

Miele, M. (2008) 'CittàSlow: producing slowness against the fast life', *Space and Polity*, 12(1): 135–56.

Mill, R.C. and Morrison, A.M. (1985) *The Tourism System: An Introductory Text*, Englewood Cliffs, NJ: Prentice-Hall.

Milmo, D. (2007) 'EasyJet calls for tax on planes, not passengers', *The Guardian*, 19 September.

—— (2008) 'Desperate times for airlines as oil price pushes losses towards $6bn', *The Guardian*, 3 June, http://www.guardian.co.uk/business/2008/jun/03/theairlineindustry.oil

Ministry of Economic Development (2007) *SMEs in New Zealand: Structure and Dynamics*, Wellington: Ministry of Economic Development.

MKG Consulting (2005) The 10 Largest Hotel Groups and the 10 Largest Hotel Brands in the European Union, http://www.hotel-online.com/News/PR2005_1st/Feb05_MKG Ranking.html (accessed 28 February 2008).

Monbiot, G. (2006) *Heat: How to Stop the Planet Burning*, London: Allen Lane.

Moore, G.E. (1965) 'Cramming more components onto integrated circuits', *Electronics* 38(8, April 19): no pages, ftp://download.intel.com/museum/Moores_Law/Articles-Press_Releases/Gordon_Moore_1965_Article.pdf (accessed 15 November 2008).

Morales-Moreno, I. (2004) 'Postsovereign governance in a globalizing and fragmenting world: the case of Mexico', *Review of Policy Research*, 21(1): 107–17.

More, T.A., Bulmer, S., Henzel, L. and Mates, A.E. (2003) *Extending the Recreation Opportunity Spectrum to Nonfederal Lands in the Northeast: An Implementation Guide*, Delaware, OH: USDA (United States Department of Agriculture Forest Service, Northeastern Research Station General Technical Report NE-309, August 2003), http://www.fs.fed.us/ne/newtown_square/publications/technical_reports/pdfs/2003/gtrne309.pdf (accessed 8 May 2008).

Morgan, N. (2004) 'Problematizing place promotion', in A.A. Lew, C.M. Hall and A.M. Williams (eds) *Companion to Tourism Geography*, London: Blackwell.

Morrison, A.J., Rimmington, M. and Williams, C. (1999) *Entrepreneurship in the Hospitality, Tourism and Leisure Industries*, Oxford: Butterworth-Heinemann.

Morse, S.S. (1993) 'AIDS and beyond: defining the rules for viral traffic', in E. Fee and D.M. Fox (eds) *AIDS: The Making of a Chronic Disease*, Berkeley: University of California Press.

Mose, I. and Weixlbaumer, N. (2007) 'A new paradigm for protected areas in Europe?' in I. Mose (ed.) *Protected Areas and Regional Development in Europe: Toward a New Model for the 21st Century*, Aldershot: Ashgate.

Muir, J. (1914) *The Yosemite*, Boston, MA: Houghton Mifflin.

Murphy, K. (2007) 'Costello brothers at loggerheads over Kyoto Protocol', *The Age*, 5 November.

Myers, N. (1988) 'Threatened biotas: "hotspots" in tropical forests', *The Environmentalist*, 8: 1–20.

——— (2003) 'Biodiversity hotspots revisited', *BioScience*, 53: 916–17.

Myers, N. and Gaia Ltd staff (1984) *Gaia: An Atlas of Planet Management*, New York: Anchor/Doubleday.

Myers, N., Mittermeier, R.A., Mittermeier, C.G., da Fonseca, G.A.B. and Kent, J. (2000) 'Biodiversity hotspots for conservation priorities', *Nature* 403: 853–8.

Nash, R. (1963) 'The American wilderness in historical perspective', *Journal of Forest History*, 6(4): 2–13.

——— (1967) *Wilderness and the American Mind*, New Haven, CT: Yale University Press.

Naughton-Treves, L., Holland, M.B. and Brandon, K. (2005) 'The role of protected areas in conserving biodiversity and sustaining livelihoods', *Annual Review of Environment and Resources*, 30: 219–52.

Nauwelaers, C. and Reid, A. (1995) *Innovative Regions? A Comparative Review of Methods of Evaluating Regional Innovation Potential. European Innovation Monitoring System (EIMS)*, Publication No. 21, Luxembourg: European Commission, Directorate General XIII.

Ndou, V. and Petti, C. (2006) 'Approaching tourism as a complex dynamic system: implications and insights', in M. Hitz, M. Sigala and J. Murphy (eds) *Information and Communication Technologies in Tourism 2006, Proceedings of the International Conference in Lausanne, Switzerland*, Vienna: Springer.

Nepal, S.K. (2008) 'Residents' attitudes to tourism in Central British Columbia, Canada', *Tourism Geographies,* 10(1): 42–65.

Nicholls, R.J., Wong, P.P., Burkett, V.R., Codignotto, J.O., Hay, J.E., Mclean, R.F. et al. (2007) 'Coastal systems and low-lying areas', in M.L. Parry, O.F. Canziani, J.P. Palutikof, P.J. van der Linden and C.E. Hanson (eds) *Climate Change 2007: Impacts, Adaptation and Vulnerability*, Cambridge: Cambridge University Press.

Nordin, S. (2003) *Tourism Clustering and Innovation*, U 2003: 14, Östersund: ETour.

Norkunas, M.K. (1993) *The Politics of Memory: Tourism, History, and Ethnicity in Monterey, California*, Albany: State University of New York Press.

Novacek, M.J. and Cleland, E.E. (2001) 'The current biodiversity extinction event: scenarios for mitigation and recovery', *PNAS,* 98(10): 5466–70.

Novelli, M., Schmitz, B. and Spencer, T. (2006) 'Networks, clusters and innovation in tourism: a UK experience', *Tourism Management,* 27(6): 1141–52.

O'Brien, K. (2006) 'Are we missing the point? Global environmental change as an issue of human security'*, Global Environmental Change,* 16(1): 1–3.

Office of Travel and Tourism Industries (OTTI) (2008) About OTTI, http://www.tinet.ita. doc.gov/about/index.html#TD (accessed 13 May 2008).

Olds, K. (1988) 'Planning for the housing impacts of a hallmark event: a case study of Expo 1986', unpublished MA thesis, Vancouver: School of Community and Regional Planning, University of British Columbia.

—— (1998) 'Urban mega-events, evictions and housing rights: the Canadian case', *Current Issues in Tourism*, 1(1): 2–46.

Oppermann, M. (1995) 'Travel life cycle', *Annals of Tourism Research,* 22(3): 535–52.

Organization for Economic Cooperation and Development (OECD) and Statistical Office of the European Communities (2005) *Oslo Manual: Guidelines for Collecting and Interpreting Innovation Data*, 3rd edn, Paris: OECD.

Osterwalder, A., Pigneur, Y. and Tucci, C.L. (2005) 'Clarifying business models: origins, present, and future of the concept', *Communications of the Association for Information Systems*, 15 (preprint).

Ostrom, E., Burger, J., Field, C.B., Norgaard, R.B. and Policansky, D. (1999) 'Revisiting the commons: local lessons, global challenges', *Science*, 284(9 April): 278–82.

Pacific Asia Travel Association (PATA) (2002) PATA Traveller's Code, http://www.pata. org/patasite/index.php?id=419 (accessed 11 May 2008).

—— (2004) 'WTO tourism negotiations: steady does it', *PATA Issues and Trends*, December.

Page, S. and Hall, C.M. (2003) *Managing Urban Tourism*, Harlow: Prentice-Hall.

Parliamentary Assembly, Council of Europe (2003) *Erosion of the Mediterranean Coastline: Implications for Tourism*, Doc. 9981 16 October, Report Committee on Economic Affairs and Development, http://assembly.coe.int/Documents/Working Docs/doc03/EDOC9981.htm (accessed 25 January 2005).

Parry, M.L., Canziani, O.F., Palutikof, J.P., Van der Linden, P.J. and Hanson, C.E. (eds) (2007) *Climate Change 2007: Impacts, Adaptation and Vulnerability. Contribution of Working Group II to the Fourth Assessment Report of the Intergovernmental Panel on Climate Change*, Cambridge: Cambridge University Press.

Pearce, P.L. (1988) *The Ulysses Factor: Evaluating Visitors in Tourist Settings*, New York: Springer-Verlag.

—— (2005) *Tourist Behaviour: Themes and Conceptual Schemes*, Clevedon: Channel View Publications.

Pearce, P.L. and Lee, U-I. (2005) 'Developing the travel career approach to tourist motivation', *Journal of Travel Research*, 43(3): 226–37.

Peeters, P., Szimba, E. and Duijnisveld, M. (2007) 'Major environmental impacts of European tourist transport', *Journal of Transport Geography*, 15: 83–93.

Philo, C. and Kearns, G. (1993) 'Culture, history, capital: a critical introduction to the selling of places', in G. Kearns and C. Philo (eds) *Selling Places: The City as Cultural Capital, Past and Present*, Oxford: Pergamon Press.

Pimentel, D., McNair, S., Janecka, J., Wightman, J., Simmonds, C., O'Connell, C. et al. (2001) 'Economic and environmental threats of alien plant, animal, and microbe invasions', *Agriculture, Ecosystems and Environment*, 84: 1–20.

Pinchot, G. (1968 [1910]) 'Ends and means', in R. Nash (ed.) *The American Environment: Readings in the History of Conservation*, Reading: Addison-Wesley.

Prahalad, C.K. and Hammond, A. (2004) 'The co-creation of value', *Journal of Marketing*, 69: 23.

Prahalad, C.K. and Ramaswamy, V. (2002) 'The co-creation connection', *Strategy and Competition*, 27: 1–12 (reprint).

Prahalad, C.K. and Ramaswamy, V. (2004) 'Co-creation experiences: the next practice in value creation', *Journal of Interactive Marketing,* 18(3): 5–14.

Preobrazhensky, V.S., Yu, A., Vedenin, A., Zorin, I.V. and Mukhina, L.I. (1976) *Current Problems of Recreational Geography*, XXIII International Geographical Congress, Moscow.

Pride, W.M. and Ferrell, O.C. (2003) *Marketing: Concepts and Strategies*, 12th edn, Boston, MA: Houghton Mifflin.

Productivity Commission (2005) *Assistance to Tourism: Exploratory Estimates*, Commission Research Paper, Canberra: Productivity Commission.

Ramesh, R. (2008) 'Paradise almost lost: Maldives seek to buy a new homeland', *The Guardian*, 10 November, http://www.guardian.co.uk/environment/2008/nov/10/maldives-climate-change (accessed 10 November, 2008).

Raptopoulou-Gigi, M. (2003) 'Severe acute respiratory syndrome (SARS): a new emerging disease in the 21st century', *Hippokratia,* 7(2): 81–3.

Ratz, T. (2006) 'The socio-cultural impacts of tourism', Budapest University of Economic Sciences, http://www.ratztamara.com/impacts.html (accessed 16 February 2008).

Rayner, S. (2006) 'What drives environmental policy?' *Global Environmental Change*, 16(1): 4–6.

Rein, S. (2007) 'Investing in China's tourism industry', Seeking Alpha (blog), http://seekingalpha.com/article/28889-investing-in-china-s-tourism-industry (accessed 27 February 2008).

Reisinger, Y. and Steiner, C.J. (2006) 'Reconceptulizing object authenticity', *Annals of Tourism Research*, 33(1): 65–86.

Relph, E. (1976) *Place and Placelessness*, London: Pion.

Ricciardi, A. (2007) 'Are modern biological invasions an unprecedented form of global change?' *Conservation Biology,* 21(2): 329–36.

Richardson, E.R. (1962) *The Politics of Conservation: Crusades and Controversies 1897–1913*, Berkeley: University of California Press.

Riley, M. (2004) 'Labor mobility and market structure in tourism', in A.A. Lew, C.M. Hall and A.M. Williams (eds) *Companion to Tourism*, London: Blackwell.

Riley, M., Ladkin, A. and Szivas, E. (2002) *Tourism Employment: Analysis and Planning*, Clevedon: Channel View.

Rittel, H.W.J. and Webber, M.M. (1973) 'Dilemmas in a general theory of planning', *Policy Sciences,* 4: 155–69.

Robertson, D. (2008) 'Airline industry faces "year of hell"', *TimeOnline* (9 October), http://business.timesonline.co.uk/tol/business/industry_sectors/transport/article4916436.ece (accessed 31 October 2008).

Robins, K. (1999) 'Tradition and translation: national culture in its global context', in D. Boswell and J. Evans (eds) *Representing the Nation: A Reader*, London: Routledge.

Robinson, J. (2003) *Work to Live*, New York: Penguin.

Rodrigues, A.S.L. and Gaston, K.J. (2001) 'How large do reserve networks need to be?' *Ecological Letters*, 4: 602–9.

Rodrigues, A.S.L. and Gaston, K.J. (2002) 'Rarity and conservation planning across geopolitical units', *Conservation Biology*, 16: 674–82.

Roehl, W. (1998) 'The tourism production system: the logic of industrial classification', in D. Ioannides and K.G. Debbage (eds) *The Economic Geography of the Tourist Industry: A Supply-side Analysis*, London: Routledge.

Rogers, E., Medina, U.E., Rivera, M.A. and Wiley, C.J. (2005) 'Complex adaptive systems and the diffusion of innovations', *Innovation Journal: The Public Sector Innovation Journal*, 10(3): www.innovation.cc/volumes-issues/rogers-adaptivesystem7final.pdf

Ropke, I. and Godsaken, M. (2007) 'Leisure activities, time and environment', *International Journal of Innovation and Sustainable Development,* 2(2): 155–74.

Rother, L. (1993) 'Fearful of tourism decline, Florida offers assurances on safety', *New York Times* (16 September), http://query.nytimes.com/gst/fullpage.html?res=9F0CE 5DE1F3AF935A2575AC0A965958260 (accessed 31 October 2008).

Rothman, H.K. (1998) *Devil's Bargains: Tourism in the Twentieth-Century American West*, Lawrence: University Press of Kansas.

Royal Watch News (2008) 'Prince Philip's tourism rant', MandC Royal Watch, 27 October, www.monstersandcritics.com/people/royalwatch/news/article_1439296.php/ (accessed 28 October 2008).

Runte, A. (1990) *Yosemite: The Embattled Wilderness*, Lincoln: University of Nebraska Press.

—— (1997) *National Parks: The American Experience*, 3rd edn, Lincoln: University of Nebraska Press.

Rushdie, S. (1991) *Imaginary Homelands*, London: Granta.

Ryan, R. and Stewart, E. (2007) 'Benn announces "stronger" climate change bill', *The Guardian*, 29 October.

Saarinen, J. and Tervo, K. (2006) 'Perceptions and adaptation strategies of the tourism industry to climate change: the case of Finnish nature-based tourism entrepreneurs', *International Journal of Innovation and Sustainable Development,* 1(3): 213–28.

Sangpikul, A. (2007) 'Travel motivations of Japanese senior travellers to Thailand', *International Journal of Tourism Research,* 10: 81–94.

Santarelli, E. (1998) 'Start-up size and post-entry performance: the case of tourism services in Italy', *Applied Economics*, 30(2): 157–63.

Sayre, N.F. (2008) 'The genesis, history, and limits of carrying capacity', *Annals of the Association of American Geographers,* 98(1): 120–34.

Schafer, A. (2000) 'Regularities in travel demand: an international perspective', *Journal of Transportation and Statistics,* 3(3): 1–31.

Schafer, A. and Victor, D. (2000) 'The future mobility of the world population', *Transportation Research A,* 34(3): 171–205.

Schenk, M. (2008) 'Oil falls to 21-month low on forecasts of lower global demand', Bloomber.com (12 November), http://www.bloomberg.com/apps/news?pid=2060 1081&sid=aozlmcJdgJ00 (accessed 14 November 2008).

Scott, D., McBoyle, G. and Mills, B. (2003) 'Climate change and the skiing industry in Southern Ontario (Canada): exploring the importance of snowmaking as a technical adaptation', *Climate Research*, 23: 171–81.

Scott, D., McBoyle. G. and Minogue, A. (2007) 'Climate change and Québec's ski industry', *Global Environmental Change*, 17(2): 181–90.

Scott, N. and Laws, E. (2005) 'Tourism crises and disasters: enhancing understanding of system effects', *Journal of Travel and Tourism Marketing,* 19(2/3): 149–58.

Scottish Wildlife Trust (SWT) (2008) They will be back: licence granted to bring back beavers, Press Release, 25 May, http://www.swt.org.uk/SingleNewsPage.aspx?News Id=33

Scoullos, M.J. (2003) 'Impact of anthropogenic activities in the coastal region of the Mediterranean Sea', in *International Conference on the Sustainable Development of the Mediterranean and Black Sea Environment*, May, Thessaloniki, Greece.

Secretariat of the Convention on Biological Diversity (SCBD) (2006) *Global Biodiversity Outlook 2*, Montreal: SCBD.

Shackley, M (ed.) (1998) *Visitor Management: Case Studies from World Heritage Sites*, Oxford: Butterworth Heinemann.

Shafer, E.L., Moeller, G.H. and Getty, R.E. (1974) *Future Leisure Environments*, Forest Research Paper NE-301, Upper Darby, PA: US Department of Agriculture, Forest Service, Northeastern Forest Experiment Station.

Sharkey, J. (2008) 'The credit squeeze compresses travel, too', *International Herald Tribune* (21 October), http://www.iht.com/articles/2008/10/21/business/21road.php (accessed 31 October 2008).

Shaw, B. and Shaw, G. (1999) '"Sun, sand and sales": enclave tourism and local entrepreneurship in Indonesia', *Current Issues in Tourism*, 2(1): 68–81.

Shaw, G. (2004) 'Entrepreneurial cultures and small business enterprises in tourism', in A.A. Lew, C.M. Hall and A.M. Williams (eds) *Companion to Tourism*, London: Blackwell.

Shaw, G. and Williams, A.M. (1998) 'Entrepreneurship, small business culture and tourism development', in D. Ioannides and K. Debbage (eds) *The Economic Geography of the Tourism Industry*, London: Routledge.

Shaw, G. and Williams, A.M. (2004) *Tourism and Tourism Spaces,* London: Sage.

Sinclair, M.T., Blake, A. and Sugiyarto, G. (2003) 'The economics of tourism', in C. Cooper (ed.) *Classic Reviews in Tourism*, Clevedon: Channel View Publications.

Singh, J.S. (2002) 'The biodiversity crisis: a multifaceted review', *Current Science*, 82(6): 638–47.

Sirgy, M.J. (1986) 'A quality-of-life theory derived from Maslow's developmental perspective: "quality" is related to progressive satisfaction of a hierarchy of needs, lower order and higher', *American Journal of Economics and Sociology*, 45(3): 329–42.

Skumanich, M. and Silbernagel, M. (1997) *Foresighting Around the World: A Review of Seven Best-In-Kind Programs*, Seattle, WA: Battelle Seattle Research Center.

Smith, S.L.J. (1998) 'Tourism as an industry: debates and concepts', in D. Ioannides and K.G. Debbage (eds) *The Economic Geography of the Tourist Industry: A Supply-side Analysis*, London: Routledge.

—— (2004) 'The measurement of global tourism: old debates, new consensus, and continuing challenges', in A.A. Lew, C.M. Hall and A.M. Williams (eds) *A Companion to Tourism*, Oxford: Blackwell.

—— (2007) 'Duelling definitions: challenges and implications of conflicting international concepts of tourism', in D. Airey and J. Tribe (eds) *Progress in Tourism Research*, Oxford: Elsevier.

Solomon, S., Qin, D., Manning, M., Chen, Z., Marquis, M., Averyt, K.B. et al. (eds) (2007) *Climate Change 2007: The Physical Science Basis. Contribution of Working Group I to the Fourth Assessment Report of the Intergovernmental Panel on Climate Change*, Cambridge: Cambridge University Press

Squires, G.D. and Hartman, C. (eds) (2006) *There Is No Such Thing as a Natural Disaster: Race, Class, and Hurricane Katrina*, New York: Routledge.

Stanway, D. (2008) 'Tibet under strain as visitors surpass locals', *The Guardian*, 14 January.

Statistics New Zealand (2006) *Business Operations Survey 2005*, Wellington: Statistics New Zealand.

—— (2007) *Innovation in New Zealand 2005*, Wellington: Statistics New Zealand.

Statistics Norway (2006) *Tourism Satellite Accounts, Final Figures for 2004 and Preliminary Figures for 2005: Total Tourism Consumption NOK 90 billion*, Oslo: Statistics Norway, http://www.ssb.no/turismesat_en/main.html

Steiner, C.J. and Reisinger, Y. (2006) 'Understanding existential authenticity', *Annals of Tourism Research,* 33(2): 299–318.

Stern, N. (2006) *The Economics of Climate Change: The Stern Review*, London: HM Treasury.

Stewart, E.J., Glen, M.H., Daly, K and O'Sullivan, D. (2001) 'To centralize or disperse – a question for interpretation: a case study of interpretive planning in the Breckes', *Journal of Sustainable Tourism*, 9(4): 342–55.

Stone, J.J. (2008) 'Europe adds flights to its emission trading system', Red Green and Blue (26 October), http://redgreenandblue.org/2008/10/26/europe-adds-flights-to-its-emission-trading-system/ (accessed 14 November 2008).

Stynes, D.J. (1997) *Economic Impacts of Tourism*, Illinois Bureau of Tourism, Department of Commerce.

—— (1999) 'Approaches to estimating the economic impacts of tourism, some examples', Michican State University, http://web4.msue.msu.edu/mgm2/econ/pdf/ecimpvol2.pdf (accessed 28 February 2008).

—— (2007) *National Park Visitor Spending and Payroll Impacts 2006*, NPS Social Science Program, Washington, DC: National Park Service.

Stynes, D.J. and Sen, Y-Y. (2005) *Economic Impacts of Grand Canyon National Park Visitor Spending on the Local Economy, 2003*, East Lansing: National Park Service Social Science Program / Department of Community, Agriculture, Recreation and Resource Studies, Michigan State University.

Sulaiman, Y. (2007) 'UNESCO ponders if World Heritage Sites endangered by mass tourism', eTurboNews Asia/Pacific, http://www.travelindustryreview.com/news/5672 (accessed 18 February 2008).

Sundbo, J., Orfila-Sintes, F. and Sørensen, F. (2007) 'The innovative behaviour of tourism firms', *Research Policy*, 36(1): 88–106.

Tanrivermis, H. (2003) 'Agricultural land use change and sustainable use of land resources in the Mediterranean region of Turkey', *Journal of Arid Environments,* 54(3): 553–64.

Tatem, A.J. and Hay, S.I. (2007) 'Climatic similarity and biological exchange in the worldwide airline transportation network', *Proceedings of the Royal Society B,* 274: 1489–96.

Tausie, V. (1981) *Art in the New Pacific*, Suva: Institute of South Pacific Studies.

Telfer, D.J. and Wall, G. (2000) 'Strengthening backward economic linkages: local food purchasing by three Indonesian hotels', *Tourism Geographies*, 2: 421–47.

Teo, P. and Leong, S. (2006) 'A post-colonial analysis of backpacking', *Annals of Tourism Research,* 33(1): 109–31.

Thomas, L. and Middleton, J. (2003) *Guidelines for Management Planning of Protected Areas*, Gland, Switzerland: IUCN – The World Conservation Union.

Tiebout, C.M. (1962) *The Community Economic Base Study*, New York: Committee for Economic Development.

Timothy, D. (2005) *Shopping Tourism, Retailing and Leisure*, Clevedon: Channel View.

Tödtling, F. and Kaufmann, A. (1998) 'Innovation systems in regions of Europe: a comparative perspective', paper presented to the 38th Congress of the European Regional Science Association, Vienna, August–September.

Torres, R. (2003) 'Linkages between tourism and agriculture in Quintana Roo, Mexico', *Annals of Tourism Research*, 30(3): 546–66.

Torres, R. and Momsen, J.H. (2004) 'Linking tourism and agriculture to achieve pro-poor tourism objectives', *Progress in Development Studies,* 4(4): 294–318.

Tourism Concern (2008) *Strategic Business Plan 2008–11*, London: Tourism Concern.

Towner, J. (1996) *An Historical Geography of Recreation and Tourism in the Western World 1540–1940*, London: Routledge.

Travel Industry Association (TIA) (2008) 'TIA lauds federal advisory panel recommendation to launch US travel promotion campaign' (16 January 2008), http://www.tia.org/pressmedia/pressrec.asp?Item=865 (accessed 28 February 2008).

Turner, B.L., Clark, W.C., Kates, R.W., Richards, J.F., Mathews, J.Y. and Meyer, W.B. (eds) (1990) *The Earth as Transformed by Human Action*, Cambridge: Cambridge University Press.

United Airlines (2008) 'United environmental flight reduces carbon emissions by nearly 33,000 pounds', *Yahoo Finance* (14 November), http://biz.yahoo.com/prnews/081 114/aqf062.html (accessed 15 November 2008).

United Nations (UN) (1994) *Recommendations on Tourism Statistics*, New York: United Nations.

United Nations and United Nations World Tourism Organization (UN and UNWTO) (2007) *International Recommendations on Tourism Statistics (IRTS) Provisional Draft*, New York/Madrid: UN and UNWTO.

United Nations Conference on Trade and Development (UNCTAD) (2002) Largest Transnational Corporations, http://www.unctad.org/Templates/Page.asp?intItemID=2443&lang=1 (accessed 28 February 2008).

United Nations Educational, Scientific, and Cultural Organization (UNESCO) (2002) *UNESCO Universal Declaration on Cultural Diversity*, http://unesdoc.unesco.org/images/0012/001271/127160m.pdf (accessed 16 February 2008; and http://www.unesco.org/education/imld_2002/unversal_decla.shtml (posted 21 February 2002; accessed 30 January 2008).

United Nations Environment Programme (UNEP) (2000) 'Tools and institutional networks for disseminating environmental information in support of the Aarhus Convention', discussion paper by Dorte Bennedbaek Information Counsellor, Ministry for the Environment and Energy, Copenhagen, Denmark. INFOTERRA 2000 – Global Conference on Access to Environmental Information, Dublin, Ireland, 11–15 September, UNEP/INF2000/WP/13.

—— (2001a) 'Economic impacts of tourism', http://www.unep.fr/scp/tourism/sustain/impacts/economic/development.htm (accessed 20 March 2009).

—— (2001b) 'Socio-cultural impacts of tourism', http://www.unep.fr/scp/tourism/sustain/impacts/ (accessed 20 March 2009).

United Nations Environment Programme – Mediterranean Action Plan – Priority Actions Programme (UNEP/MAP/PAP) (2001) *White Paper: Coastal Zone Management in the Mediterranean, Priority Actions Programme*, Split: UNEP/MAP/PAP.

United Nations Environment Programme, United Nations World Tourism Organization and World Meteorological Organization (UNEP, UNWTO and WMO) (2008) *Climate Change Adaptation and Mitigation in the Tourism Sector: Frameworks, Tools and Practice* (M. Simpson, S. Gössling, D. Scott, C.M. Hall, and E. Gladin), Paris: UNEP, University of Oxford: UNWTO, WMO.

United Nations, European Commission, International Monetary Fund, Organization for Economic Co-operation and Development, United Nations Conference on Trade and Development, and World Trade Organization (2002) *Manual on Statistics of International Trade in Services*, Department of Economic and Social Affairs Statistics Division, Series M No.86. New York: United Nations.

United Nations World Tourism Organization (UNWTO) (1991) *Guidelines for the Collection and Presentation of Domestic and International Tourism Statistics*, Madrid: UNWTO.

——— (1994) *Recommendations on Tourism Statistics*, Madrid: UNWTO.

——— (1997) *Tourism 2020 Vision*, Madrid: UNWTO.

——— (1999) Global Code of Ethics for Tourism, http://www.unwto.org/code_ethics/eng/principles.htm, and http://www.unwto.org/code_ethics/eng/brochure.htm (accessed 3 May 2008).

——— (2001) *Tourism 2020 Vision – Global Forecasts and Profiles of Market Segments*, Madrid: UNWTO.

——— (2006) *International Tourist Arrivals, Tourism Market Trends, 2006 Edition –* Annex, Madrid: UNWTO.

——— (2008a) 'Facts and figures. information, analysis and know how: Tourism 2020 Vision', http://www.unwto.org/facts/eng/vision.htm (accessed 1 June 2008).

——— (2008b) World tourism exceeds expectations in 2007 – arrivals grow from 800 million to 900 million in two years, Press Release, Madrid 29 January, http://www.unwto.org/media/news/en/press_det.php?id=1665

United Nations World Tourism Organization and United Nations Environment Programme (UNWTO and UNEP) (2008) *Climate Change and Tourism: Responding to Global Challenges* (prepared by Scott, D., Amelung, B., Becken, S., Ceron, J.P., Dubois, G., Gossling, S., Peeters, P. and Simpson, M.), Madrid/Paris: UNWTO/UNEP.

United Nations World Tourism Organization, United Nations Environment Programme and World Meteorological Organization (UNWTO, UNEP and WMO) (2007) *Climate Change and Tourism: Responding to Global Challenges: Summary*, (prepared by Scott, D., Amelung, B., Becken, S., Ceron, J.P., Dubois, G., Gossling, S., Peeters, P. and Simpson, M.), Madrid: UNWTO, and Paris: UNEP.

——— (2008) 'Climate change and tourism: responding to global challenges, technical report' (draft), (prepared by Scott, D., Amelung, B., Becken, S., Ceron, J.P., Dubois, G., Gossling, S., Peeters, P. and Simpson, M.), Madrid: UNWTO, and Paris: UNEP.

United States Department of Transportation (USDOT) (2008) 'Aviation consumer protection division, organization and functions', http://airconsumer.ost.dot.gov/org.htm (accessed 13 May 2008).

United States Environmental Protection Agency (USEPA) (2008) 'Environmental Management System', http://www.epa.gov/ems/ (accessed 13 June.

Urry, J. (2002) *The Tourist Gaze*, 2nd edn, London: Sage.

Uyarra, M.C., Cote, I., Gill, J., Tinch, R., Viner, D. and Watkinson, A. (2005) 'Island-specific preferences of tourists for environmental features: implications of climate change for tourism-dependent states', *Environmental Conservation,* 32: 11–19.

van der Veen, T. (1995) 'Historical changes in Pacific arts', *Oceania* (11/12; Radboud University Nijmegen, Netherlands), http://cps.ruhosting.nl/12/ (accessed 3 November 2008).

Vargo, S.L. and Lusch, R.F. (2004) 'Evolving to a new dominant logic of marketing', *Journal of Marketing*, 68(1): 1–17.

Votolo, G. (2007) *Transport Design: A Travel History*, London: Reaktion Books.

Waldrop, M.M. (1992) *Complexity: The Emerging Science at the Edge of Order and Chaos*, New York: Touchstone.

Walker, P.A., Greiner, R., McDonald, D. and Lyne, V. (1998) 'The tourism futures simulator: a systems thinking approach', *Environmental Modelling and Software*, 14(1): 59–67.

Wall, G. (1997) 'Tourism attractions: points, lines and areas', *Annals of Tourism Research*, 24(1): 240–3.

Walton, J. (2003) *Storied Land: Community and Memory in Monterey*, San Francisco: University of California Press.

Wang, N. (2000) *Tourism and Modernity: A Sociological Analysis*, Amsterdam: Pergamon.

Warnes, A. (1992) 'Migration and the life course', in A. Champion and A. Fielding (eds) *Migration Processes and Patterns*. Vol. 1: *Research Progress and Prospects*, London: Belhaven.

Warnken, J. (2000) 'Monitoring diffuse impacts: Australian tourism developments', *Environmental Management*, 25(4): 453–61.

Warnken, J. and Buckley, R. (1998) 'Monitoring diffuse impacts: Australian tourism developments. Scientific quality of tourism environmental impact assessment', *Journal of Applied Ecology*, 35(1): 1–8.

Warnken, J., Thompson, D. and Zakus, D.H. (2002) 'Golf course development in a major tourist destination: implications for planning and management', *Environmental Management*, 27(5): 681–96.

Watson, M. (2003) 'Environmental Impact Assessment and European Community law', XIV International Conference, Danube – River of Cooperation, Beograd, November.

Western Australian Tourism Commission (WATC) (1997) *Tourism Research Brief on Daytripping*, Perth: Western Australian Tourism Commission.

Whitson, D. and Horne, J. (2006) 'Underestimated costs and overestimated benefits? Comparing the outcomes of sports mega-events in Canada and Japan', *Sociological Review Monograph, Sports Mega-events: Social Scientific Analyses of a Global Phenomenon*, 54(s2): 73–89.

Wild Ennerdale Partnership (2006) *Stewardship Plan Text 2006*. Wild Ennerdale Partnership.

Wiley, D.N., Moller, J.C., Pace III, R.M. and Carlson, C. (2008) 'Effectiveness of voluntary conservation agreements: case study of endangered whales and commercial whale watching', *Conservation Biology*, 22(2): 450–7.

Wilkinson, J. (1994) *The Olympic Games: Past History and Present Expectations*, Sydney: NSW Parliamentary Library.

Williams, A. (2004) 'Towards a political economy of tourism', in A.A. Lew, C.M. Hall and A.M. Williams (eds) *Companion to Tourism,* London: Blackwell.

Williams, A.M. and Montanari, A. (1999) 'Sustainability and self-regulation: critical perspectives', *Tourism Geographies,* 1(1): 26–40.

Wilpert, C.B. (1985) *Südsee Souvenirs*, Hamburg: Hamburgisches Museum für Völkerkunde.

Wincott, D. (2003) 'Slippery concepts, shifting context: (national) states and welfare in the Veit-Wilson/Atherton debate', *Social Policy and Administration*, 37(3): 305–15.

Winter, T. (2007) *Post-Conflict Heritage, Postcolonial Tourism: Tourism, Politics and Development at Angkor*, London: Routledge.

—— (2008) 'An overview of the issue', *Post-Conflict Heritage Blog* (17 October), http://www.postconflictheritage.com/home/pch_blog/pch_blog.html (accessed 3 November 2008).

Wintour, P. and Elliott, L. (2007) 'Smash and grab: how Labour stole the Tories' big ideas', *The Guardian*, 10 October.

Woodside, A.G. and Dubelaar, C. (2002) 'A general theory of tourism consumption systems: a conceptual framework and an empirical exploration', *Journal of Travel Research*, 41(2): 120–32.

World Commission on Environment and Development (WCED) (the Brundtland Report) (1987) *Our Common Future*, Oxford: Oxford University Press.

World Health Organization (WHO) (2008) *International Travel and Health*, Geneva: WHO.

World Trade Organization (WTO) (1998) *Annual Report 1998*, Geneva: WTO.

—— (2000) Communication from the United States: Tourism and Hotels, Council for Trade in Services, Special Session, 18 December (00–5572), World Trade Organization S/CSS/W/31.

—— (2001a) *International Trade Statistics 2001*, Geneva: WTO.

—— (2001b) Communication from Columbia: Tourism and Travel-Related Services, Council for Trade in Services, Special Session, 27 November (01–6056), World Trade Organization S/CSS/W/122.

—— (2006) *International Trade Statistics 2006*, Geneva: WTO.

—— (2007a) 'Merchandise trade by product', http://www.wto.org/english/res_e/statis_e/its2007_e/its07_merch_trade_product_e.htm (accessed 16 February 2008).

—— (2007b) 'Trade in commercial services by category', http://www.wto.org/english/res_e/statis_e/its2007_e/its07_trade_category_e.htm (accessed 16 February 2008).

World Travel and Tourism Council (WTTC) (2008a) Continued growth signalled for travel and tourism industry, Press Release, 6 March, Berlin, http://www.wttc.org/eng/Tourism_News/Press_Releases/Press_Releases_2008/Continued_growth_signalled_f or_Travel_and_Tourism_Industry/

—— (2008b) Travel and tourism leaders forecast continued growth for 2008, Press Release, 21 January, Dubai, http://www.wttc.org/eng/Tourism_News/Press_Releases/Press_Releases_2008/Travel_and_Tourism_leaders_forecast_continued_growth_for_2008_/

—— (2008c) *The 2008 Travel and Tourism Economic Research Executive Summary*, London: WTTC.

World Wide Fund for Nature (WWF) (2001) *Tourism Threats in the Mediterranean*, Rome: WWF Mediterranean Programme.

Worster, D. (1977) *Nature's Economy: A History of Ecological Ideas*, Cambridge: Cambridge University Press.

Wu, W.P. (2004) 'Cultural strategies in Shanghai: regenerating cosmopolitanism in an era of globalization', *Progress in Planning*, 61(3): 159–80.

Xie, P.F. and Lane, B. (2006) 'A life cycle model for aboriginal arts performance in tourism: perspectives on authenticity', *Journal of Sustainable Tourism*, 14(6): 545–61.

Xie, P.F. and Lew, A.A. (2009) 'Podcasting and tourism: an exploratory study of types, approaches and content', *Journal of Information Technology and Tourism*, 10(2): 173–80.

Xinhua (2007) 'Lhasa to build mini Potala Palace amid tourist boom', *People's Daily Online* (14 March), http://english.peopledaily.com.cn/200703/14/eng20070314_357 519.html (accessed 3 November 2008).

Yeoh, B.S.A., Tan, E.S., Wang, J. and Wong, T. (2001) 'Tourism in Singapore: an overview of policies and issues', in E.S. Tan, B.S.A. Yeoh and J. Wang (eds) *Tourism Management and Policy: Perspectives from Singapore*, Singapore: World Scientific Publishing.

Yeoman, I. and McMahon-Beattie, U. (2005) 'Developing a scenario planning process using a blank piece of paper', *Tourism and Hospitality Research*, 5(3): 273–85.

Yeoman, I., Munro, C. and McMahon-Beattie, U. (2006) 'Tomorrow's: World, consumer and tourist', *Journal of Vacation Marketing*, 12(2): 173–90.

Zhang, J. and Marcussen, C. (2007) 'Tourist motivation, market segmentation and marketing strategies', paper presented at 5th Biannual Symposium of the International Society of Culture, Tourism, and Hospitality Research, June, Charleston, SC, USA, http://www.crt.dk/media/Tourism_Motivation_and_Marketing_Strategies_Denmark_ Jie_Zhang_Carl_Henrik_Marcussen_CRT_2007.pdf (accessed 8 May 2008).

Zhang, Y. and Lew, A.A. (1997) 'The People's Republic of China: two decades of tourism', *Pacific Tourism Review,* 1(2): 161–72.

Zimmermann, E.W. (1951) *World Resources and Industries*, rev. edn, New York: Harper.

Index

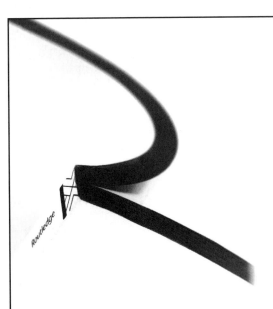

Routledge
Paperbacks Direct

Bringing you the cream of our hardback publishing at paperback prices

This exciting new initiative makes the best of our hardback publishing available in paperback format for authors and individual customers.

Routledge Paperbacks Direct is an ever-evolving programme with new titles being added regularly.

To take a look at the titles available, visit our website.

www.routledgepaperbacksdirect.com

Routledge
Taylor & Francis Group

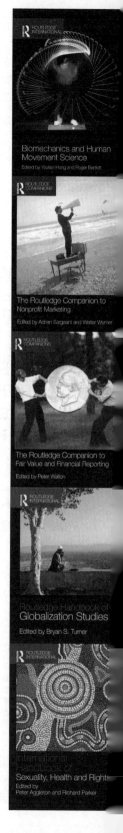